MODERN ASPECTS OF ELECTROCHEMISTRY

No. 54

Series Editors:
Ralph E. White
Department of Chemical Engineering
University of South Carolina
Columbia, SC 29208

Constantinos G. Vayenas
Department of Chemical Engineering
University of Patras
Patras 265 00
Greece

Managing Editor:
Maria E. Gamboa-Aldeco
1107 Raymer Lane
Superior, CO 80027

For further volumes:
http://www.springer.com/series/6251

Previously from Modern Aspects of Electrochemistry

Modern Aspects of Electrochemistry No. 52

Applications of Electrochemistry and Nanotechnology in Biology and Medicine I

Edited by Noam Eliaz, Professor of Engineering at Tel-Aviv University

Topics in Number 52 include:

- Monitoring of cellular dynamics with electrochemical detection techniques
- Fundamental studies of long- and short-range electron exchange mechanisms between electrodes and proteins
- Microbial fuel cell scalability and applications in robotics
- Electrochemical coating of medical implants
- Electrochemical techniques for obtaining biofunctional materials
- Preparation and properties of bioactive metals prepared by surface modification

Modern Aspects of Electrochemistry No. 53

Applications of Electrochemistry and Nanotechnology in Biology and Medicine II

Edited by Noam Eliaz, Professor of Engineering at Tel-Aviv University

Topics in Number 53 include:

- Fundamental studies of electron tunneling between electrodes and proteins
- Electron transfer kinetics at oxide films on metallic biomaterials
- How adsorption of organic molecules and ions depends on surface crystallography of the metal electrode
- Studying and modifying biomaterial surfaces with high resolution using the scanning electrochemical microscope
- Electrochemical method for high-throughput screening of enzymatic activity

Stojan S. Djokić
Editor

Electrochemical Production of Metal Powders

Springer

Editor
Stojan S. Djokić
Elchem Consulting Ltd.
Edmonton, AB, Canada

ISSN 0076-9924
ISBN 978-1-4614-2379-9 ISBN 978-1-4614-2380-5 (eBook)
DOI 10.1007/978-1-4614-2380-5
Springer New York Heidelberg Dordrecht London

Library of Congress Control Number: 2012934056

Printed on acid-free paper

Springer is part of Springer Science+Business Media (www.springer.com)

Preface

The present volume of *Modern Aspects of Electrochemistry* brings readers the newest developments and achievements in the production of metallic powders by electrochemical and electroless methods from aqueous solutions. Although the deposition of metallic powders from aqueous solutions was intensively studied for years, the last summarized results (Calusaru) on this topic were published in 1979.

Electrochemically and chemically produced metal powders from aqueous solutions are of high purity. These powders find applications in metallurgy, automotive, aerospace, energy device, electronics, and biomedical industries. Disperse deposits and electrochemically produced metal powders are also very suitable for use as catalytic surfaces in chemical industry.

This volume of *Modern Aspects of Electrochemistry* reviews the electrochemical aspects of the latest developments in the deposition of metal powders. Distinguished international contributors have written chapters devoted to this fine area which may impact significant technological advancement in the future. Following is a brief description of chapters in this volume of *Modern Aspects of Electrochemistry*.

Popov and Nikolić in Chapter 1 discuss the fundamental aspects of disperse metals electrodeposition. The shapes of polarization curves in relation to the deposition process parameters are analyzed. Disperse metal deposits are formed with a nonuniform current density distribution over the surface of the macroelectrode. Adherent granular disperse deposits are produced in an electrodeposition process characterized by a large exchange current density, due to the formation of nucleation exclusion zones around growing grains on the inert substrate. Nonadherent dendritic or spongy deposits are formed in the dominant diffusion control on the level of the macroelectrode and an activation control on the tips of microelectrodes placed inside the diffusion layer of the macroelectrode. Nonadherent honeycomb-like deposit is formed in the presence of strong hydrogen codeposition. All the above cases are discussed in detail and explained using appropriate mathematical models. It is also shown that the formation of dendritic deposits at low level of coarseness strongly increases the apparent exchange current density for the deposition process, producing electrocatalytic effect.

Chapter 2, by Jović et al., presents the results of morphology investigation of different metal powders, e.g., Ag, Pd, Pb, Cd, Fe, Ni, and Co. It is shown that the morphology is different for each metal. The conditions for deposition of each powder are specified. Diffusion control, based on the descriptions in this chapter, is necessary for the formation of powders in accordance with the current theory. Presented results correspond to either disperse deposits on the electrode or powders spontaneously detached or removed by tapping from the electrode.

In Chapter 3, by Nikolić and Popov, types, properties, and modeling of copper powders are presented. Powdered copper deposits are formed at overpotentials and current densities belonging to the plateau of the limiting diffusion current density and/or at higher, where the simultaneous hydrogen evolution reaction occurs. The effect of periodically changing regimes of electrolysis, such as pulsating current, reversing current, and pulsating overpotential, on the formation of disperse copper deposits is analyzed. It is shown that the effects on morphology of electrodeposited copper with an application of square-waves pulsating current are equivalent to those attained by electrodepositions in the constant regimes of electrolysis from solutions of different $CuSO_4$ and H_2SO_4 concentrations.

Chapter 4, by Nikolić, discusses the formation of open and porous electrodes by the constant and periodically changing regimes of electrolysis. The formation of these electrodes in both potentiostatic and galvanostatic electrodeposition is presented. Three dimensional foam or honeycomb-like copper electrodes are formed by electrochemical deposition at high current densities and overpotentials where parallel to copper electrodeposition, hydrogen evolution occurs. Hydrogen evolution enabling the formation of these electrodes is vigorous enough to cause such stirring of the copper solution which leads to the decrease of the cathode diffusion layer thickness and to the increase in the limiting diffusion current density and hence to the change of the hydrodynamic conditions in the near-electrode layer. The phenomenology of the formation of the honeycomb-like structures by potentiostatic electrodeposition, as well as parameters affecting number, size, and distribution holes in the honeycomb-like structures, is analyzed.

Jović et al. in Chapter 5 discuss morphology, chemical and phase composition of electrodeposited Co–Ni, Fe–Ni, and Mo–Ni–O powders. The processes of Co–Ni, Fe–Ni, and Mo–Ni–O powders electrodeposition were investigated by polarization measurements compensated for IR drop. All polarization curves were characterized with two inflection points. The first one, positioned at a less negative potential reflecting the onset of electrodeposition, is seen as a sudden increase in the current density and the second one, at more negative potential, is characterized by a decrease of the slope on the polarization curves, representing the stage when the electrodeposition becomes controlled by the rate of hydrogen bubbles formation. Powder samples for the investigation of morphology, chemical, and phase composition are electrodeposited at current density slightly lower than that corresponding to the second inflection point.

Chapter 6, by Magagnin and Cojocaru, is a review of recent progress in the electrochemical synthesis of dispersed nanoparticles, including the sonoelectrochemical approach. Results on the synthesis of silver and gold particles with size from the nanoscale to the mesoscale in sulfite-based solutions are reported. The electrochemical behavior of the electrolytes used in the electrodeposition is studied on different substrates such as glassy carbon, Ti, and indium tin oxide. Silver particles below 50 nm were easily obtained on glassy carbon substrate by potential-controlled deposition achieving a high nucleation density. Silver particle deposition on titanium showed low nucleation density and a strong tendency to form large particles, clusters, and agglomerates, mostly in connection with surface irregularities. Gold particles were successfully deposited by either a potential pulse or a potential sweeping technique, achieving good results in terms of nucleation density. This was observed on titanium substrate, using a single potential pulse technique for the deposition of Au particles. The preparation of dispersed nanoparticles supported on silicon by galvanic displacement reactions in microemulsions is also presented. Examples of gold and palladium particles are included, discussing the mechanism of formation and the coalescence behavior of the nanostructures.

Finally, in Chapter 7 Djokić discusses the deposition of metallic powders from aqueous solutions without an external current source. Metallic powders can be successfully produced via galvanic displacement reaction or by electroless deposition from homogenous aqueous solutions or slurries. The formation of various metallic powders without an external current source e.g., Cu, Ni, Co, Ag, Pd, and Au, using appropriate reducing agents is presented. The mechanistic aspects of electroless deposition of powders are also discussed. It is shown that the hydrolysis of metallic ions is the most important factor leading to the deposition of metal powders from aqueous solutions.

This new volume of *Modern Aspects of Electrochemistry* brings to scientists, engineers, and students new concepts and summarized results in the fields of electrochemical and chemical deposition, which may have significant influence for future practical applications.

Edmonton, AB, Canada

Stojan S. Djokić

Contents

Contributors

Paula Cojocaru Dip. Chimica, Materiali e Ing., Chimica G. Natta, Politecnico di Milano, Milano, Italy

Stojan S. Djokić Elchem Consulting Ltd., Edmonton, AB, Canada

Borka M. Jović Department of Materials Science, Institute for Multidisciplinary Research, University of Belgrade, Belgrade, Serbia

Vladimir D. Jović Department of Materials Science, Institute for Multidisciplinary Research, University of Belgrade, Belgrade, Serbia

Uroš Č. Lačnjevac Department of Materials Science, Institute for Multidisciplinary Research, University of Belgrade, Belgrade, Serbia

Luca Magagnin Dip. Chimica, Materiali e Ing., Chimica G. Natta, Politecnico di Milano, Milano, Italy

Nebojša D. Nikolić ICTM-Institute of Electrochemistry, University of Belgrade, Belgrade, Serbia

Konstantin I. Popov ICTM-Institute of Electrochemistry, University of Belgrade, Belgrade, Serbia

Faculty of Technology and Metallurgy, University of Belgrade, Belgrade, Serbia

Chapter 1
General Theory of Disperse Metal Electrodeposits Formation

Konstantin I. Popov and Nebojša D. Nikolić

1.1 Introduction

The most frequently used form of the cathodic polarization curve equation for flat or large spherical electrode of massive metal is given by

$$j = \frac{j_0(f_c - f_a)}{1 + \frac{j_0 f_c}{j_L}}, \tag{1.1}$$

where j, j_0 and j_L, are the current density, exchange current density, and limiting diffusion current density, respectively, and

$$f_c = 10^{\frac{\eta}{b_c}}, \tag{1.2}$$

K.I. Popov (✉)
ICTM-Institute of Electrochemistry, University of Belgrade, Njegoševa 12, P.O.B. 473,11001 Belgrade, Serbia

Faculty of Technology and Metallurgy, University of Belgrade, Karnegijeva 4, P.O.B. 3503,11001 Belgrade, Serbia
e-mail: kosta@tmf.bg.ac.rs

N.D. Nikolić
ICTM-Institute of Electrochemistry, University of Belgrade, Njegoševa 12, P.O.B. 473,11001 Belgrade, Serbia
e-mail: nnikolic@tmf.bg.ac.rs

Stojan S. Djokić (ed.), *Electrochemical Production of Metal Powders*, Modern Aspects of Electrochemistry 54, DOI 10.1007/978-1-4614-2380-5_1,

$$f_a = 10^{-\frac{\eta}{b_a}}, \qquad (1.3)$$

where b_c and b_a are the cathodic and anodic Tafel slopes and η is the overpotential. Equation (1.1) is modified for use in electrodeposition of metals by taking cathodic current density and overpotential as positive. Derivation of Eq. (1.1) is performed under assumption that the concentration dependence of j_0 can be neglected [1–4].

It is known [3] that electrochemical processes on microelectrodes in bulk solution can be under activation control at overpotentials which correspond to the limiting diffusion current density plateau of the macroelectrode. The cathodic limiting diffusion current density for steady-state spherical diffusion, $j_{L,Sp}$ is given by

$$j_{L,Sp} = \frac{nFDC_0}{r}, \qquad (1.4)$$

and for steady-state linear diffusion, j_L, it is given by

$$j_L = \frac{nFDC_0}{\delta}, \qquad (1.5)$$

where n is the number of transferred electrons, F is the Faraday constant, D and C_0 are the diffusion coefficient and bulk concentration of the depositing ion, respectively, r is the radius of the spherical microelectrode, and δ is the diffusion layer thickness of the macroelectrode. It follows from Eqs. (1.4) and (1.5) that

$$\frac{j_{L,Sp}}{j_L} = \frac{\delta}{r}. \qquad (1.6)$$

An electrode around which the hydrodynamic diffusion layer can be established, being considerably lower than dimensions of it, could be considered as a macroelectrode. An electrode, mainly spherical, whose diffusion layer is equal to the radius of it, satisfying

$$\delta \gg r, \qquad (1.7)$$

can be considered as a microelectrode [5].

According to Eq. (1.1) for

$$f_c \gg f_a \text{ and } \frac{j_0 f_c}{j_L} \gg 1, \tag{1.8}$$

the cathodic process on the macroelectrode enters full diffusion control, i.e.,

$$j \cong j_L. \tag{1.9}$$

Simultaneously, the cathodic current density on the spherical microelectrode, j_{Sp}, is given by

$$j_{Sp} = \frac{j_0(f_c - f_a)}{1 + \frac{j_0 f_c}{j_{L,Sp}}} \tag{1.10}$$

or, because of Eq. (1.6),

$$j_{Sp} = \frac{j_0(f_c - f_a)}{1 + \frac{j_0}{j_L} \cdot \frac{r}{\delta} \cdot f_c} \tag{1.11}$$

and, if condition (1.8) is also valid, but

$$\frac{r}{\delta} \to 0. \tag{1.12}$$

Equation (1.11) can be rewritten in the form

$$j = j_0 f_c. \tag{1.13}$$

This means that the process on the microelectrode in the bulk solution can be under complete activation control at the same overpotential at which the same process on the macroelectrode is simultaneously under full diffusion control.[1]

[1] The reversible potential of a surface with radius of curvature r_{cur} would depart from that of a planar surface by the quantity $\Delta E_r = 2\sigma V/(nFr_{cur})$, where σ is the interfacial energy between metal and solution, and V is the molar volume of metal [5]. It is valid at extremely low r_{cur}, being of the order of few millivolts, and it can be neglected except in some special cases, like the stability of the shape of the tips of dendrites [5].

The different behavior of macroelectrodes and microelectrodes under the same conditions of electrodeposition causes the disperse deposits formation.

Since the paper of Barton and Bockris [5] on the growth of silver dendrites, a lot of papers, chapters, and even books, dealing with electrodeposition of disperse metals were published. The aim of this chapter is to unite the basic statements of the previous contributions in a general all-inclusive theory.

1.2 Active Microelectrodes Placed the Inside Diffusion Layer of the Active Macroelectrode

1.2.1 Basic Facts

Naturally, the microelectrodes can be placed on the macroelectrodes inside their diffusion layers. Let us consider the model of surface irregularities shown in Fig. 1.1. The electrode surface irregularities

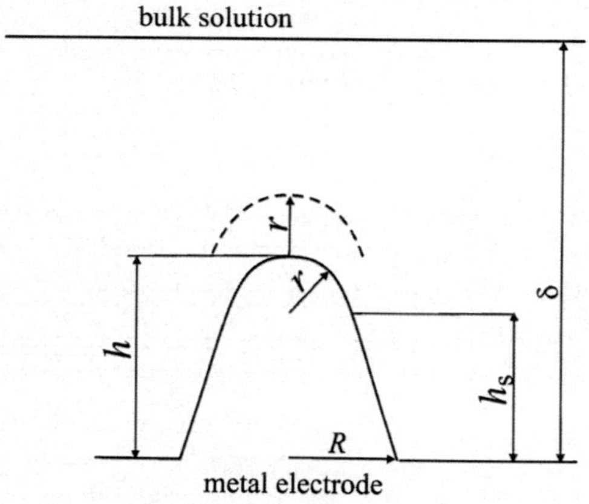

Fig. 1.1 Model of a paraboloidal surface protrusion: h is the height of the protrusion relative to the flat portion of the surface, h_s is the corresponding local side elongation, r is the radius of the protrusion tip, R is the radius of the protrusion base, δ is the thickness of the diffusion layer, and $\delta \gg h$ (Reprinted from [1] with permission from Springer and [6] with permission from Elsevier.)

are buried deep in the diffusion layer, which is characterized by a steady linear diffusion to the flat portion of the surface [1, 6, 7].

At the side of an irregularity, the limiting diffusion current density, $j_{L,S}$, is given as

$$j_{L,S} = \frac{nFDC_0}{\delta - h_s} = j_L \frac{\delta}{\delta - h_s}. \quad (1.14)$$

Obviously, this is valid if the protrusion height does not affect the outer limit of the diffusion layer and that a possible lateral diffusion flux supplying the reacting ions can be neglected. At the tip of an irregularity, the lateral flux cannot be neglected and the situation can be approximated by assuming a spherical diffusion current density, $j_{L,tip}$, given by [7]

$$j_{L,tip} = \frac{nFDC^*}{r}, \quad (1.15)$$

where C^* is the concentration of the diffusing species at a distance r from the tip, assuming that around the tip a spherical diffusion layer having a thickness equal to the radius of the protrusion tip is formed [5]. Obviously, if $R > \delta$ the spherical diffusion layer around the tips of protrusion cannot be formed and Eq. (1.16) is valid:

$$j_{L,tip} = \frac{nFDC_0}{\delta - h}. \quad (1.16)$$

If deposition to the macroelectrode is under full diffusion control, the distribution of the concentration C inside the linear diffusion layer is given by [3]

$$C = C_0 \frac{h}{\delta}, \quad (1.17)$$

where $0 \leq h \leq \delta$. Hence,

$$C^* = C_0 \frac{h + r}{\delta} \quad (1.18)$$

and

$$j_{\mathrm{L,tip}} = j_{\mathrm{L}}\left(1 + \frac{h}{r}\right) \tag{1.19}$$

because of Eqs. (1.5), (1.15), and (1.18).

The tip radius of the paraboloidal protrusion is given by [3, 5, 8]

$$r = \frac{R^2}{2h}, \tag{1.20}$$

and substitution of r from Eq. (1.20) in Eq. (1.19) gives

$$j_{\mathrm{L,tip}} = j_{\mathrm{L}}\left(1 + \frac{2h^2}{R}\right) \tag{1.21}$$

or

$$j_{\mathrm{L,tip}} = j_{\mathrm{L}}(1 + 2k^2), \tag{1.22}$$

where

$$k = \frac{h}{R}. \tag{1.23}$$

Hence for a hemispherical protrusion,
If $h = R$, $k = 1$

$$j_{\mathrm{L,tip}} = 3j_{\mathrm{L}}, \tag{1.24}$$

if $h << R$, $k \to 0$

$$j_{\mathrm{L,tip}} \to j_{\mathrm{L}}, \tag{1.25}$$

and if $R << h$, $k \to \infty$

$$j_{\mathrm{L,tip}} \to \infty. \tag{1.26}$$

Substituting $j_{L,tip}$ from Eq. (1.22) instead of j_L in Eq. (1.1) and further rearranging gives

$$j_{tip} = \frac{j_{0,tip}(f_c - f_a)}{1 + \frac{j_{0,\,tip}}{j_L} \cdot \frac{1}{1+2k^2} f_c}, \tag{1.27}$$

if j_0 around the tip is $j_{0,\,tip}$ and if the surface energy term [3, 5] can be neglected. The current density on the tip of a protrusion, j_{tip}, is determined by k, hence by the shape of the protrusion. If $k \to 0$, $j_{tip} \to j$ (see Eq. (1.1)) and if $k \to \infty$, $j_{tip} \to j_{0,\,tip}\,(f_c - f_a) >> j$. The electrochemical process on the tip of a sharp needle-like protrusion can be under pure activation control outside the diffusion layer of the macroelectrode. Inside it, the process on the tip of a protrusion is under mixed control, regardless it is under complete diffusion control on the flat part of the electrode for $k \to 0$. If $k = 1$, hence for hemispherical protrusion, j_{tip} will be somewhat larger than j, but the kind of control will not be changed. It is important to note that the current density to the tip of hemispherical protrusion does not depend on the size of it if $k = 1$. This makes a substantial difference between spherical microelectrodes in bulk solution [9] and microelectrodes inside diffusion layer of the macroelectrode [3]. In the first case the limiting diffusion current density depends strongly on the radius of the microelectrode.

1.2.2 *Physical Illustration*

1.2.2.1 General Observation

Activation-controlled deposition of copper produces large grains with relatively well-defined crystal shapes. This can be explained by the fact that the values of the exchange current densities on different crystal planes are quite different, whereas the reversible potential is approximately the same for all planes [10, 11]. This can lead to preferential growth of some crystal planes, because the rate of deposition depends only on the orientation, which leads to the formation of a large-grained rough deposit. However, even at low

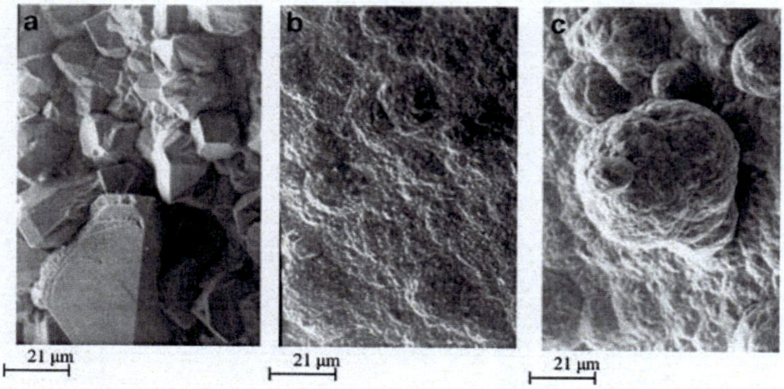

Fig. 1.2 Copper deposits obtained from 0.10 M $CuSO_4$ in 0.50 M H_2SO_4. Quantity of electricity, Q: 20 mAh cm^{-2}. (**a**) Activation-controlled deposition: deposition overpotential, η: 90 mV, initial current density: 3.3 mA cm^{-2}; (**b**) electrodeposition under mixed activation–diffusion control: $\eta = 140$ mV, initial current density: 4.2 mA cm^{-2}, and (**c**) electrodeposition under dominant diffusion control: $\eta = 210$ mV, initial current density 6.5 mA cm^{-2} (Reprinted from [7, 10] with permission from Springer and [12] with permission from Elsevier.)

degrees of diffusion control, the formation of large, well-defined grains is not to be expected, because of irregular growth caused by mass transport limitations. Hence, the current density which corresponds to the very beginning of mixed control (a little larger than this at the end of the Tafel linearity) will be the optimum one for compact metal deposition [12].

All the above facts are illustrated in Fig. 1.2 [12].

1.2.2.2 Cauliflower-Like Forms

It can be seen from Fig. 1.2c that the surface protrusions are globular and cauliflower-like. If the initial electrode surface protrusions are ellipsoidal shape, they can be characterized by the base radius R_0 and the height h as shown in Fig. 1.3a.

The tip radius is then given by

$$r = \frac{R_0^2}{h}. \tag{1.28}$$

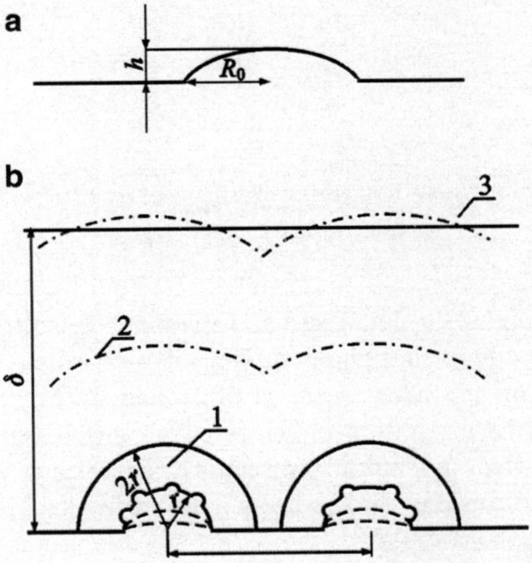

Fig. 1.3 Schematic representation of (**a**) the initial electrode surface protrusion and (**b**) the establishment of spherical diffusion layers around independently growing protrusions. (1) $r < (\delta - h)$ and $r < 1/4\ l$, spherical diffusion zones are formed; (2) $r < (\delta - h)$ and $r > 1/4\ l$, spherical diffusion zones overlap; (3) $r > (\delta - h)$, spherical diffusion zones are not formed (Reprinted from [13] with permission from the Serbian Chemical Society and [7, 10] with permission from Springer.)

The initial electrode surface protrusion is characterized by $h \to 0$ and $r \to \infty$ if $R_0 \neq 0$. In this situation, a spherical diffusion layer cannot be formed around the tip of the protrusion if $r < \delta - h$, and linear diffusion control occurs, leading to an increase in the height of the protrusion relative to the flat surface.

The rate of growth of the tip of a protrusion for $r > \delta$ is equal to the rate of motion of the tip relative to the rate of motion of the flat surface. Hence,

$$\frac{\mathrm{d}h}{\mathrm{d}t} = \frac{V}{nF}\left(j_{\mathrm{L,\,tip}} - j_{\mathrm{L}}\right). \tag{1.29}$$

Substituting $j_{\mathrm{L,tip}}$ from Eq. (1.16) and j_{L} from Eq. (1.5) in Eq. (1.29) and further rearranging gives

$$\frac{\mathrm{d}h}{\mathrm{d}t} = \frac{VDC_0}{\delta} h \tag{1.30}$$

or

$$h = h_0 \exp\left(\frac{VDC_0}{\delta^2} t\right). \tag{1.31}$$

When h increases, r decreases, and spherical diffusion control can be operative around the whole surface of protrusion, if it is sufficiently far from the other ones, as illustrated by Fig. 1.3b. In this situation, second-generation protrusions can grow inside the diffusion layer of first-generation protrusions in the same way as first-generation protrusions grow inside the diffusion layer of the macroelectrode and so on.

A cauliflower-like deposit is formed under such conditions, as is shown in Fig. 1.4. It can be seen from Fig. 1.4a that the distance between the cauliflower-like grains is sufficiently large to permit the formation of spherical diffusion zones around each of them. Simultaneously, second-generation protrusions grow in all directions, as shown in Fig. 1.4b, c. This confirms the assumption that the deposition takes place in a spherically symmetric fashion.

To a first approximation, the rate of propagation can be taken to be practically the same in all directions, meaning that the cauliflower-type deposit formed by spherically symmetric growth inside the diffusion layer of the macroelectrode will be hemispherical, as is illustrated in Fig. 1.4a–c.

This type of protrusion is much larger than that formed by linearly symmetric growth inside the diffusion layer of the macroelectrode (Fig. 1.4a–c).

This is because a spherical diffusion layer cannot be formed around closely packed protrusions, their diffusion fields overlap and they grow in the diffusion layer of the macroelectrode.

If spherical diffusion layer can be established around the tip of a protrusion the limiting diffusion current to the tip is given by Eq. (1.19) or by

$$j_{\mathrm{L,tip}} = j_{\mathrm{L}} \frac{h}{r} \tag{1.32}$$

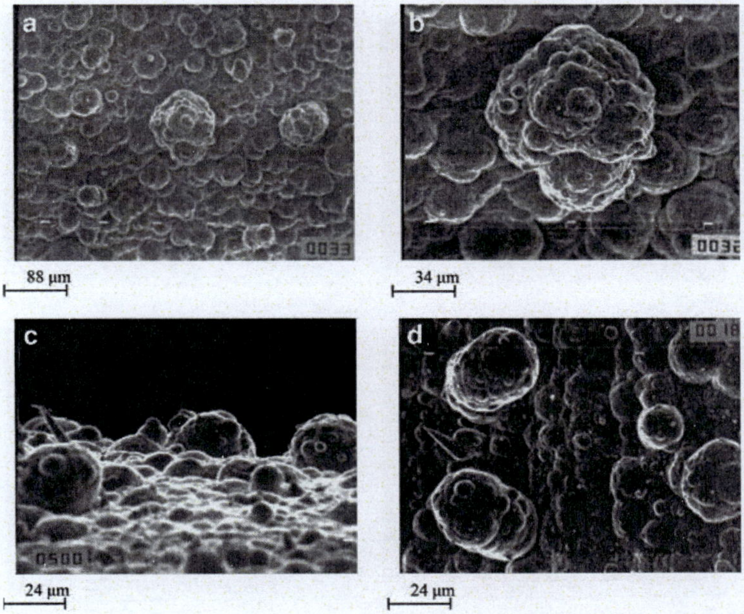

Fig. 1.4 Copper deposits obtained from 0.30 M $CuSO_4$ in 0.50 M H_2SO_4 by electrodeposition under mixed activation–diffusion control. Deposition overpotential: 220 mV (**a**) Quantity of electricity: 40 mAh cm^{-2}; (**b**) The same as in (**a**), and (**c**) and (**d**) quantity of electricity: 20 mAh cm^{-2} (Reprinted from [7, 10] with permission from Springer and [13] with permission from the Serbian Chemical Society.)

for

$$\frac{h}{r} \gg 1. \qquad (1.33)$$

1.2.2.3 Carrot-Like Forms

It can also be seen from Figs. 1.4c, d and 1.5 that the growth of such protrusions produces carrot-like forms, another typical form obtained in copper deposition under mixed activation–diffusion control. This happens under the condition $k << 1$, when spherical diffusion

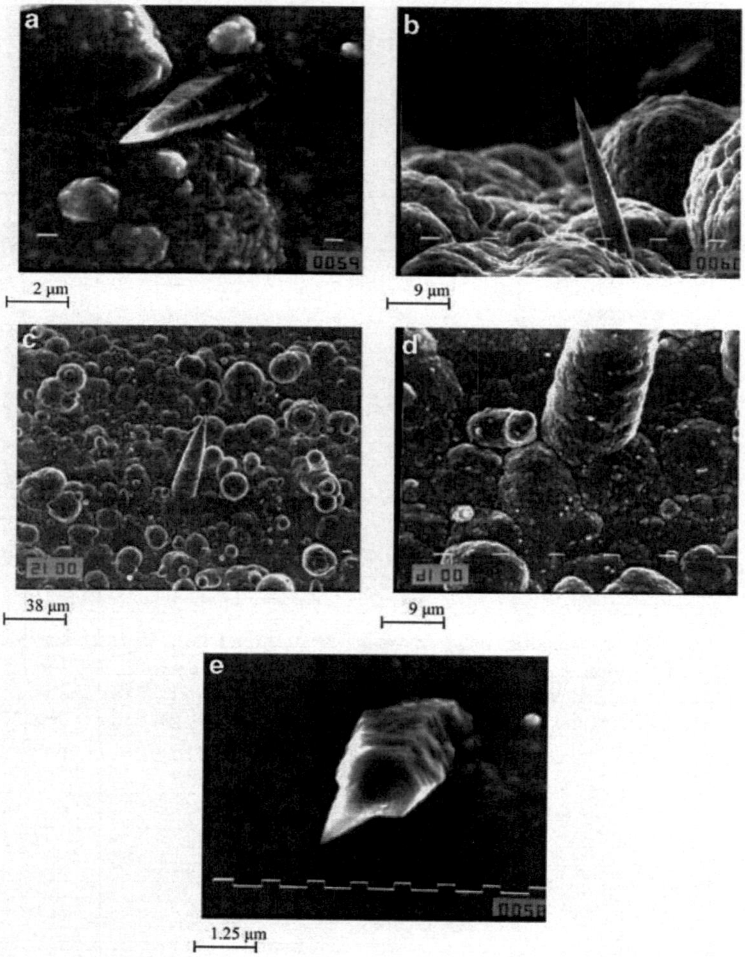

Fig. 1.5 Copper deposits obtained from 0.30 M $CuSO_4$ in 0.50 M H_2SO_4 by electrodeposition under mixed activation–diffusion control. Deposition overpotential: 220 mV. Quantity of electricity (**a**) 10 mAh cm^{-2}; (**b**) 40 mAh cm^{-2}; (**c**) 20 mAh cm^{-2}; (**d**) the root of the carrot from (**c**); and (**e**) 10 mAh cm^{-2} (Reprinted from [7, 10] with permission from Springer and [14] with permission from the Serbian Chemical Society.)

control takes place only around the tip of the protrusion, as is illustrated in Fig. 1.5. In this case, Eq. (1.27) can be rewritten in the form:

$$j_{\mathrm{tip}} = j_{0,\mathrm{tip}}(f_{\mathrm{c}} - f_{\mathrm{a}}), \tag{1.34}$$

meaning that deposition on the protrusion tip can be under pure activation control at overpotentials lower than the critical one for the initiation of dendritic growth.

This happens if the nuclei have a shape like that in Fig. 1.5. The assumption that the protrusion tip grows under activation control is confirmed by the regular crystallographic shape of the tip [14] just as in the case of grains growing on the macroclectrode under activation control (see Fig. 1.2a).

The maximum growth rate at a given overpotential corresponds to activation-controlled deposition. As a result, the propagation rate at the tip will be many times larger than that in other directions, resulting in protrusions like that in Fig. 1.5b. The final form of the carrot-like protrusion is shown in Fig. 1.5c. It can be concluded from the parabolic shape that such protrusions grow as moving paraboloids in accordance with the Barton–Bockris theory [5], the tip radius remaining constant because of the surface energy effect. It can be concluded from Fig. 1.5d that thickening of such a protrusion is under mixed activation–diffusion control because the deposit is seen to be of the same quality as that on the surrounding macroelectrode surface. It can be seen from Fig. 1.5e that activation control takes place only at the very tip of the protrusion.

1.2.3 The Essence of Dendritic Deposits Formation

Two phenomena seem to distinguish dendritic from carrot-like growth [15–17]:

1. A certain well-defined critical overpotential value appears to exist below which dendrites do not grow.
2. Dendrites exhibit a highly ordered structure and grow and branch in well-defined directions. According to Wranglen [18], a dendrite is a skeleton of a monocrystal and consists of a stalk and branches, thereby resembling a tree.

It is known that dendritic growth occurs selectively at three types of growth sites [16]:

1. Dendritic growth occurs at screw dislocations. Sword-like dendrites with pyramidal tips are formed by this process [3, 16].
2. Many investigations of the crystallographic properties of dendrites have reported the existence of twin structures [19–21]. In the twinning process, a so-called indestructible reentrant groove is formed. Repeated one-dimensional nucleation in the groove is sufficient to provide for growth extending in the direction defined by the bisector of the angle between the twin plants [16].
3. It is a particular feature of a hexagonal close-packed lattice that growth along a high-index axis does not lead to the formation of low index planes. Grooves containing planes are perpetuated and so is the chance for extended growth by the one-dimensional nucleation mechanism [22].

In all the above cases, the adatoms are incorporated into the lattice by repeated one-dimensional nucleation. On the other hand, deposition to the tip of screw dislocations can be theoretically considered as deposition to a point; in the other two cases, the deposition is to a line.

From the electrochemical point of view, a dendrite can be defined as an electrode surface protrusion that grows under activation or mixed control, while deposition to the flat part of the electrode surface is under complete diffusion control [3, 4, 8, 15].

Considering the model of surface irregularities shown in Fig. 1.1, the surface irregularities are buried deep in the diffusion layer, which is characterized by a steady linear diffusion to the flat portion of completely active surface.

If the protrusion does not affect the outer limit of the diffusion layer, i.e., if $\delta \gg h$, the limiting diffusion current density to the tip of the protrusion from Fig. 1.1, $j_{\mathrm{L,tip}}$, is given by

$$j_{\mathrm{L,tip}} = j_{\mathrm{L}}\left(1 + \frac{h}{r}\right). \tag{1.19}$$

Substitution of $j_{L,tip}$ from Eq. (1.19) into Eq. (1.1) produces for $h/r >> 1$:

$$j_{tip} = j_{0,tip}(f_c - f_a), \quad (1.34)$$

where $j_{0,tip}$ is the exchange current density at the tip of a protrusion.

Obviously, deposition to the tip of such protrusion inside the diffusion layer is activation-controlled relative to the surrounding electrolyte, but it is under mixed activation–diffusion control relative to the bulk solution.

If deposition to the flat part of electrode is a diffusion-controlled process and assuming a linear concentration distribution inside diffusion layer, the concentration C_{tip} at the tip of a protrusion can be given by modified Eq. (1.17) [3]

$$C_{tip} = C_0 \frac{h}{\delta}. \quad (1.17a)$$

According to Newman [23] the exchange current density at the tip of a protrusion is given by

$$j_{0,tip} = j_0 \left(\frac{C_{tip}}{C_0} \right)^{\xi}, \quad (1.35)$$

where

$$\xi = \frac{d \log j_0}{d \log C_0} \quad (1.36)$$

and j_0 is the exchange current density for a surface concentration C_0 equal to that in the bulk,

or

$$j_{0,tip} = j_0 \left(\frac{h}{\delta} \right)^{\xi} \quad (1.37)$$

because of Eq. (1.17a).

Taking into account Eq. (1.34), the current density to the tip of a protrusion is then given by

$$j_{\rm tip} = j_0 \left(\frac{h}{\delta}\right)^{\xi} (f_{\rm c} - f_{\rm a}) \tag{1.38}$$

being under mixed control due to the $(h/\delta)^{\xi}$ term, which takes into account the concentration dependence of $j_{0,\rm tip}$, expressing in this way a mixed-controlled electrodeposition process.

Outside the diffusion layer $h \geq \delta$, Eq. (1.38) becomes

$$j_{\rm tip} = j_0 (f_{\rm c} - f_{\rm a}), \tag{1.39}$$

indicating pure activation control, as the $(h/\delta)^{\xi}$ term is absent.

For the dendrite growth, the current density to the tip of a protrusion formed on the flat part of the electrode surface growing inside the diffusion layer should be larger than the corresponding limiting diffusion current density [24]. Hence,

$$j_{\rm L} < j_{\rm tip}, \tag{1.40}$$

the protrusion grows as a dendrite.

In accordance with Eq. (1.40), instantaneous dendrite growth is possible at overpotentials larger than some critical value, $\eta_{\rm c}$, which can be derived from Eq. (1.38) as shown in [15, 17]

$$\eta_{\rm c} = \frac{b_{\rm c}}{2.3} \ln \frac{j_{\rm L}}{j_0} \left(\frac{\delta}{h}\right)^{\xi}, \tag{1.41}$$

and minimum overpotential at which dendritic growth is still possible, $\eta_{\rm i}$ is given by

$$\eta_{\rm i} = \frac{b_{\rm c}}{2.3} \ln \frac{j_{\rm L}}{j_0} \tag{1.42}$$

for $f_{\rm c} \gg f_{\rm a}$, where h and δ are the protrusion height and the diffusion layer thickness, respectively. For very fast processes, when $j_0/j_{\rm L} \gg 1$, and if $f_{\rm c} \approx f_{\rm a}$ but $f_{\rm c} > f_{\rm a}$, Eq. (1.41) becomes

$$\eta_{\rm c} = \frac{RT}{nF} \frac{j_{\rm L}}{j_0} \left(\frac{\delta}{h}\right)^{\xi} \tag{1.43}$$

and Eq. (1.42)

$$\eta_{\rm i} = \frac{\rm RT}{nF}\frac{j_{\rm L}}{j_0}, \tag{1.44}$$

meaning that in the case of ohmic-controlled reactions, dendritic growth can be expected at very low overpotentials, or better to say, if $j_0 \to \infty$, instantaneous dendritic growth is possible at all overpotentials if only mass transfer limitations are taken into consideration.

In fact, dendrite propagation under such conditions is under diffusion and surface energy control, and $\eta_{\rm c}$ is then given by [5, 24]

$$\eta_{\rm c} = \frac{8\sigma V}{nFh} \tag{1.45}$$

where σ is the interfacial energy between metal and solution and V is the molar volume of the metal, and minimum overpotential at which dendritic growth is still possible, $\eta_{\rm i}$ is given by

$$\eta_{\rm i} = \frac{8\sigma V}{nF\delta}. \tag{1.46}$$

Hence, a critical overpotential for initiation dendritic growth is also expected in such cases, being of the order of few millivolts [15, 17, 24].

1.3 Polarization Curves

1.3.1 The Polarization Curve Equation for Partially Covered Inert Electrode

A mathematical model can be derived under the assumption that the electrochemical process on the microelectrodes inside the diffusion layer of a partially covered inert macroelectrode is under activation control, despite the overall rate being controlled by the diffusion layer of the macroelectrode [6, 25]. The process on the microelectrodes decreases the concentration of the electrochemically active ions on the surfaces of the microelectrodes inside the diffusion layer of the macroelectrode, and the zones of decreased concentration around them overlap, giving way to linear mass transfer to an effectively planar surface [26]. Assuming that the surface concentration is the same on the total area of the electrode surface, under steady-state conditions, the current density on the whole electrode surface, j, is given by

$$j = \frac{nFD(C_0 - C_S)}{\delta}, \tag{1.47}$$

where n is the number of transferred electrons, F is the Faraday constant, and D is the diffusion coefficient of the reacting ion. Obviously, the current density from Eq. (1.47) is due to the difference in the bulk, C_0, and surface concentration, C_S, of the reactive ion. The concentration dependence of the exchange current density [23] is expressed as

$$j_{0,S} = \left(\frac{C_S}{C_0}\right)^{\xi} j_0, \tag{1.48}$$

where $j_{0,S}$ is the exchange current density for a surface concentration C_S.

The current density on the macroelectrode can also be written as

$$j = S_w j_0 \left(\frac{C_S}{C_0}\right)^{\xi} (f_c - f_a), \tag{1.49}$$

assuming a reversible activation-controlled electrode process on the hemispherical active microelectrodes on an inert substrate, where S_w is the active surface per square centimeter of the macroelectrode, and j_0 is the exchange current density on the massive active electrode, standardized to the apparent electrode surface.

The current densities given by Eqs. (1.47) and (1.49) are mutually equal and substitution of C_S/C_0 from Eq. (1.47) into Eq. (1.49) gives

$$j = S_w j_0 \left(1 - \frac{j}{j_L}\right)^{\xi} (f_c - f_a). \tag{1.50}$$

Different forms of polarization curve equation were discussed in detail [25] and this form was chosen for digital simulation. The use of any other form of the polarization curve equation will give some similar results.

1.3.2 Calculated Polarization Curves When Ohmic Potential Drop Is Not Included

The shape of polarization curves can be estimated by digital simulation [6]. It will be performed for example for one-electron transfer process and $\beta = 0.5$ and $\xi = 0.5$. In all cases the apparent current density is standardized to the apparent surface of modified electrode.

Using Eq. (1.50) with $\xi = 0.5$ and $j_0/j_L = 100$, 1, and 0.01, $S_w = 0.05$, 0.1, 0.25, 0.5, 0.75, and 1, and $f_c = 10^{\frac{\eta}{120}}$ and $f_a = 10^{-\frac{\eta}{120}}$, the diagrams presented in Fig. 1.6 are obtained. The current density–overpotential dependence above each set of polarization curves corresponds to $S_w = 1$. It follows from Fig. 1.6 that for large values of $j_{0,\text{eff}}/j_L$, electrochemical polarization can probably be neglected and that complete ohmic control of the deposition process can be expected, for $j_{0,\text{eff}}/j_L \geq 100$ up to a current density about 0.95 j_L and for $j_{0,\text{eff}}/j_L = 0.5$ for current densities lower than 0.3 j_L.

As told earlier, the shape of polarization curves does not depend strongly on S_w at large j_0/j_L ratios. At lower ones the important effect arises.

1.3.3 Calculated Polarization Curves with Included Ohmic Potential Drop

The polarization curves for the electrodeposition process which include the ohmic voltage drop can be obtained as follows, assuming $S_w = 1$ in all cases [25, 27]. This will be performed for a one-electron transfer process and $\beta = 0.5$, meaning $\xi = 0.5$ [6].

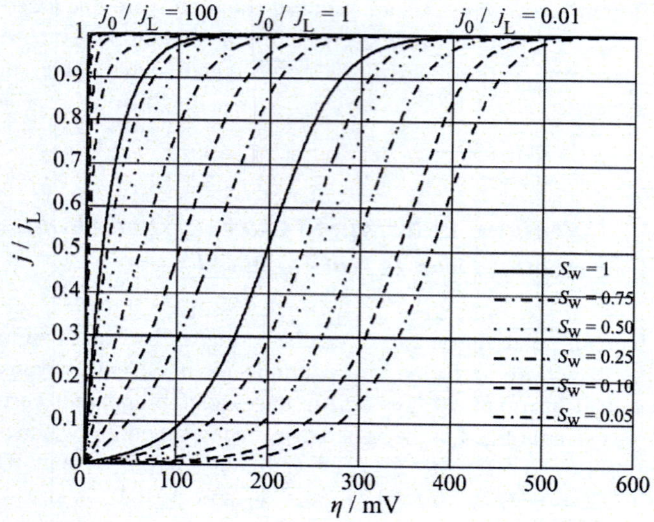

Fig. 1.6 Dependences $j/j_L - \eta$ calculated from Eq. (1.50), using $j_0/j_L = 100$, 1, and 0.01, $S_w = 0.05$, 0.1, 0.25, 0.5, 0.75, and 1, $f_c = 10^{\frac{\eta}{120}}$, $f_a = 10^{-\frac{\eta}{120}}$, and $\xi = 0.5$ (Reprinted from [6] with permission from Elsevier and [25] with permission from Springer.)

Using Eq. (1.50) with $\xi = 0.5$ and $j_0/j_L = 100$, 10, 1, and 0.01, $f_c = 10^{\frac{\eta}{120}}$, $f_a = 10^{-\frac{\eta}{120}}$, and $j_L = 50$ mA cm^{-2}, the dependences presented by the dashed line in Figs. 1.7 and 1.8 are obtained. The ohmic potential drop is not included in the calculated polarization curves depicted in Figs. 1.7 and 1.8 by the dashed line. It follows from Figs. 1.7 and 1.8 that for large values of j_0/j_L, electrochemical polarization can probably be neglected but mass transfer limitations are present in all cases, which can also be shown by differentiation of Eq. (1.1).

On the other hand, the measured value of overpotential, η_m, is given by

$$\eta_m = \eta + j\frac{L}{\kappa} \tag{1.51}$$

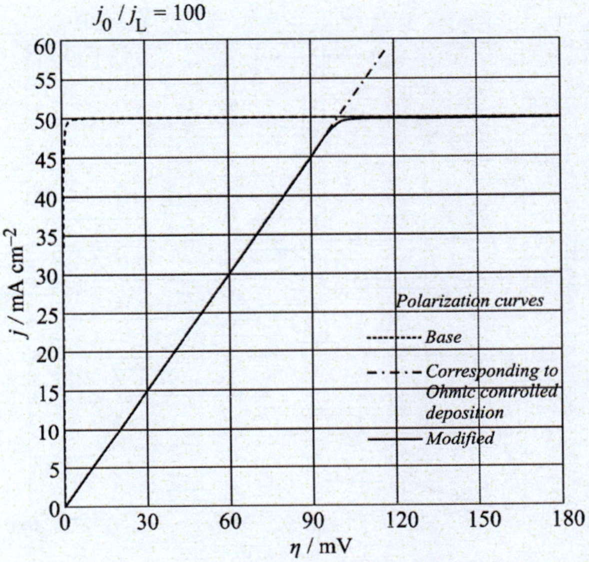

Fig. 1.7 The dependence $j - \eta$ calculated using Eq. (1.50), $j_0/j_L = 100$, $f_c = 10^{\frac{\eta}{120}}$, $f_a = 10^{-\frac{\eta}{120}}$, $\xi = 0.5$, $S_w = 1$, and $j_L = 50$ mA cm^{-2}, and one modified using Eq. (1.51), $L = 0.2$ cm, $\kappa = 0.1$ S cm^{-1} (Reprinted from [27] with permission from Elsevier and [25] with permission from Springer.)

due to the *IR* error [28], where L is the length of the electrolyte column between the tip of a liquid capillary and the electrode and κ is the specific conductivity of the electrolyte.

For a 1 M solution of a typical fully dissociated electrolyte, the value of κ is around 0.1 S cm^{-1}, L can be taken as 0.2 cm and $j_L = 50$ mA cm^{-2}. Using these given values, as well as $\kappa = 0.033$ S cm^{-1}, Eq. (1.51), and the diagrams presented in Figs. 1.7 and 1.8 by the dashed line, polarization curves including the ohmic potential drop can be obtained, as shown in Figs. 1.7 and 1.8 by the solid line.

In the case under consideration, complete ohmic control of the deposition process can be expected for $j_0/j_L \geq 100$ up to a current density about $0.95j_L$ (Fig. 1.7). It is obvious from Figs. 1.7 and 1.8 that, regardless of the shape of the polarization curve which depends

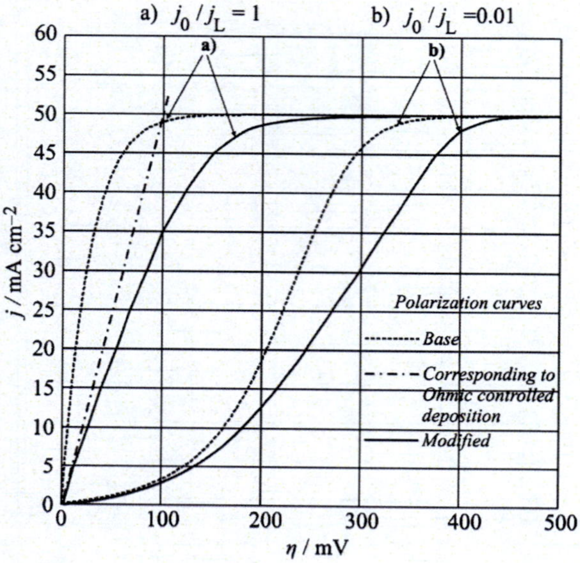

Fig. 1.8 The dependences $j - \eta$ calculated using Eq. (1.50), $j_0/j_L = 1$ and 0.01, $f_c = 10^{\frac{\eta}{120}}$, $f_a = 10^{-\frac{\eta}{120}}$, $\xi = 0.5$, $S_w = 1$, and $j_L = 50$ mA cm^{-2}, and ones modified using Eq. (1.51), $L = 0.2$ cm, $\kappa = 0.1$ S cm^{-1} (Reprinted from [27] with permission from Elsevier and [25] with permission from Springer.)

on the j_0/j_L ratio and κ, a limiting diffusion current density plateau is present in all cases.

It can be noticed that before the increase of the current density, over the value of the limiting diffusion one, the first part of the polarization curve for silver deposition from nitrate solution [6] has practically the same shape as that from Fig. 1.7 and that those from Fig. 1.8 are very similar to the ones for Cd and Cu deposition [29]. The value of j_0 for Ag deposition is very large [30]. In the cases of both Cd [31] and Cu [32] deposition, j_0 is considerably lower than in the case of Ag deposition.

The increase in the current density over the limiting diffusion current in the absence of some other electrochemical process indicates a decrease of the mass transport limitations, due to initiation of growth of dendrites and further dendritic growth.

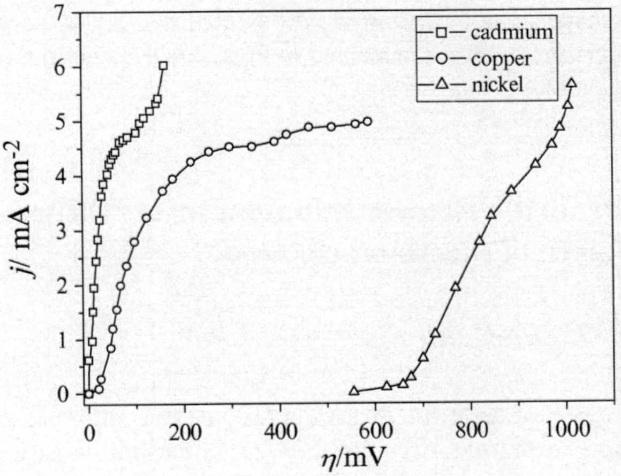

Fig. 1.9 Polarization curve for cadmium (*open square*), copper (*open circle*) and nickel (*open triangle*) depositions (Reprinted from [33] with permission from the Serbian Chemical Society.)

Table 1.1 The exchange current density and j_L/j_0 ratios for Cd, Cu, and Ni deposition processes

Metal	j_0 (A cm^{-2})	j_L/j_0
Cadmium	1.5×10^{-3}	3.0
Copper	3.2×10^{-4}	14.4
Nickel	1.6×10^{-9}	2.9×10^{6}

Reprinted from [33] with permission from Serbian Chemical Society and [15] with permission from Springer

1.3.4 Polarization Curves Measured for Different j_0/j_L Ratios

The polarization curves for nickel, copper, and cadmium deposition are shown in Fig. 1.9, while corresponding Tafel plots and the results of linear polarization experiments are given in [33]. The limiting diffusion currents in all cases are practically the same, but the exchange current densities (given in Table 1.1) are very different.

The shape of polarization curves is qualitatively in accordance with polarization curves presented in Figs. 1.6–1.8 due to the different j_0/j_L ratios.

1.4 Dendritic Growth Initiation Inside Diffusion Layer of the Macroelectrode

1.4.1 Ohmic-Controlled Deposition

The initiation of dendritic growth is followed by an increase of the deposition current density, and the overall current density will be larger than the limiting diffusion current on a flat active electrode. Based on the above discussion, the polarization curve equation in the ohmic-controlled electrodeposition of metals can be determined now by [27]

$$j = \frac{\kappa\eta}{L} \quad \text{for} \quad 0 \leq \eta < j_L \frac{L}{\kappa}, \tag{1.52a}$$

$$j = j_L \quad \text{for} \quad j_L \frac{L}{\kappa} \leq \eta < \eta_c + j_L \frac{L}{\kappa}, \tag{1.52b}$$

$$j = j_L\theta + (1-\theta)j_0 \frac{(f_c - f_a)}{N} \sum_{i=1}^{i=N} \left(\frac{h_i}{\delta}\right)^{\xi}, \quad \text{for} \quad \eta_c \leq \eta, \tag{1.52c}$$

where $N = N(t)$ is the number of dendrites and $\theta = \theta(t) \leq 1$, where θ is the flat part of the electrode surface.

Equation (1.52a) describes the linear part of the polarization curves for tin [34], silver [6] and lead [35] deposition, and Eq. (1.52b) foresees the inflection point in the cases when η_c is low and the resistance of the electrolyte is large. Finally, Eq. (1.52c) describes the part of the polarization curve after initiation of dendrite growth.

The exchange current density of the silver reaction in nitrate electrolytes is sufficiently large to permit ohmic-controlled deposition, as well as dendritic growth at low overpotentials [30].

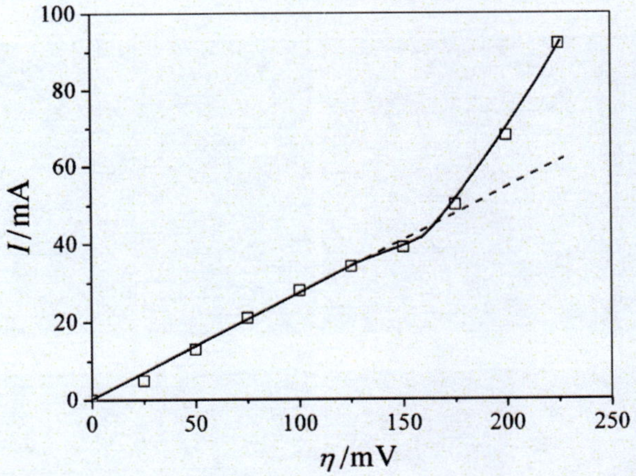

Fig. 1.10 The polarization curve for silver electrodeposition from nitrate solution on a graphite electrode (Reprinted from [27] with permission from Elsevier and [36] with permission from Springer.)

After a linear increase of the deposition current density with increasing overpotential, an exponential increase after the inflection point appears, meaning the elimination of mass transfer limitations due to the initiation of dendritic growth. Thus, instead of a limiting diffusion current density plateau, a curve inflection point or a short inclined plateau can be expected on the polarization curve in ohmic-controlled electrodeposition of metals, as observed.

The polarization curve for silver electrodeposition from nitrate solution, 0.50 M $AgNO_3$ in 0.20 M HNO_3, onto a graphite electrode is shown in Fig. 1.10. The overpotential was increased from the initial to the final value and held for 30 s before measurement in all cases during the polarization measurements. The polarization curve in Fig. 1.10 means that mass transfer limitations were decreased or even eliminated. The SEM photomicrographs of the deposit corresponding to the points from Fig. 1.10 are shown in Fig. 1.11.

It can be seen from Figs. 1.10 and 1.11a that at an overpotential of 100 mV, only grains [27] can be seen, which means that the deposition was not under diffusion control. It follows from Figs. 1.10 and 1.11b that deposition at an overpotential of 125 mV is still out of

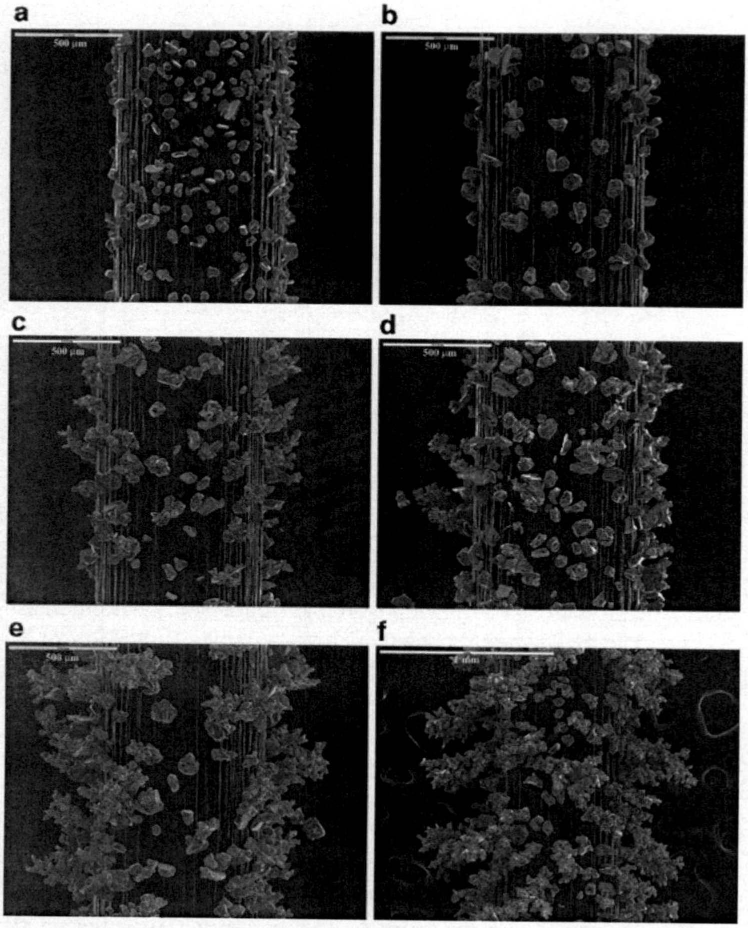

Fig. 1.11 The SEM photomicrographs of the silver deposit obtained on a graphite electrode obtained after the recording of the current at different overpotentials in polarization measurements (**a**) 100 mV; (**b**) 125 mV; (**c**) 150 mV; (**d**) 175 mV; (**e**) 200 mV; and (**f**) 225 mV (Reprinted from [27] with permission from Elsevier and [36] with permission from Springer.)

diffusion control. At 150 mV, the current density is somewhat lower than that which could be expected from the linear dependence of current on overpotential. This indicates the initiation of diffusion control of the deposition process, but also the initiation of dendrite growth, which compensates the mass transfer limitations, as can be seen from Figs. 1.10 and 1.11c. The point corresponding to an overpotential of 150 mV can be considered as the inflection point of the polarization curve in Fig. 1.10.

At overpotentials larger than 175 mV, the current density is considerably larger than the one expected from the linear dependence of current on overpotential. The formation of dendritic deposits (Fig. 1.11d–f) confirms that the deposition was dominantly under activation control. Thus, the elimination of mass transport limitations in the ohmic-controlled electrodeposition of metals is due to the initiation of dendritic growth at overpotentials close to that at which complete diffusion control of the process on the flat part of the electrode surface occurs.

It is necessary to note that the silver deposits shown in Fig. 1.11d–f are not similar to ideal silver dendrites [18], but they behave as dendritic ones in regard to their electrochemical properties. Hence, they can be considered as degenerate dendritic deposits.

Occasionally, the needle-like dendrites can also be formed.

1.4.2 Ohmic-Diffusion and Activation–Diffusion Controlled Deposition

In these cases the dendritic growth starts at overpotentials larger than the one which corresponds to the beginning of the limiting diffusion current density plateau [15, 17].

There is an induction period before the initiation of dendritic growth [5]. During this induction period, dendrite precursors are formed and become sufficiently high to satisfy Eq. (1.41) at a given overpotential, as illustrated in Figs. 1.12 and 1.13. The cross-like grains seen in Fig. 1.12a, b further develop into dendrite precursors (Fig. 1.12a, c).

The propagation of this structure by branching (Fig. 1.12d) produces dendrites as shown in Fig. 1.12e.

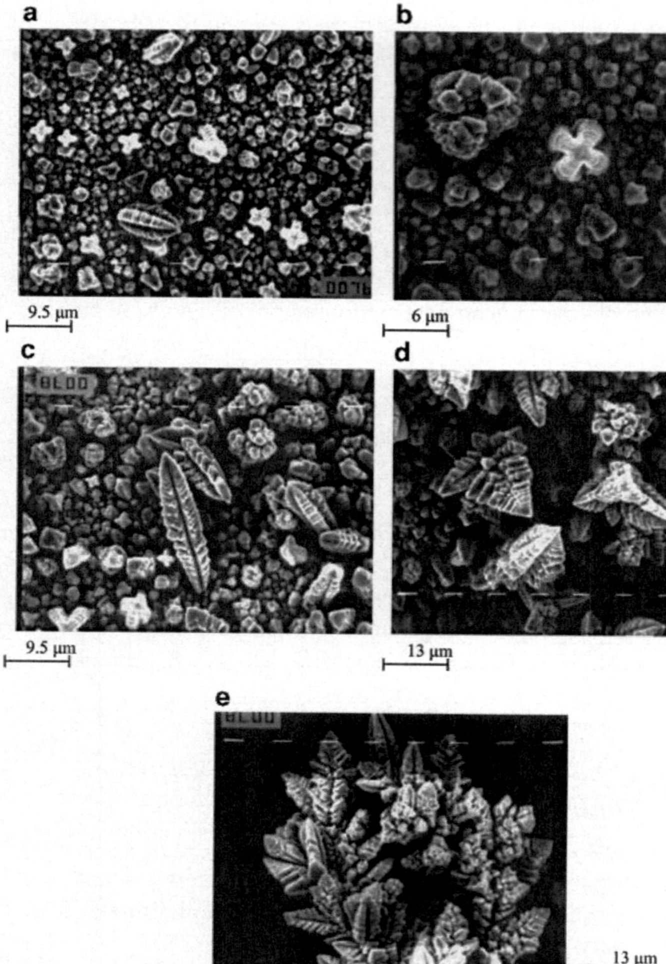

Fig. 1.12 SEM micrographs of copper deposits obtained by deposition from 0.30 M $CuSO_4$ in 0.50 M H_2SO_4 onto a copper wire electrode. Deposition overpotential: 550 mV. Quantity of electricity: (**a**) 2 mAh cm^{-2}; (**b**) 2 mAh cm^{-2}; (**c**) 5 mAh cm^{-2}; (**d**) 10 mAh cm^{-2}, and (**e**) 10 mAh cm^{-2} (Reprinted from [37] with permission from the Serbian Chemical Society and [15, 17] with permission from Springer.)

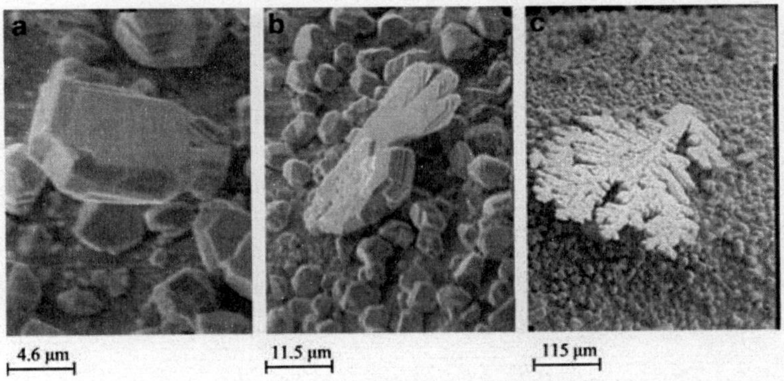

Fig. 1.13 The cadmium deposits obtained by deposition from 0.10 M $CdSO_4$ in 0.50 M H_2SO_4 onto a cadmium electrode. Deposition overpotential: 50 mV. Deposition times (**a**) 2 min; (**b**) 2 min, and (**c**) 10 min (Reprinted from [38] with permission from Elsevier and [39] with permission from Springer.)

On the other hand, lighter-like precursors from Fig. 1.13a develop in 2D dendrites, as shown in Fig. 1.13c.

The initiation of dendritic growth is followed by a change in the slope of the current density–time curves [15–17], indicating a change in the growth mechanism of the deposit.

The slopes of these dependences are similar to one another and independent of the deposition overpotential during the nondendritic amplification of the surface-coarsening according to Eq. (1.31).

The change of the slope of the current–time dependences due to the dendritic growth initiation will be treated here in somewhat simplified way.

The limiting diffusion current density to the elevated points of a surface protrusion, $j_{L,e}$, is given by

$$j_{L,e} = \frac{nFDC_0}{\delta - h} \tag{1.53}$$

and

$$j_L = \frac{nFDC_0}{\delta} \tag{1.5}$$

to the flat part of the electrode.

The limiting diffusion current density will then be given by

$$j = \theta \frac{nFDC_0}{\delta} + \frac{1-\theta}{N} \sum_{i=1}^{i=N} \frac{nFDC_0}{\delta - h_i}, \tag{1.54}$$

where θ is the flat part of electrode surface, N is the number of elevated points on the electrode surface, and h_i changes with time according to Eq. (1.55):

$$h_i = h_{0,i} \exp\left(\frac{VDC_0}{\delta^2} t\right), \tag{1.55}$$

which is somewhat modified Eq. (1.31).

It is obvious that $\mathrm{d}j_L/\mathrm{d}t$ does not depend on overpotential.

After initiation of dendritic growth, the slopes become dependent on the overpotential. A dendrite is a surface protrusion growing under mixed or activation control, while deposition to the flat part of the electrode surface is under complete diffusion control. The overpotential η and current density j_{tip} on the tip of a dendrite are related by

$$j_{tip} = j_0 \frac{h}{\delta} f_c \tag{1.56}$$

for $f_c >> f_a$, and Eq. (1.52c) can be rewritten in the form

$$j = j_L \theta + (1-\theta) j_0 \frac{f_c}{N} \sum_{i=1}^{i=N} \frac{h_i}{\delta}, \tag{1.57}$$

and $\frac{\mathrm{d}j}{\mathrm{d}t}$ in this case depends on overpotential.

Hence, the maximum overpotential at which the slope of the apparent current density–time dependence remains constant and equal to that in nondendritic amplification of the surface roughness corresponds to η_i. The minimum overpotential at which this slope cannot be recorded corresponds to η_c.

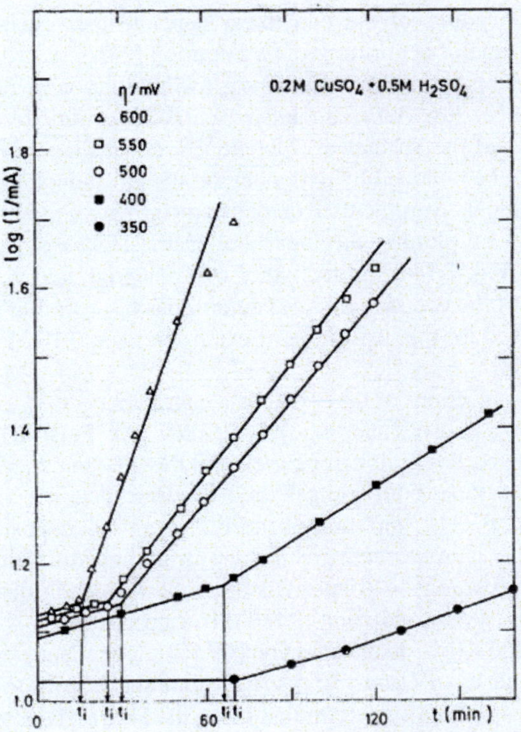

Fig. 1.14 log *I* as a function of time for copper deposition (Reprinted from [15, 24, 40] with permission from Springer.)

In this way η_i and η_c can be estimated. It is known that the *j*–*t* dependences are different from case to case owing to different mechanisms of dendritic growth initiation and dendritic growth [15]. As a result of this, the analytical approach to the determination of η_i and η_c must be specific for each system under consideration; the procedure for one particular case is as follows.

Typical log (current)–time dependences obtained for copper deposition from 0.20 M $CuSO_4$ in 0.50 M H_2SO_4 at overpotentials belonging to the limiting diffusion current plateau are shown in Fig. 1.14. According to the above discussion, it is clear that the intersection points of the two linear dependencies determines the induction time of dendritic growth initiation [24].

The induction times for dendritic growth initiation extracted from the graphs in Fig. 1.14 can be presented as a function of overpotential, and the critical overpotential for instantaneous dendritic growth can be obtained by extrapolation to zero induction time.

The critical overpotential of dendritic growth initiation can be determined by plotting the logarithm of the slopes of the straight lines from Fig. 1.14 as a function of overpotential, and the intersection point of the two straight lines determines η_i. A similar procedure was followed for the deposition of cadmium from 0.10 M $CdSO_4$ in 0.50 M H_2SO_4.

The cross sections of the copper and cadmium deposits obtained at $\eta < \eta_i$, $\eta_i < \eta < \eta_c$, and $\eta > \eta_c$ are shown in Figs. 1.15a–c and 1.16a–c, respectively. It can be seen that there is no dendrite formation when $\eta < \eta_i$, both compact and dendritic deposits are formed when $\eta_i < \eta < \eta_c$ and only dendritic metal is deposited when $\eta > \eta_c$. This is in perfect agreement with findings of Calusaru [41] for the morphology of deposits of the same metals deposited at overpotentials corresponding to full diffusion control.

The η_i and η_c of 260 mV and 660 mV for copper deposition (lower j_0 value) and 27 mV and 110 mV for cadmium deposition (larger j_0 value) are successfully determined using the above given procedure, being in perfect agreement with experimental findings as can be seen from Figs. 1.15 and 1.16 [24, 29, 40].

The shapes of the polarization curves presented in Figs. 1.15 and 1.16 are in accordance with values of the exchange current density to the limiting diffusion current density ratios.

It is known [40] that, apart from decreasing the concentration of the depositing ion, the formation of a dendritic deposit can also be enhanced by increasing the concentration of the supporting electrolyte, increasing the viscosity of the solution, decreasing the temperature, and decreasing the velocity of motion of the solution. Practically, all the above facts can be explained by Eqs. (1.42) and (1.44), assuming that a decrease in η_i means enhanced dendrite formation because of the lower electrical work required to produce the dendrites. The possibility of obtaining dendrites of Pb [42] and Sn [43] from aqueous solutions at lower overpotentials than

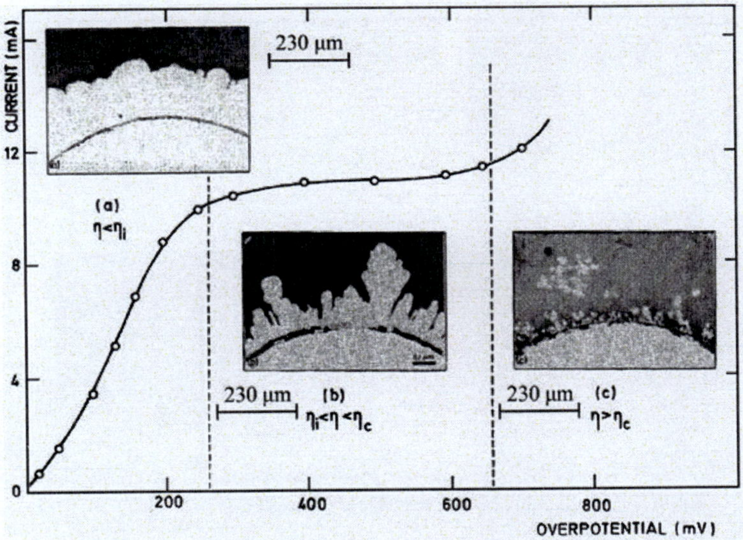

Fig. 1.15 Polarization curve for the potentiostatic deposition of copper from 0.20 M $CuSO_4$ in 0.50 M H_2SO_4 and the cross sections of copper deposits obtained on copper wire electrodes previously plated with nickel (**a**) overpotential: 200 mV, deposition time: 6 h; (**b**) overpotential: 300 mV, deposition time: 5 h, and (**c**) overpotential: 700 mV, deposition time: 2 min (Reprinted from [24, 29, 40] with permission from Springer and copied by permission from the "Electrochemistry Encyclopedia" (http://electrochem.cwru.edu/ed/encycl/) on 04/25/2007. The original material is subject to periodical changes and updates.)

required for the formation of dendrites of Ag from aqueous solutions can also be explained by Eq. (1.46) owing to the much lower melting points of these metals, i.e., their lower surface energy at room temperature. Dendrites of silver can be obtained from molten salts at overpotentials of a few millivolts [5], as in the case of Pb and Sn deposition from aqueous solutions [42, 43], because the difference between the melting point of silver and the working temperature for deposition from molten salts is not very different from the difference between the melting point of lead or tin and room temperature. On the other hand, dendrites grow from screw

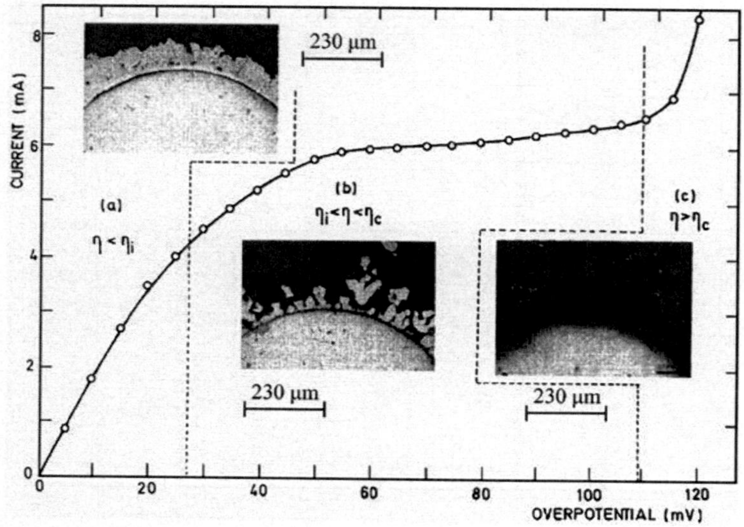

Fig. 1.16 Polarization curve for the potentiostatic deposition of cadmium from 0.10 M $CdSO_4$ in 0.50 M H_2SO_4 and the cross sections of cadmium deposits obtained on copper wire electrode (**a**) overpotential: 20 mV, deposition time: 8 h; (**b**) overpotential: 40 mV, deposition time: 2 h, and (**c**) overpotential: 120 mV, deposition time: 9 min (Reprinted from [24, 29, 40] with permission from Springer.)

dislocation and nuclei of higher indices or twinned ones only [15–17]. The probability of formation of such nuclei increases with increasing overpotential [44], and η_i can also be defined as the overpotential at which they are formed. Regardless of this, Eqs. (1.42), (1.44), and (1.46) illustrate well the effect of different parameters on the initiation of dendritic growth.

It is obvious that the electrochemical conditions, as well as the crystallographic ones, under which dendritic deposits are formed can be precisely determined. One problem that still seems to remain unsolved is the question what causes the dendrite precursors to appear at regularly spaced locations along the dendrite stem. Further investigations in this direction are necessary.

1.5 Inhibition of Dendritic Growth by Vigorous Hydrogen Codeposition (Formation of the Honeycomb-Like Structures)

Honeycomb-like structures are formed by electrochemical deposition processes at high current densities and overpotentials, where parallel to copper electrodeposition hydrogen evolution reaction occurs. Hydrogen evolution responsible for the formation of this type of structure is vigorous enough to cause such stirring of the solution leading to the change of the hydrodynamic conditions in the near-electrode layer [45].

Figure 1.17 shows the polarization curve for electrodeposition of copper from 0.15 M $CuSO_4$ in 0.50 M H_2SO_4. The plateau of the limiting diffusion current density corresponds to the range of overpotentials between 350 and 750 mV. Hydrogen evolution, as the second reaction, commences at some overpotential belonging to the plateau of the limiting diffusion current density, and increasing overpotential intensifies hydrogen evolution reaction. For this copper solution, hydrogen evolution commences at an overpotential of 680 mV [45]. At some overpotential outside the plateau of the

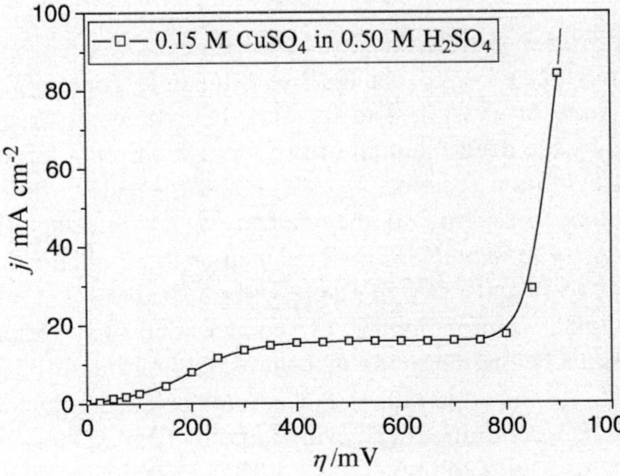

Fig. 1.17 Polarization curve for the cathodic process of copper deposition from 0.15 M $CuSO_4$ in 0.50 M H_2SO_4. Temperature: 18.0 ± 1.0°C (Reprinted from [45] with permission from Elsevier and [46] with permission from Springer.)

Table 1.2 The average current efficiency of hydrogen evolution, $\eta_{I,av}(H_2)$ (%), obtained at overpotentials of 550, 700, 800, and 1,000 mV

Overpotential, η (mV)	550	700	800	1,000
The average current efficiency of hydrogen evolution, $\eta_{I,av}(H_2)$ (%)	0	1.97	10.8	30.0

Reprinted from [45] with permission from Elsevier and [46] with permission from Springer

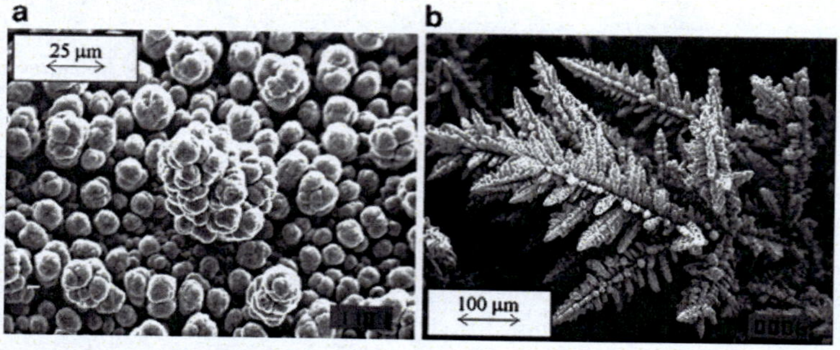

Fig. 1.18 Copper deposits electrodeposited from 0.15 M $CuSO_4$ in 0.50 M H_2SO_4 at overpotentials of (**a**) 550 mV and (**b**) 700 mV. The quantity of electricity: 10 mAh cm^{-2} (Reprinted from [45, 47] with permission from Elsevier and [46] with permission from Springer.)

limiting diffusion current density, hydrogen evolution becomes vigorous enough leading to change hydrodynamic conditions in the near-electrode layer [45]. The quantity of evolved hydrogen was quantified by the determination of the average current efficiency for hydrogen evolution reaction, $\eta_{I,av}(H_2)$, and the values obtained at overpotentials belonging to the plateau of the limiting diffusion current density (550 and 700 mV), as well as those obtained outside this plateau (800 and 1,000 mV), are given in Table 1.2.

The typical morphologies of copper deposits obtained at overpotentials belonging to the plateau of the limiting diffusion current density are shown in Fig. 1.18. Cauliflower-like agglomerates of copper grains were formed at an overpotential of 550 mV, where there was no hydrogen evolution (Fig. 1.18a). Very branchy copper dendrites were formed at an overpotential of 700 mV, where hydrogen evolution was very small, and corresponded to the average current efficiency of hydrogen evolution of about 2.0% (Fig. 1.18b) [45].

Fig. 1.19 Copper deposits electrodeposited from 0.15 M $CuSO_4$ in 0.50 M H_2SO_4 at overpotentials of (**a**) and (**b**) 800 mV and (**c**) and (**d**) 1,000 mV. The quantity of electricity: 10 mAh cm^{-2} (Reprinted from [45, 47] with permission from Elsevier and [46] with permission from Springer.)

Morphologies of copper deposits obtained at overpotentials of 800 and 1,000 mV which were about 50 and 250 mV outside the plateau of the limiting diffusion current density are shown in Fig. 1.19. In both cases holes formed by attached hydrogen bubbles (Fig. 1.19a, c) surrounded by agglomerates of copper grains (Fig. 1.19b, d) were obtained. It is necessary to note that the number of holes formed at an overpotential of 1,000 mV (Fig. 1.19c) was larger than the number of holes formed at an overpotential of 800 mV (Fig. 1.19a). It is understandably due to more vigorous hydrogen evolution at 1,000 mV ($\eta_{I,av}(H_2) = 30.0\%$) than at 800 mV ($\eta_{I,av}(H_2) = 10.8\%$) [45]. These copper deposits are the typical honeycomb-like structures, and the concept of "effective overpotential" was proposed to explain their formation.

It is known that the hydrogen evolution effects onto the hydrodynamic conditions inside the electrochemical cell [48–50]. The increase in hydrogen evolution rate leads to the decrease of the diffusion layer thickness and hence to the increase of limiting diffusion current density of electrode processes. It was shown [48] that if the rate of gas evolution at the electrode is larger than 100 cm^3/cm^2 min (> 5 A/cm^2), the diffusion layer becomes only a few micrometers thick. It is also shown [47] that a coverage of an electrode surface with gas bubbles can be about 30%. If the thickness of the diffusion layer in conditions of natural convection is ~5 × 10^{-2} cm and in strongly stirred electrolyte ~5×10^{-3} cm [51], it is clear that gas evolution is the most effective way of the decrease of mass transport limitations for electrochemical processes in mixed activation–diffusion control.

For electrochemical process in mixed activation–diffusion control, the overpotential η and the current density j are related by Eq. (1.58) [45]:

$$\eta = \frac{b_c}{2.3}\ln\frac{j}{j_0} + \frac{b_c}{2.3}\ln\frac{1}{1-\frac{j}{j_L}}. \tag{1.58}$$

The first term in Eq. (1.58) corresponds to the activation part of deposition overpotential and the second one is due to the mass transfer limitations. If one and the same process occurs under two different hydrodynamic conditions, characterized by two different values of the limiting diffusion current densities $j_{L,1}$ and $j_{L,2}$, Eq. (1.58) can be rewritten in the forms

$$\eta_1 = \frac{b_c}{2.3}\ln\frac{j_1}{j_0} + \frac{b_c}{2.3}\ln\frac{1}{1-\frac{j_1}{j_{L,1}}} \tag{1.59}$$

and

$$\eta_2 = \frac{b_c}{2.3}\ln\frac{j_2}{j_0} + \frac{b_c}{2.3}\ln\frac{1}{1-\frac{j_2}{j_{L,2}}}, \tag{1.60}$$

where η_1 and η_2 and j_1 and j_2 are the corresponding values of overpotentials and current densities. The same degree of diffusion control is obtained if

$$\frac{j_1}{j_{L,1}} = \frac{j_2}{j_{L,2}} \tag{1.61}$$

or

$$j_2 = j_1 \frac{j_{L,2}}{j_{L,1}} \tag{1.62}$$

and substitution of j_2 from Eq. (1.62) in Eq. (1.61) and further rearanging give

$$\eta_2 = \frac{b_c}{2.3} \ln\frac{j_1}{j_0} + \frac{b_c}{2.3} \ln\frac{1}{1 - \frac{j_1}{j_{L,1}}} + \frac{b_c}{2.3} \ln\frac{j_{L,2}}{j_{L,1}} \tag{1.63}$$

and if Eq. (1.59) is taken into account

$$\eta_2 = \eta_1 + \frac{b_c}{2.3} \ln\frac{j_{L,2}}{j_{L,1}}. \tag{1.64}$$

Hence, if

$$j_{L,2} > j_{L,1} \tag{1.65}$$

to obtain the same degree of diffusion control in two hydrodynamic conditions, Eq. (1.64) must be satisfaied, meaning that

$$\eta_2 > \eta_1. \tag{1.66}$$

In the absence of strong hydrogen evolution, the diffusion layer is due to the natural convection and does not depend on the overpotential of electrodeposition. The vigorous hydrogen evolution changes the hydrodynamic conditions and decreases the degree of diffusion control. Hence, Eq. (1.64) should be rewritten in the form

$$\eta_1 = \eta_2 - \frac{b_c}{2.3} \ln \frac{j_{L,2}}{j_{L,1}}, \tag{1.67}$$

where η_1 becomes the effective overpotential, $\eta_1 = \eta_{eff}$, related to conditions of natural convection at which there is the same degree of diffusion control as at overpotential η_2 with the hydrogen codeposition. Hence, the dendritic growth can be delayed or completely avoided, as can be seen from Fig. 1.19b, d, meaning that there is a really lower degree of diffusion control at overpotentials of 800 and 1,000 mV with the hydrogen codeposition than at an overpotential of 700 mV where the hydrogen codeposition is very small.

Anyway, the concept of "effective overpotential" can be summarized as follows: when hydrogen evolution is vigorous enough to change hydrodynamic conditions in the near-electrode layer, then electrodeposition process occurs at some overpotential which is effectively lower than the specified one. This overpotential is denoted by "effective overpotential" of electrodeposition process. From morphological point of view, it means that morphologies of metal deposits become similar to those obtained at some lower overpotentials where there is no hydrogen evolution or it is very small. More about the formation of the honeycomb-like structure and the concept of "effective overpotential" can be found in [46, 52–61].

The dendritic growth in this system at larger overpotentials is possible by the application of the appropriate pulsating overpotential (PO) regime. For example, the well-developed dendrites were formed with an overpotential amplitude of 1,000 mV, a deposition pulse, t_c, of 10 ms, and pause, t_p, of 100 ms (the pause to pulse ratio, p, where $p = t_p/t_c = 10$), as shown in Fig. 1.20. Dendrites are formed during the overpotential pulses with the average current density of hydrogen evolution in pulsating conditions which was not vigorous enough to cause the change of hydrodynamic conditions in the near-electrode layer [62, 63].

In the systems characterized by the strong hydrogen evolution which cause the change of hydrodynamic conditions in the near-electrode layer, the formation of dendrites mainly occurs in sheltered parts of the surface area, such as the bottom of holes (Fig. 1.21) [64].

Fig. 1.20 The copper dendrites formed by the pulsating overpotential (PO) regime: deposition pulse of 10 ms, pause duration of 100 ms; deposition time: 18 min; the amplitude overpotential used was 1,000 mV (Reprinted from [62] with permission from Elsevier and [63] from permission from Springer.)

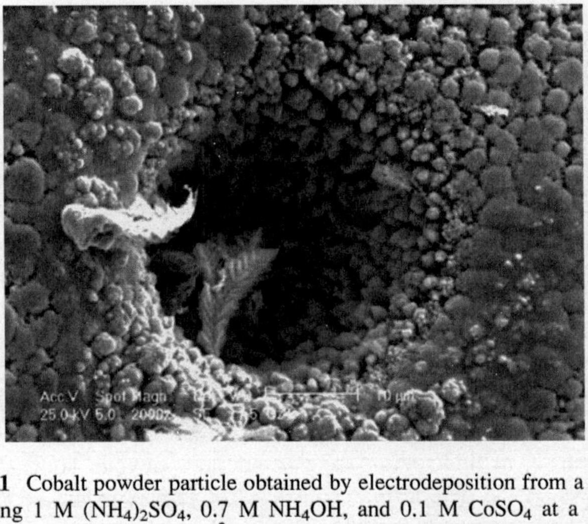

Fig. 1.21 Cobalt powder particle obtained by electrodeposition from a solution containing 1 M $(NH_4)_2SO_4$, 0.7 M NH_4OH, and 0.1 M $CoSO_4$ at a constant current density of 0.5 A cm^{-2} (Reprinted from [64] with permission from Springer.)

1.6 Granular Electrodeposits Formation

Metal electrodeposition on inert electrodes begins with the formation of separate growth centers until a continuous or disperse deposit is produced. Once a nucleus of the depositing metal has been formed, the current flowing causes a local deformation of the electric field in the vicinity of the growing center. As a result, an ohmic potential drop occurs along the nucleus-anode direction. Considering the high dependence of the nucleation rate on the overpotential, new nuclei would be expected to form only outside the spatial region around the initial nucleus. In that region the potential difference between the cathode and the electrolyte surpasses some critical value η_{cr}. Using simple mathematics, one obtains for the radius of the screening zone, r_{sz}, in an ohmic-controlled deposition:

$$r_{sz} = f \frac{U_\Omega}{U_\Omega - \eta_{cr}} r_N, \qquad (1.68)$$

where η_{cr} is the critical overpotential for nucleation to occur, U_Ω is the ohmic drop between the anode and cathode, f is a numerical factor, and r_N is the radius of the nucleus. The radius of the screening zone depends on the value of both U_Ω and η_{cr}. At a constant η_{cr}, an increase in U_Ω leads to a decrease in the radius of the screening zone; the same is true if η_{cr} decreases at constant U_Ω [65].

The radius of a nucleation exclusion zone can be calculated on the basis of the following discussion, taking into account the charge transfer overpotential also. If there is a half-spherical nucleus on a flat electrode, the extent of the deviation in the shape of the equipotential surfaces which occurs around it depends on the crystallization overpotential, current density, resistivity of the solution, and radius of the nucleus r_N. If the distance from the flat part of the substrate surface to the equipotential surface which corresponds to the critical nucleation overpotential, η_n, is l, then this changes around defect to the extent kr_N, as is presented in Fig. 1.22.

Therefore, in this region the current lines deviate from straight lines towards the defect, thus causing an increase in the deposition

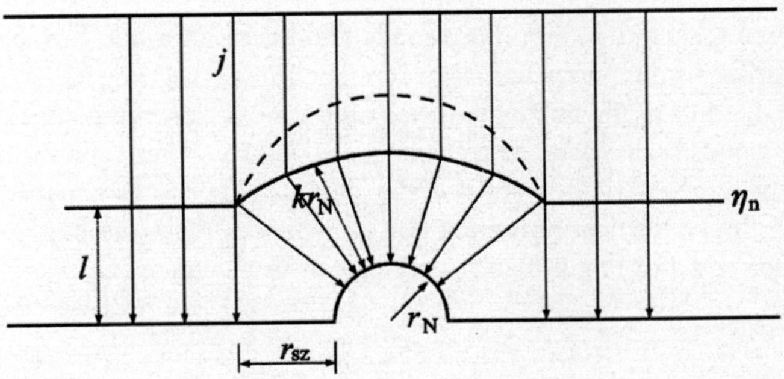

Fig. 1.22 A schematic representation of the deformation of the current field around a defect or a grain grown on a foreign substrate. For an explanation of the symbols see the text (Reprinted from [66] with permission from the Serbian Chemical Society and [67] with permission of Springer.)

rate, while in the surrounding region nucleation does not occur, i.e., a nucleation exclusion zone is formed. The voltage drop between the point from which the deviation occurs and the nucleus surface consists of the ohmic drop between these points and the charge transfer overpotential at the nucleus solution interface. The nucleation overpotential includes both the crystallization and charge transfer (deposition) overpotential:

$$\eta_n = \eta_{cryst} + \eta_d. \tag{1.69}$$

Hence, at the moment when kr_N become equal to l

$$kr_N j\rho = \eta_{cr}, \tag{1.70}$$

where j is the current density along the current lines and ρ is the electrolyte resistivity. Hence, when the ohmic drop between the deviation point and nucleus surface becomes equal to the crystallization overpotential, a new nucleation becomes possible on inert substrate assuming in both the cases the same charge transfer overpotential and the same value of the current density between the two symmetrical points on the anode and inert cathode surface and between the same point on the anode and the point at the surface of the earlier formed nucleus.

The radius of the nucleation exclusion zone or screening zone, r_{sz}, corresponds to the distance between the edge of a nucleus and the first current line which does not deviate (when kr_N becomes equal to l). Accordingly, nucleation will occur at distances from the edge of a nucleus equal or larger than r_{sz}, which can be calculated as

$$r_{sz} = r_N\left(\sqrt{2k+1} - 1\right). \tag{1.71}$$

If Eq. (1.70) is taken into account, one obtains

$$r_{sz} = r_N\left(\sqrt{\frac{2\eta_{cr}}{r_N \rho j} + 1} - 1\right). \tag{1.72}$$

According to Eq. (1.72), a new nucleation is possible in the vicinity of a nucleus if $\eta_{cr} \to 0$ or $j \to \infty$ or $\rho \to \infty$.

During the cathodic process at low j/j_0 the crystallization overpotential is considerably high; with increasing j/j_0, however, it decreases rapidly [68]. Hence, for $j_0 \to 0$, it follows that $r_{sz} \to 0$.

Electrodeposits of cadmium, copper, and nickel are shown in Figs. 1.23–1.25, respectively. In the cadmium deposition, boulders were formed by the independent growth of formed nuclei inside zones of zero nucleation. As a result of the high value of j_0 the deposition overpotential is low and the crystallization overpotential is relatively large and so the screening zone, according to Eq. (1.72), is relatively large. On the other hand, the nucleation rate is low. This results in the deposits shown in Fig. 1.23. These types of granular electrodeposits are mainly considered as disperse ones. In this chapter, only deposits based on dendritic and spongy growth will be treated as disperse ones in more details.

In the case of copper, a surface film is practically formed by a smaller quantity of electricity, as seen in Fig. 1.24, due to the lower exchange current density. The value of the deposition overpotential is larger than in the case of cadmium and the crystallization overpotential is lower, resulting in a decrease in the zero nucleation

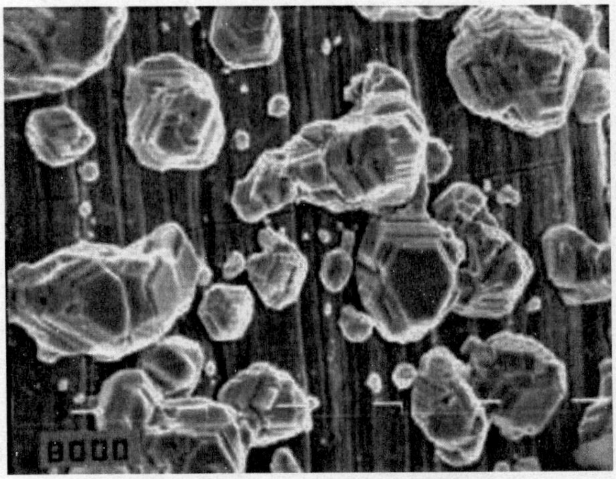

Fig. 1.23 Cadmium deposit on a copper substrate obtained at a current density of 1 mA cm^{-2} from 0.07 M $CdSO_4$ in 0.5 M H_2SO_4. Deposition overpotential: 15 mV. Deposition time 1,200 s. Magnification: ×2000. (Reprinted from [33] with permission from the Serbian Chemical Society and [67] with permission of Springer.)

Fig. 1.24 SEM microphotograph of copper deposits on a silver substrate obtained at a current density of 1 mA cm^{-2} from 0.07 M $CuSO_4$ in 0.5 M H_2SO_4. Deposition overpotential: 60 mV. Deposition time 300 s. Magnification: ×5000 (Reprinted from [33] with permission from the Serbian Chemical Society and [67] with permission of Springer.)

Fig. 1.25 SEM microphotograph of nickel deposits on a copper substrate obtained at a current density of 1 mA cm^{-2} from 0.07 M $NiSO_4$ in 0.5 M Na_2SO_4 + 30 g/l H_3BO_3. pH = 4. Deposition overpotential: 715 mV. Deposition time: 120 s. Magnification: ×5000 (Reprinted from [33] with permission from the Serbian Chemical Society and [67] with permission of Springer.)

zone radius, and hence a considerably larger nucleation rate. A further decrease in the exchange current density value, as in the case of Ni, leads to the situation shown in Fig. 1.25. A surface film is formed, but it is porous, probably due to hydrogen codeposition.

On the other hand, the classical expression for the steady-state nucleation rate, J, is given by [69–71]

$$J = K_1 \exp\left(-\frac{K_2}{\eta^2}\right), \tag{1.73}$$

where K_1 and K_2 are practically overpotential-independent constants. Equation (1.73) is valid for a number of systems regardless of the value of the exchange current density for the deposition process [69, 71]. At one and the same deposition current density, j, decreasing j_0 leads to an increasing nucleation rate and decreasing nucleation exclusion zones radii. Hence, the limiting case for nucleation exclusion zones can be expected when $j/j_0 \to 0$, and the limiting case for active centers when $j/j_0 \to \infty$.

The saturation nucleus density, i.e., the exchange current density of the deposition process, strongly affects the morphology of metal deposits. At high exchange current densities, the radii of the screening zones are large and the saturation nucleus density is low. This permits the formation of large, well-defined crystal grains and granular growth of the deposit. At low exchange current densities, the screening zones radii are low, or equal to zero, the nucleation rate is large, and a thin surface film can be easily formed. The saturation nucleus density depends also on the deposition overpotential.

The nucleation law can be written as [72]

$$N = N_0[1 - \exp(-At)], \tag{1.74}$$

where

$$A = K_1 j_0 \exp\left(-\frac{K_2}{\eta^2}\right) \tag{1.75}$$

and N_0 is the saturation nucleus surface density (nuclei cm^{-2}), being dependent on the exchange current density of deposition process and the deposition overpotential.

The overpotential and the current density in activation-controlled deposition inside the Tafel region are related by

$$\eta = \frac{b_c}{2.3} \ln \frac{j}{j_0}. \tag{1.76}$$

Therefore, increasing b_c and decreasing j_0 leads to an increase in the deposition overpotential. According to Eq. (1.75), the value of A increases with increasing overpotential and decreases with decreasing exchange current density. It follows from all available data that the former effect is more pronounced resulting in deposits with a finer grain size with decreasing value of the exchange current density.

1.7 Spongy and Spongy–Dendritic Growth Initiation Inside Diffusion Layer of Microelectrodes

1.7.1 Spongy Deposits

According to Barton and Bockris [5], if the electrodeposition process on the microelectrode with

$$r \leq 55 \ \mu\text{m} \tag{1.77}$$

is under complete diffusion control, a spherical diffusion layer, having a thickness equal to the radius of microelectrode is formed around it [5]. Equation (1.77) is always satisfied if

$$r \ll \delta. \tag{1.78}$$

Hence, it can be expected that the diffusion layer of hemispherical active particles on the inert substrate will not overlap if the distance between centers of the particles is larger than 4r, as illustrated in Fig. 1.26a. The common diffusion layer of the macroelectrode will be formed at larger times.

In the case presented in Fig. 1.26b the diffusion layers of the microelectrodes are not formed and the diffusion layer of the macroelectrode

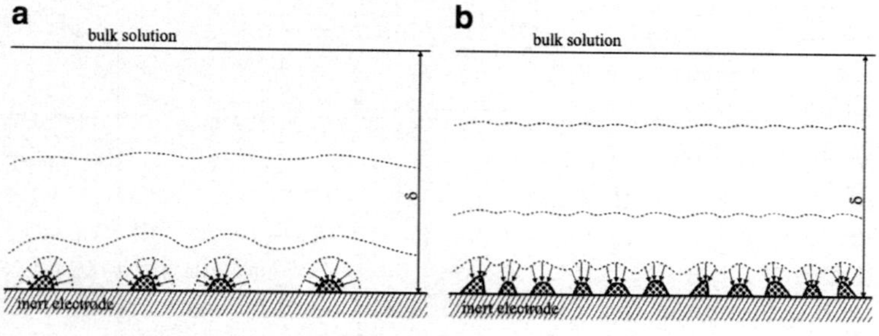

Fig. 1.26 Schematic presentation of formation of diffusion layer on the inert electrode covered with active grains. (**a**) The diffusion layers of macroelectrodes do not overlap and (**b**) the diffusion layers of the macroelectrodes overlap

is formed as on the massive electrode of active metal. Naturally, the initial electrodeposition on the grain from Fig. 1.26a will be performed in a spherical symmetry.

It follows from the previous discussion that this effect can be registered in the systems with the large j_0 values.

It follows from Eq. (1.1)

$$j = \frac{j_0(f_c - f_a)}{1 + \frac{j_0 f_c}{j_L}}, \tag{1.1}$$

that deposition in systems with low exchange current densities comes under full diffusion control at sufficiently large overpotentials. On the other hand, if:

$$\frac{j_0}{j_L} \gg 1 \tag{1.79}$$

deposition will be under complete diffusion control at all overpotentials if some other kind of control does not occur (e.g., for silver deposition on a well-defined silver crystal grains at a silver electrode at low overpotentials, two-dimensional nucleation is the rate-determining step) [73].

At low overpotentials a small number of nuclei are formed and they can grow independently. The limiting diffusion current density to the growing nucleus $j_{L,N}$ is given by

$$j_{L,N} = \frac{nFDC_0}{r_N}, \tag{1.80}$$

if

$$r_N < \delta, \tag{1.81}$$

where r_N is the radius of the nucleus. Hence, if $r_N \to 0$, the condition given by Eq. (1.79) is not satisfied and deposition is under activation or mixed control. Pure activation-controlled deposition is, thus, possible even at $j_0 \gg j_L$ on very small electrodes such as nuclei on an inert substrate.

An increase in r_N leads to a decrease of $j_{L,N}$, and, at sufficiently large r_N, the deposition comes under mixed activation–diffusion control, i.e., when

$$r_N > r_c, \tag{1.82}$$

where r_c is the radius of a growing nucleus where the process comes under mixed control [7, 74].

Under mixed control of the deposition, amplification of the surface irregularities on the growing nucleus occurs, leading to the formation of a spherical agglomerate of filaments. Thereby a spongy deposit is formed. The above reasoning is valid if spherical diffusion control can occur around growing grains, as in the case of cauliflower-like deposit growth. Assuming that around each grain with radius r_N, growing under spherical diffusion control, a diffusion layer of the same thickness is formed, then the initiation of spongy growth is possible if the number of nuclei per square centimeter, N, satisfies the condition

$$N \leq \frac{1}{(4r_c)^2}. \tag{1.83a}$$

Typical spongy electrodeposits are formed during zinc and cadmium electrodeposition at low overpotentials [7, 74]. Scanning electron microscopy images of zinc deposited at an overpotential of 20 mV onto a copper electrode from an alkaline zincate solution are shown in Fig. 1.27.

The increase in the number of nuclei formed with increasing deposition time can be seen in Fig. 1.27a, b, and a spongy deposit is formed as can be seen in Fig. 1.27b. The spongy growth takes place on a relatively small number of nuclei, as is shown in Fig. 1.27b, c.

The initiation of spongy growth at a fixed overpotential is possible if the condition $r_N > r_c$ (Eq. (1.82)) is satisfied, which is the case after some time. On the other hand, increasing the deposition time leads to the formation of a larger number of nuclei, and so the condition given by Eq. (1.83a) is not satisfied over a large part of the electrode surface. Regardless of this, the coverage of the electrode surface by spongy deposits increases with increasing deposition time up to full coverage, as can be seen in Fig. 1.27d.

Spongy growth can start on the growing nucleus if the conditions given by Eqs. (1.82) and (1.83a) are both satisfied simultaneously.

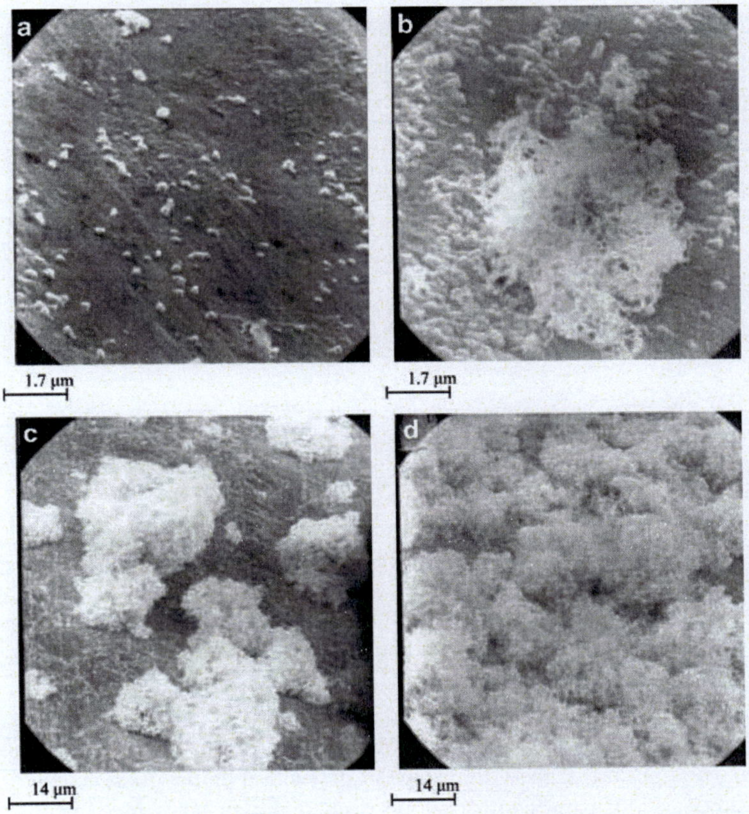

Fig. 1.27 Zinc deposits obtained by deposition at 20 mV from 0.1 M zincate and 1.0 M KOH solution. Deposition time (**a**) 10 min; (**b**) 20 min; (**c**) 30 min; and (**d**) 60 min (Reprinted from [7, 74, 75] with permission from Springer.)

In the first stage of deposition, the formation of nuclei having a regular crystal shape can be expected because the deposition is activation-controlled. After r_c is reached, the system comes under mixed control, producing polycrystalline grains like those shown in Fig. 1.28a, just as in the case of mixed control of copper deposition [12], Fig. 1.2c. In this situation, amplification of the surface irregularities on the growing grains occurs and spongy growth is initiated.

An ideal spongy nucleus obtained in a real system is shown in Fig. 1.28b which illustrates the above discussion. The agglomerate of

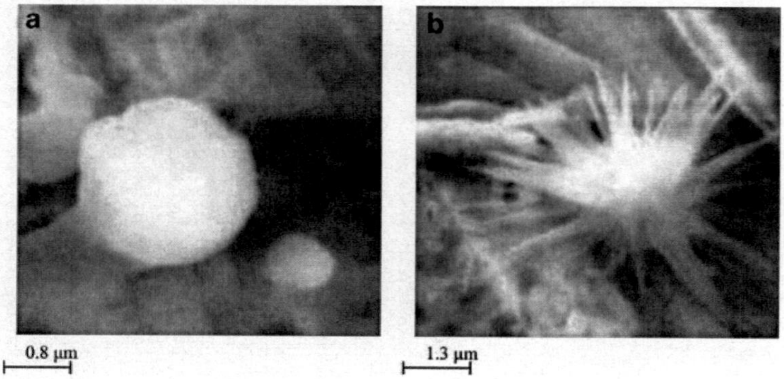

Fig. 1.28 Zinc deposits obtained by deposition at 35 mV from 0.1 M zincate solution in 1.0 M KOH solution. Deposition time (**a**) 7 min and (**b**) 15 min. The substrate is a copper plane electrode (Reprinted from [76] with permission from the Serbian Chemical Society and [7, 75] with permission from Springer.)

filaments in Fig. 1.27b is obviously formed by further growth of nuclei like that in Fig. 1.28b.

Hence, it can be concluded that at low overpotentials the initiation of spongy growth is due to the amplification of surface protrusions directly inside the spherical diffusion layer formed around each independently growing grain, as in the case of the formation of cauliflower-like deposits. The growth of protrusions in all directions is good proof that the initial stage of deposition on the grain is under spherical diffusion control, while further growth takes place in the diffusion layer of the macroelectrode. In less ideal situations, non-ideal spongy nuclei are formed, which, however, after further deposition result in a macroelectrode with the same appearance.

1.7.2 Spongy-Dendritic Deposits

The limiting diffusion current to the growing nucleus, $j_{L,N}$, can be related to j_L using Eqs. (1.5) and (1.80) by

$$j_{L,N} = \frac{\delta}{r_N} j_L, \tag{1.84}$$

where δ is the diffusion layer thickness of the macroelectrode of massive metal, and r_N is the radius of the growing nucleus. Equation (1.11) for the growing nucleus can be rewritten in the form

$$j_N = \frac{j_0 f_c}{1 + \frac{j_0 f_c}{j_L} \frac{r_N}{\delta}}, \tag{1.85}$$

where j_N is the current density to the growing nucleus. It is obvious from Eqs. (1.11) and (1.85) that deposition process on the macro-electrode can be under complete diffusion control if

$$\frac{j_0 f_c}{j_L} \gg 1 \tag{1.86}$$

and that at the same overpotential, process on the growing nucleus can be under pure activation control if

$$\frac{j_0 f_c}{j_L} \frac{r_N}{\delta} \ll 1 \tag{1.87}$$

or at

$$\frac{r_N}{\delta} \to 0. \tag{1.88}$$

Equations (1.87) and (1.88) are fulfilled in the initial stage of electrodeposition to the nuclei of metal formed on the inert substrate [77]. In this case the nuclei behave as microelectrodes, because of their complete independent growth well before the formation of the diffusion layer of the macroelectrode. The radius r_0 of the initial stable nucleus at overpotential η is given by [78]

$$r_0 = \frac{2\sigma V}{nF}, \tag{1.89}$$

where σ is the interfacial energy between metal and solution and V is the molar volume of the metal. The radius of the growing nucleus will vary with time according to [79]

$$r_N = r_0 + \frac{V}{nF} j_0 f_c t \tag{1.90}$$

or

$$r_N \approx \frac{V}{nF} j_0 f_c t \tag{1.91}$$

because r_0 is extremely low.

Obviously, Eqs. (1.85), (1.87), (1.90), and (1.91) are only the approximation on growth times, because the effect of surface energy has not been taken into consideration. At larger deposition times they are valid, because the surface energy term at higher value can be neglected [5].

An increase in r_N leads to a decrease of $j_{L,N}$, and at sufficiently large r_N deposition comes under mixed activation–diffusion control. It can be assumed that this happens at

$$j_N > \upsilon j_{L,N}, \tag{1.92}$$

where $0 < \upsilon < 1$. By combining Eqs. (1.85) and (1.92) one obtains

$$r_{c,N} \approx \frac{j_L \upsilon \delta}{j_0 f_c (1 - \kappa)}, \tag{1.93}$$

where $r_{c,N}$ is the radius of the growing nucleus when the process comes under mixed or spherical diffusion control. According to Barton and Bockris [5] the diffusion layer around such grain forms very fast. The further combination of Eqs. (1.91) and (1.93) gives the corresponding induction time, t_i, given by

$$t_i = \frac{j_L}{j_0^2} \frac{\upsilon F \delta}{V f_c^2 (1 - \upsilon)}. \tag{1.94}$$

For sufficiently high overpotentials Eqs. (1.93) and (1.94) can be rewritten in the forms [80]

$$r_{c,N} \approx \frac{j_L}{j_0} \frac{\delta}{4 f_c} \tag{1.95}$$

and

$$t_{\mathrm{i}} = \frac{j_{\mathrm{L}}}{j_0^2} \frac{nF\delta}{4Vf_{\mathrm{c}}^2} \tag{1.96}$$

for $\upsilon = 0.2$.

At $r < r_{\mathrm{c,N}}$ and $t < t_{\mathrm{i}}$ the deposition on the growing grain is under activation control.

Hence, if $r_{\mathrm{N}} > r_{\mathrm{c,N}}$, the spherical diffusion layer around microelectrode can be formed. This is the condition for deposition in spherical diffusion control.

The nucleus of spongy deposit, i.e., hedgehog-like particle, appears when amplification of surface coarseness on the nucleus in spherical diffusion control starts growing. It was shown earlier [80] that this amplification is very fast so the induction time when growing nucleus enters mixed control can be taken also as induction time of spongy formation. It follows from Eqs. (1.95) and (1.96) that $r_{\mathrm{c,N}}$ and t_{i} decrease with increasing overpotential.

On the other hand, it was also shown [74] that spongy deposit can be formed only if around each grain with radius $r_{\mathrm{c,N}}$, growing under spherical diffusion control, a diffusion layer of the same thickness is formed, as illustrated earlier. This condition is fulfilled if

$$N \leq \frac{1}{(4r_{\mathrm{c,N}})^2}, \tag{1.83b}$$

where N is the number of grain per square centimeter of the macroelectode. Hence, deposition in spherical diffusion control on the growing grain is possible if both Eqs. (1.93) and (1.83b) are satisfied in the same time the nucleation law can be written in the form [72]

$$N = N_0\left[1 - \mathrm{e}^{-At}\right], \tag{1.74}$$

where

$$A = K_1 j_0 \mathrm{e}^{-\frac{K_2}{\eta^2}}. \tag{1.75}$$

N_0 is the maximum number of active sites for selected value of overpotential and K_1 and K_2 are constants.

Spongy deposits formation is possible if

$$N_0\left[1 - \mathrm{e}^{-At_\mathrm{i}}\right] < \frac{1}{4r_{\mathrm{c,N}}^2} \tag{1.97}$$

and

$$At_\mathrm{i} \approx 0 \tag{1.98}$$

which happens at sufficiently high overpotentials where $K_2/\eta^2 \rightarrow 0$, $A \rightarrow K_1\, j_0$ and $t_\mathrm{i} \rightarrow 0$. Hence, the spongy deposit formation at high overpotentials starts at very low deposition times, when the spherical diffusion layer formed around grains do not overlap. The critical overpotential of spongy formation can be obtained by substitution of $r_{\mathrm{c,N}}$ from Eq. (1.95) and t_i from Eq. (1.96) in Eq. (1.97) and further calculation. if this overpotential is larger than critical one for instantaneous dendritic growth the dendrite spongy nuclei can be formed over inert substrate.

The experimental verification of the above discussion is given by the consideration of the morphology of electrodeposited silver from 0.50 M $AgNO_3$ in 0.20 M HNO_3 on the graphite electrode at different overpotentials of deposition and with different deposition times [81]. In Fig. 1.29 the deposit obtained at an overpotential of 100 mV during 180 s is shown.

As expected, the boulders are obtained. In Fig. 1.30, the deposits obtained at 200 mV during 1 and 10 s are presented. At 1 s, the boulders are formed, but at 10 s the needle-like deposit is obtained. This means that the spherical diffusion layer around the growing grains is not formed before the formation of the diffusion layer of the macroelectrode. The electrodeposition inside the diffusion layer of the macroelectrode is confirmed by the growth of needles towards the bulk of solution.

At an overpotential of 300 mV, the conditions of the spherical diffusion control around the growing grains are fulfilled and dendritic-spongy deposit is formed, as can be seen from Fig. 1.31.

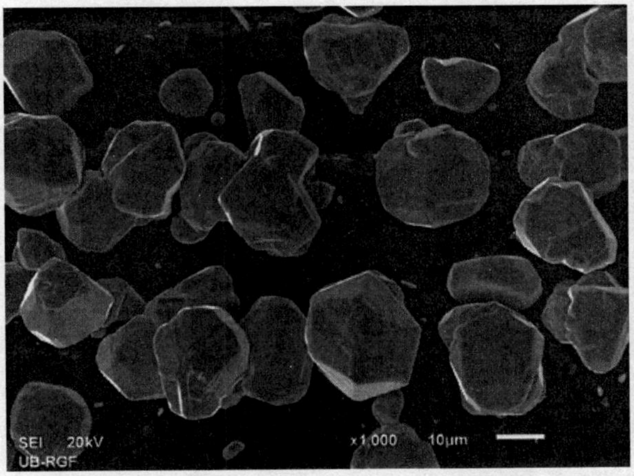

Fig. 1.29 Silver deposit obtained at an overpotential of 100 mV with electrolysis time of 180 s

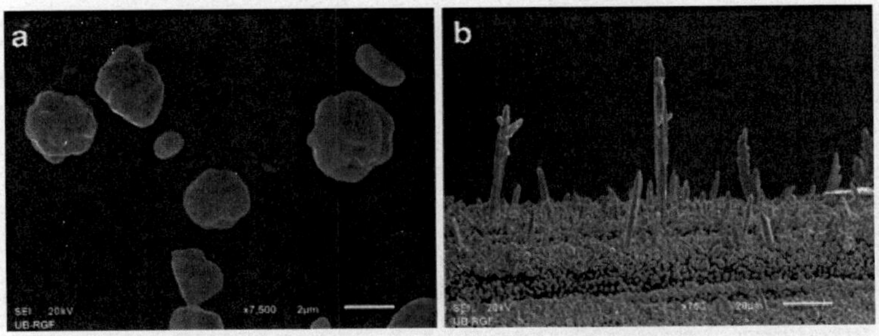

Fig. 1.30 Silver deposit obtained at an overpotential of 200 mV. Deposition time (**a**) 1 s and (**b**) 10 s

The growth of dendrites in all directions means that there is a spherical control to the growing grains in the initial stage of the electrodeposition.

Finally, the fact that $r_{c,N}$ and t_i (Eqs. (1.95) and (1.96)) decrease with the increasing overpotential can be verified by Figs. 1.31 and 1.32.

It is obvious that the semiqualitative agreement between the theory and experiments is fair. Besides, the deposits from Figs. 1.31 and 1.32 are similar to those from Fig. 1.29. Unfortunately, the ideal spongy–dendritic

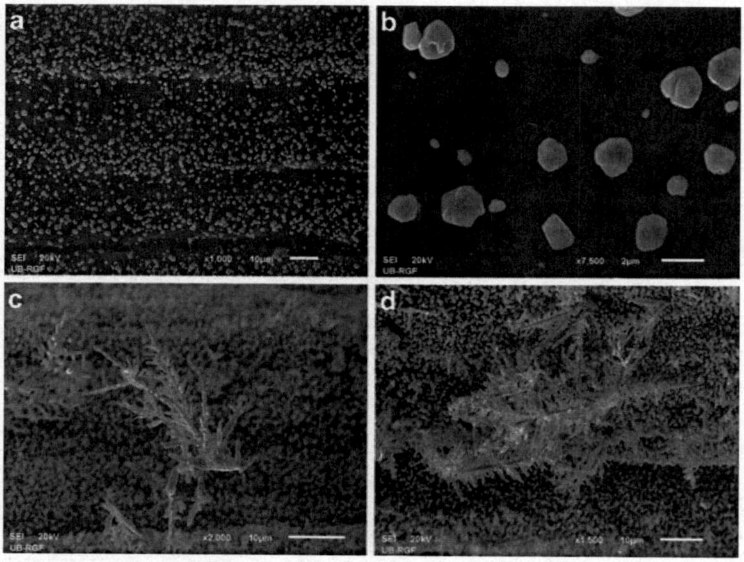

Fig. 1.31 Silver deposits obtained at an overpotential of 300 mV (**a**) 1 s; (**b**) 1 s; (**c**) 3 s, and (**d**) 5 s

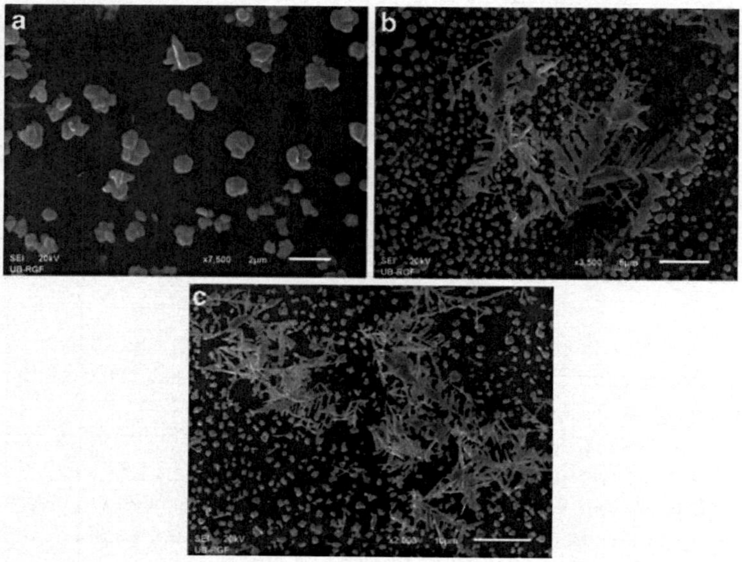

Fig. 1.32 Silver deposits obtained at an overpotential of 700 mV (**a**) 0.15 s; (**b**) 0.30 s; and (**c**) 0.5 s

nucleus similar to the one from Fig. 1.30 has not been formed so far; it will probably be formed in the future investigations.

1.8 Conclusions

The formation of disperse metal electrodeposits is discussed by the consideration of corresponding physical and mathematical models. It is shown that the mechanisms of formation of all different forms can be elucidated by the use of conclusions of a few classic works in the field of metal electrodeposition making a general theory of disperse metal electrodeposits formation.

The appearance of different forms of disperse metal electrodeposits is correlated with the properties of electrodeposited metal and deposition conditions. In this way, the theoretical basis of powdered electrodeposits formation and inert electrodes activation is formed.

Acknowledgments This chapter is based on a few classical studies and a research in the field of metal electrodeposition performed at the Department of Physical Chemistry and Electrochemistry at the Faculty of technology and Metallurgy and Institute of Electrochemistry, ICTM, University of Belgrade, Serbia. We would like to acknowledge all colleagues and students who participated in it.

The work was supported by the Ministry of Education and Science of the Republic of Serbia under the research project: "Electrochemical synthesis and characterization of nanostructured functional materials for application in new technologies" (No. 172046).

References

1. Popov KI, Živković PM, Nikolić ND (2010) The effect of morphology of activated electrodes on their electrochemical activity. In: Djokić SS (ed) Electrodeposition: theory and practice. Modern aspects of electrochemistry, vol. 48. Springer, Berlin, pp 163–213, 165
2. Bockris JO'M, Reddy AKN, Gamboa-Aldeco M (2000) Modern electrochemistry 2A, 2nd edn. Kluwer Academic/Plenum, New York, p 1248
3. Diggle JW, Despić AR, Bockris JO'M (1969) J Electrochem Soc 116:1503
4. Popov KI, Djokić SS, Grgur BN (2002) Fundamental aspects of electrometallurgy, chap 3. Kluwer Academic/Plenum, New York, p 14

5. Barton JL, Bockris JO'M (1962) Proc Roy Soc A 268:485
6. Popov KI, Živković PM, Grgur BN (2007) Electrochim Acta 52:4696
7. Popov KI, Krstajić NV, Čekerevac MI (1996) The mechanism of formation of coarse and disperse electrodeposits. In: White RE, Conway BE, Bockris JO'M (eds) Modern aspects of electrochemistry, vol 30. Plenum, New York, pp 261–312, 262
8. Despić AR, Diggle JW, Bockris JO'M (1969) J Electrochem Soc 116:507
9. Gilleadi E (1993) Electrode kinetics. VCH, New York, p 443
10. Popov KI, Djokić SS, Grgur BN (2002) Fundamental aspects of electrometallurgy, chap 3. Kluwer Academic/Plenum, New York, p 56
11. Damjanović A (1965) Plating 52:1017
12. Popov KI, Pavlović MG, Pavlović LjJ, Čekerevac MI, Remović GŽ (1988) Surf Coat Technol 34:355
13. Popov KI, Grgur BN, Pavlović MG, Radmilović V (1993) J Serb Chem Soc 58:1055
14. Popov KI, Radmilović V, Grgur BN, Pavlović MG (1994) J Serb Chem Soc 59:47
15. Popov KI, Djokić SS, Grgur BN (2002) Fundamental aspects of electrometallurgy, chap 3. Kluwer Academic/Plenum, New York, p 78
16. Despić AR, Popov KI (1972) Transport controlled deposition and dissolution of metals. In: Conway BE, Bockris JO'M (eds) Modern aspects of electrochemistry, vol 7. Plenum, New York, pp 199–313, 203, 220
17. Popov KI, Krstajić NV, Čekerevac MI (1996) The mechanism of formation of coarse and disperse electrodeposits. In: White RE, Conway BE, Bockris JO'M (eds) Modern aspects of electrochemistry, vol 30. Plenum, New York, pp 261–312, 294
18. Wranglen G (1960) Electrochim Acta 2:130
19. Bechtoldt CJ, Ogburn F, Smith J (1968) J Electrochem Soc 115:813
20. Faust JW, John HF (1963) J Electrochem Soc 110:463
21. Faust JW, John HF (1961) J Electrochem Soc 108:855
22. Justinijanović IN, Despić AR (1973) Electrochim Acta 18:709
23. Newman JS (1973) Electrochemical systems. Prentice-Hall, Engelwood Cliffs, NJ, p 177
24. Popov KI, Maksimović MD, Trnjavčev JD, Pavlović MG (1981) J Appl Electrochem 11:239
25. Popov KI, Živković PM, Nikolić ND (2010) The effect of morphology of activated electrodes on their electrochemical activity. In: Djokić SS (ed) Electrodeposition: theory and practice. Modern aspects of electrochemistry series, vol 48. Springer, Berlin, pp 163–213, 171
26. Scharifker B, Hills G (1983) Electrochim Acta 28:879
27. Popov KI, Živković PM, Krstić SB, Nikolić ND (2009) Electrochim Acta 54:2924
28. Bockris JO'M, Reddy AKN, Gamboa-Aldeco M (2000) Modern electrochemistry 2 A, 2nd edn. Kluwer Academic/Plenum, New York, p 1107
29. Popov KI, Djokić SS, Grgur BN (2002) Fundamental aspects of electrometallurgy, chap 3. Kluwer Academic/Plenum, New York, pp 87, 88

30. Price PB, Vermilyea DA (1958) J Chem Phys 28:720
31. Lorenz W (1954) Z Electrochem 58:912
32. Mattsson BE, Bockris JO'M (1959) Trans Faraday Soc 55:1586
33. Popov KI, Grgur BN, Stoiljković ER, Pavlović MG, Nikolić ND (1997) J Serb Chem Soc 62:433
34. Meibhur S, Yeager E, Kozawa A, Hovorka F (1963) J Electrochem Soc 110:190
35. Popov KI, Pavlović MG, Stojilković ER, Stevanović ZŽ (1997) Hydrometallurgy 46:321
36. Popov KI, Živković PM, Nikolić ND (2010) The effect of morphology of activated electrodes on their electrochemical activity. In: Djokić SS (ed) Electrodeposition: theory and practice. Modern aspects of electrochemistry, vol 48. Springer, Berlin, pp 163–213, 190
37. Popov KI, Radmilović V, Grgur BN, Pavlović MG (1994) J Serb Chem Soc 59:119
38. Popov KI, Čekerevac MI (1989) Surf Coat Technol 37:435
39. Popov KI, Djokić SS, Grgur BN (2002) Fundamental aspects of electrometallurgy, chap 2. Kluwer Academic/Plenum, New York, p 24
40. Popov KI, Pavlović MG (1993) Electrodeposition of metal powders with controlled grain size and morphology. In: White RE, Bockris JO'M, Conway BE (eds) Modern aspects of electrochemistry, vol 24. Plenum, New York, pp 299–391, 300
41. Calusaru A (1979) Electrodepositon of metal powders. Material science monography, vol 3. Elsevier, Amsterdam
42. Popov KI, Krstajić NV, Pantelić RM, Popov SR (1985) Surf Technol 26:177
43. Popov KI, Pavlović MG, Jovićević JN (1989) Hydrometallurgy 23:127
44. Pangarov NA (1967) Phys Stat Sol 20:371
45. Nikolić ND, Popov KI, Pavlović LjJ, Pavlović MG (2006) J Electroanal Chem 588:88
46. Nikolić ND, Popov KI (2010) Hydrogen co-deposition effects on the structure of electrodeposited copper. In: Djokić SS (ed) Electrodeposition: theory and practice. Modern aspects of electrochemistry, vol 48. Springer, Berlin, pp 1–70
47. Nikolić ND, Popov KI, Pavlović LjJ, Pavlović MG (2006) Surf Coat Technol 201:560
48. Ibl N (1961) Chemie Ing Techn 33:69
49. Ibl N (1963) Chemie Ing Techn 35:353
50. Jenssen LJ, Hoogland JG (1970) Electrochim Acta 15:1013
51. Bockris JO'M, Reddy AKN, Gamboa M (2000) Aldeco, Modern electrochemistry 2A, Fundamentals of electrodics, 2nd edn. Kluwer Academic/Plenum, New York
52. Nikolić ND, Popov KI, Pavlović LjJ, Pavlović MG (2007) J Solid State Electrochem 11:667
53. Nikolić ND, Pavlović LjJ, Pavlović MG, Popov KI (2007) Electrochim Acta 52:8096
54. Nikolić ND, Popov KI, Pavlović LjJ, Pavlović MG (2007) Sensors 7:1

55. Nikolić ND, Pavlović LjJ, Krstić SB, Pavlović MG, Popov KI (2008) Chem Eng Sci 63:2824
56. Nikolić ND, Branković G, Pavlović MG, Popov KI (2008) J Electroanal Chem 621:13
57. Nikolić ND, Pavlović LjJ, Branković G, Pavlović MG, Popov KI (2008) J Serb Chem Soc 73:753
58. Nikolić ND, Pavlović LjJ, Pavlović MG, Popov KI (2007) J Serb Chem Soc 72:1369
59. Nikolić ND, Branković G, Pavlović MG, Popov KI (2009) Electrochem Commun 11:421
60. Nikolić ND, Branković G, Maksimović VM, Pavlović MG, Popov KI (2010) J Solid State Electrochem 14:331
61. Nikolić ND, Branković G, Maksimović VM, Pavlović MG, Popov KI (2009) J Electroanal Chem 635:111
62. Popov KI, Nikolić ND, Živković PM, Branković G (2010) Electrochim Acta 55:1919
63. Popov KI, Živković PM, Nikolić ND (2010) The effect of morphology of activated electrodes on their electrochemical activity. In: Djokić SS (ed) Electrodeposition: theory and practice, vol 48, Modern aspects of electrochemistry. Springer, Berlin, pp 163–213
64. Jović VD, Maksimović V, Pavlović MG, Popov KI (2006) J Solid State Electrochem 10:373
65. Markov I, Boynov A, Toshev S (1973) Electrochim Acta 18:377
66. Štrbac S, Rakočević Z, Popov KI, Pavlović MG, Petrović R (1999) J Serb Chem Soc 64:483
67. Popov KI, Djokić SS, Grgur BN (2002) Fundamental aspects of electrometallurgy, chap 3. Kluwer Academic/Plenum, New York, p 30
68. Klapka V (1970) Collection Czechoslov Chem Comun 35:899
69. Gutzov I (1964) Izv Inst Fiz Chim Bulgar Acad Nauk 4:69
70. Erdey-Grúz T, Volmer Z (1931) Z Phys Chem 157A:165 (in German)
71. Fetter K (1967) Electrochemical kinetics. Khimiya, Moscow (in Russian)
72. Fleischmann M, Thirsk HR (1959) Electrochim Acta 1:146
73. Popov KI, Krstajić NV, Jerotijević ZĐ, Marinković SR (1985) Surf Technol 26:185
74. Popov KI, Krstajić NV (1983) J Appl Electrochem 13:775
75. Popov KI, Djokić SS, Grgur BN (2002) Fundamental aspects of electrometallurgy, chap 3. Kluwer Academic/Plenum, New York, p 72
76. Popov KI, Krstajić NV, Simičić MV, Bibić NM (1992) J Serb Chem Soc 57:927
77. Popov KI, Krstajić NV, Popov SR, Čekerevac MI (1986) J Appl Electrochem 14:771
78. Toshev S, Markov I (1967) Electrochim Acta 12:489
79. Bockris JO'M, Nagy Z, Dražić D (1973) J Electrrochem Soc 120:30
80. Popov KI, Krstajić NV, Popov SR (1985) J Appl Electrochem 15:151
81. Popov KI, Nikolić ND, Živković PM (2012) Int J Electrochem Sci 7:686

Chapter 2
Morphology of Different Electrodeposited Pure Metal Powders

V.D. Jović, N.D. Nikolić, U.Č. Lačnjevac, B.M. Jović, and K.I. Popov

2.1 Introduction

As mentioned in other chapters, metal powders obtained by electrolytic processes are mainly dendrites which can spontaneously fall off or can be removed from the electrode by tapping or other similar techniques [1]. Also, powder particles can have other morphological forms, such as flakes or needles, fibrous or spongy, and needle or

V.D. Jović (✉) • U.Č. Lačnjevac • B.M. Jović
Department of Materials Science, Institute for Multidisciplinary Research, University of Belgrade, Kneza Višeslava 1, 11030 Belgrade, P.O. Box 33, Serbia
e-mail: vladajovic@imsi.rs; uros.lacnjevac@imsi.rs; bjovic@imsi.rs

N.D. Nikolić
ICTM-Institute of Electrochemistry, University of Belgrade, Njegoseva 12, P.O.B. 473, 11001 Belgrade, Serbia
e-mail: nnikolic@tmf.bg.ac.rs

K.I. Popov
ICTM-Institute of Electrochemistry, University of Belgrade, Njegoseva 12, P.O.B. 473, 11001 Belgrade, Serbia

Faculty of Technology and Metallurgy, University of Belgrade, Karnegijeva 4, P.O.B. 3503, 11001 Belgrade, Serbia
e-mail: kosta@tmf.bg.ac.rs

Stojan S. Djokić (ed.), *Electrochemical Production of Metal Powders*, Modern Aspects of Electrochemistry 54, DOI 10.1007/978-1-4614-2380-5_2,

cauliflower-like ones, and the shape of powder particles depends on the electrodeposition conditions and the nature of the metal.

Some technologically and academically important metal powders, such as copper, silver, nickel, cobalt, lead, and cadmium, are obtained by electrolysis of aqueous solutions.

According to Winand [2], metals can be classified into three groups:

(a) Normal metals (Cd, Zn, Sn, Ag (silver nitrate solutions), Pb) which have low melting points, T_m, and high exchange current densities, j_0
(b) Intermediate metals [Au, Cu, Ag (silver ammonia complex)], which have moderate T_m and medium j_0
(c) Inert metals (Fe, Ni, Co, Pt, Cr, Mn), which have high T_m and low j_0

The aim of this chapter is morphological analysis of some of the powders from these groups. Metal powders obtained by the constant and periodically changing regimes of electrolysis are analyzed in detail.

2.2 Silver

2.2.1 *Effect of Exchange Current Density on the Morphology of Silver Powder Particles*

Electrodeposition of silver from nitrate solutions is characterized by the high exchange current densities, j_0 ($j_0 \gg j_L$, j_L is the limiting diffusion current density) [3, 4] and the typical polarization curve obtained from 0.06 M $AgNO_3$, 1.2 M $NaNO_3$, and 0.05 M HNO_3 solution is shown in Fig. 2.1 (denoted by "o"). The plateau of the limiting diffusion current density corresponds to the range of overpotentials between 75 and 175 mV. On the other hand, electrodeposition of silver from ammonium solutions is characterized by the relation $j_0 < j_L$ [4, 5], and the polarization curve for silver electrodeposition obtained from a solution containing 0.1 M $AgNO_3$, 0.5 M $(NH_4)_2SO_4$, and 0.5 M NH_3 is also shown in Fig. 2.1 (denoted by "□"). The plateau of the limiting diffusion current density was

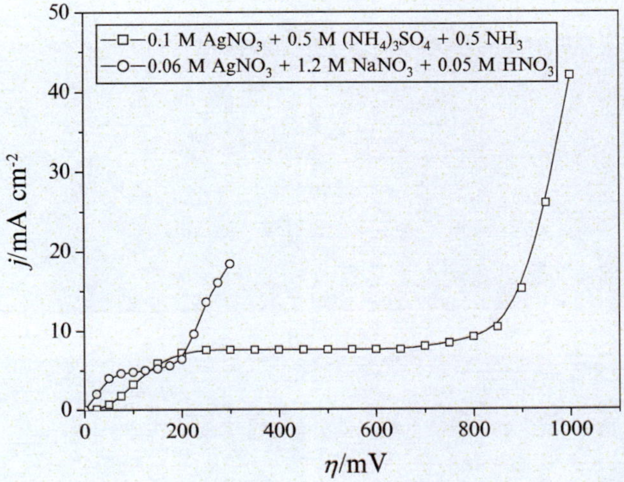

Fig. 2.1 Polarization curves for silver electrodepositions from both 0.1 M $AgNO_3$ + 0.5 M $(NH_4)_2SO_4$ + 0.5 M NH_3 and 0.06 M $AgNO_3$ + 1.2 M $NaNO_3$ + 0.05 M HNO_3 (Reprinted from [4] with permission from Electrochemical Society.)

considerably wider, corresponding to the range of overpotentials between 250 and 700 mV. For both the examined solutions, electrodeposition of silver was performed at the room temperature using vertical cylindrical graphite electrodes [4]. The processes of silver electrodeposition from these solutions were not accompanied with hydrogen evolution reaction.

Due to the large exchange current density in silver nitrate solutions [3], which is significantly higher than in the case of silver ammonium solution [5], an instantaneous growth of dendrites starts at relatively low overpotential [6]. The silver powder electrodeposited from nitrate electrolyte at an overpotential of 150 mV is shown in Fig. 2.2a. Silver particles obtained by tapping the silver deposit from electrode surface represent a mixture of different morphological forms, as illustrated in Fig. 2.2b–d. Some of the particles had the shape of two-dimensional (2D) dendrites (Fig. 2.2b). The presence of other morphological forms, such as crystals of irregular shape and needle-like particles (Fig. 2.2c, d), was also noticed.

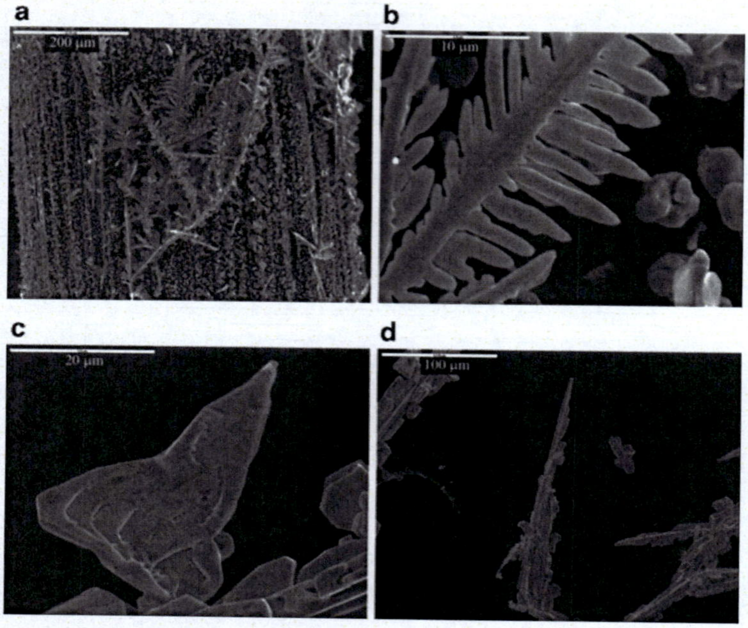

Fig. 2.2 (**a**) Macrostructure and (**b–d**) typical morphological forms of silver obtained by electrodeposition from 0.06 M $AgNO_3$ in both 1.2 M $NaNO_3$ and 0.05 M HNO_3 at an overpotential of 150 mV (Reprinted from [4] with permission from Electrochemical Society.)

A typical silver deposit obtained from the ammonium solution at an overpotential of 650 mV is shown in Fig. 2.3a. From Fig. 2.3a, it can be seen that very branchy dendrites are produced at this overpotential. Silver particles obtained by tapping the silver deposit (Fig. 2.3a) are shown in Fig. 2.3b. The dendritic character of this particle is made of the corncob-like elements as presented by the images in Fig. 2.3c, d. A further analysis of the corncob-like elements at the microlevel showed that they are composed of small agglomerates of silver grains (Fig. 2.3d). Anyway, morphologies of silver particles electrodeposited from ammonium solution were completely different than those formed during silver electrodeposition from nitrate electrolyte.

Fig. 2.3 (**a**) Macrostructure of silver powdered deposits electrodeposited at an overpotential of 650 mV from 0.1 M $AgNO_3$ in both 0.5 M $(NH_4)_2SO_4$ and 0.5 M NH_3; (**b**) dendritic particles obtained by tapping this silver deposit; (**c**) and (**d**) the corncob-like elements of which dendrites are composed (Reprinted from [4] with permission from Electrochemical Society.)

It is necessary to note that the shape of the polarization curve for silver electrodeposition from ammonium solution (Fig. 2.1) and morphologies of silver particles (Fig. 2.3) were very similar to those obtained by copper electrodeposition from sulfate solutions at overpotentials corresponding to the plateaus of the limiting diffusion current density [4, 7–11] (see also other chapters). The similarity of silver and copper dendrites was observed at both macro- and micro-levels, because both copper and silver dendrites were composed of corncob-like forms, while the corncob-like forms were built of small agglomerates of metal grains [4].

Also, the sudden increase of the current density with the increasing overpotential above 700 mV is the common characteristic of

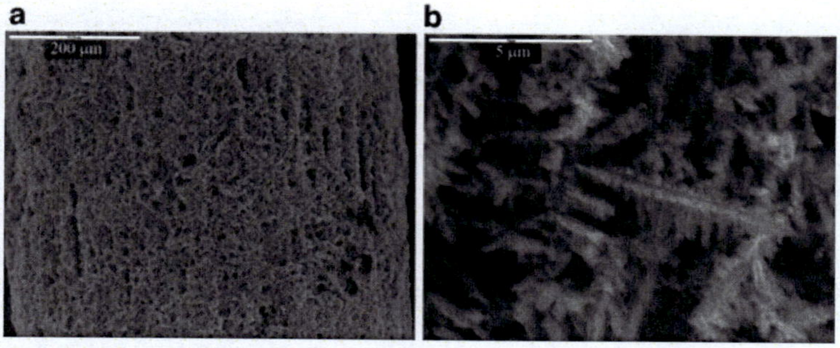

Fig. 2.4 (**a**) Macrostructure of silver powdered deposits electrodeposited at an overpotential of 1,000 mV from 0.1 M $AgNO_3$ in both 0.5 M $(NH_4)_2SO_4$ and 0.5 M NH_3 and (**b**) dendritic particles obtained by tapping silver deposit (Reprinted from [4] with permission from Electrochemical Society.)

both copper electrodeposition from sulfate solution and silver electrodeposition from ammonium solution. Silver powdered deposit obtained from the ammonium solution at an overpotential of 1,000 mV is shown in Fig. 2.4a. At the first sight, spongy-like structure can be noticed from this figure. An analysis of silver deposit produced at the overpotential of 1,000 mV with a higher magnification (Fig. 2.4b) showed that this material is dendritic in shape. Hence, the macrostructure of silver powder formed at an overpotential of 1,000 mV is similar to that of copper and silver deposited at 650 mV. The only difference is in the size of dendrites; silver dendrites electrodeposited at an overpotential of 1,000 mV were considerably smaller than those formed at overpotential of 650 mV, which is attributed to higher nucleation rate at 1,000 mV than at 650 mV [4].

Meanwhile, the macrostructure of the silver deposit produced at an overpotential of 1,000 mV was completely different than that of the copper electrodeposited at the same overpotential. As already mentioned, holes formed by attached hydrogen bubbles surrounded by cauliflower-like agglomerates of copper grains (the honeycomb-like structure; see other chapters) were formed by copper electrodeposition at an overpotential of 1,000 mV [7, 8, 10–13]. The copper powder obtained by tapping the powdered deposit consists of an aggregate of small cauliflower-like particles [14]. Similar copper structures were also observed by electrodeposition at periodically changing rate [15–20].

The observed difference in the morphology of silver and copper deposits produced at an overpotential of 1,000 mV can be ascribed to the simultaneous hydrogen evolution reaction during copper electrodeposition at high overpotentials. In the case of copper, hydrogen evolution commences at some overpotential belonging to the plateau of the limiting diffusion current density, and the increasing overpotential intensifies this reaction [7, 11]. At some overpotential outside the plateau of the limiting diffusion current density, hydrogen evolution becomes vigorous enough to cause a strong stirring of the solution leading to a change of the hydrodynamic conditions in the near-electrode layer. Copper dendrites were formed without and with the quantity of evolved hydrogen which was insufficient to lead to the change of hydrodynamic conditions in the near-electrode layer, while cauliflower-like particles (formed around holes) were obtained with the quantity of evolved hydrogen which was enough to lead to the change of the hydrodynamic conditions in the near-electrode layer [7, 11]. The growth of the current with the increasing overpotential was just a result of parallel hydrogen evolution reaction to copper electrodeposition.

In the case of silver, there is no hydrogen evolution, and the increase in the current with the rising overpotential above 700 mV can be ascribed to the instantaneous dendrite growth. The absence of hydrogen evolution during silver electrodeposition is explained by the fact that the equilibrium potential of silver electrode in silver ammonium solution is slightly more positive than the one of copper electrode in copper sulfate solution [21]. On the other hand, it is well known that the hydrogen evolution reaction on copper electrode is somewhat faster than on silver electrode [22]. Consequently, hydrogen evolution does not occur on silver electrode even at an overpotential of 1,000 mV vs. Ag reference electrode.

2.2.2 *Effect of Regime of Pulsating Overpotential on the Shape of Silver Powder Particles*

Typical deposits obtained in electrodeposition by constant overpotential are shown in Fig. 2.5. Disperse or irregular deposits

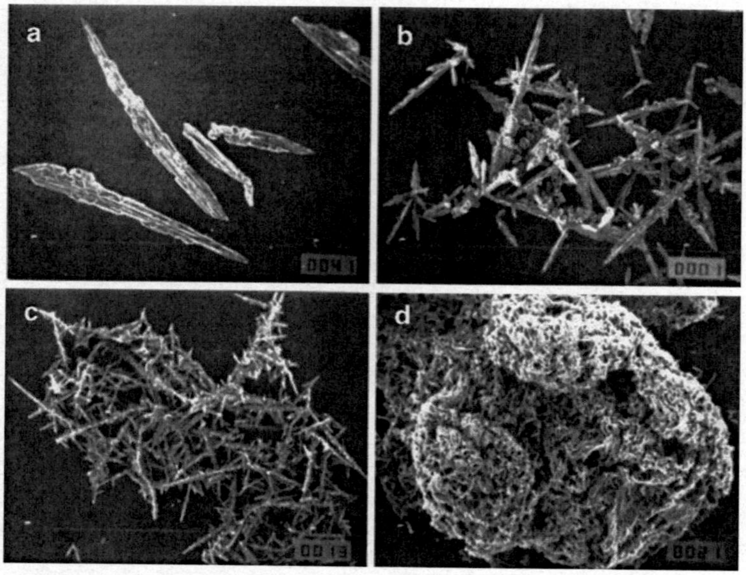

Fig. 2.5 Disperse electrodeposits of silver obtained in constant overpotential deposition at different overpotentials: (**a**) 50 mV, magnification: ×50; (**b**) 100 mV; (**c**) 150 mV; and (**d**) 200 mV; Magnification: ×100 (Reprinted from [23] with permission from the Serbian Chemical Society.)

were obtained at all overpotentials used. In this section, all presented morphologies of silver particles were obtained by electrodepositions from solution containing 0.06 M $AgNO_3$ and 1.2 M $NaNO_3$ in 0.05 M HNO_3 at the room temperature using platinum as the working electrode [23].

Single, nonbranched dendrites were obtained at 50 mV (Fig. 2.5a). Dendrites obtained at 100 mV were occasionally branched (Fig. 2.5b), and some interweaving of the growing dendrites started producing spongy-like agglomerates at 150 mV (Fig. 2.5c). The spongy-like agglomerates obtained at an overpotential of 200 mV were also dendritic, but dendrites are more than ten times shorter than those obtained at lower overpotentials (Fig. 2.6).

The shape of powder particles strongly depends on the type of working electrode used [23]. Agglomeration of silver powder

Fig. 2.6 Same as in Fig. 2.5 but (**a**) 200 mV, ×2000; (**b**) 250 mV, ×3500; and (**c**) 300 mV, ×5000 (Reprinted from [23] with permission from the Serbian Chemical Society and copied by permission from the "Electrochemistry Encyclopedia" (http://electrochem.cwru.edu/ed/encycl/) on 04/25/2007. The original material is subject to periodical changes and updates.)

particles was not observed when electrodeposition of silver was performed from the same solution in the overpotential range between 140 and 200 mV onto graphite electrodes [24].

The elimination of agglomerates of powder particles obtained at high overpotentials and the formation of single particles can be realized by treating the powder in an ultrasonic bath [23]. Figure 2.7a shows the single particles obtained by destroying the agglomerates obtained at an overpotential of 200 mV (Fig. 2.5d) in an ultrasonic bath. The detail from Fig. 2.7a is shown in Fig. 2.7b. A similar situation was observed with powder agglomerates formed at overpotentials of 250 and 300 mV [23].

The prevention of formation of agglomerates of particles and the formation of individual particles can be realized by the application of periodically changing regimes of electrolysis [23]. This can be

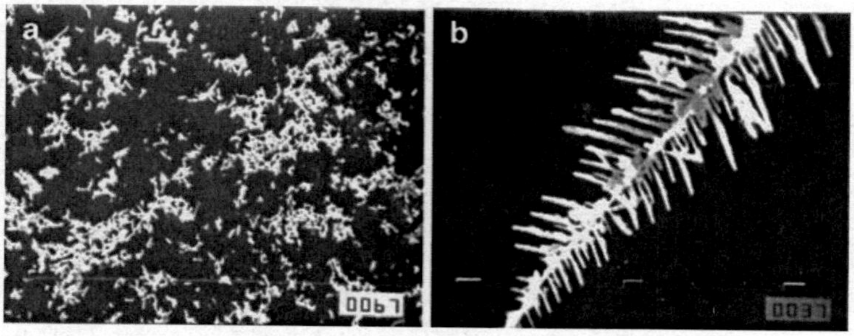

Fig. 2.7 (**a**) Same as in Fig. 2.5d but after treatment in ultrasonic bath: ×100 and (**b**) detail from Fig. 2.7a; ×5000 (Reprinted from [23] with permission from the Serbian Chemical Society.)

explained by the fact that the reversible potential of a surface with a radius of curvature r would depart from that of a planar surface by the quantity [25]:

$$\Delta E = \frac{2\sigma V}{Fr}, \qquad (2.1)$$

where σ is the interfacial energy between the metal and solution, F is the Faraday constant, V is the molar volume, and r is the protrusion tip radius. For example, this makes the equilibrium potential of spongy zinc deposits 7–10 mV more cathodic than that of zinc foil [26–28]. Obviously, the tips of dendrites characterized by small tip radii will dissolve faster than the flat surface in electrodeposition by all current or overpotential waveforms that are characterized by some anodic current flow [26–30]. In this way, the branching of dendrites is decreased and powder particles become less dendritic and more compact.

Typical deposits obtained by square-wave pulsating overpotential (PO) are shown in Figs. 2.8–2.10. It can be noticed by the analysis of these figures the decrease of both agglomeration and dendritic character of powder particles with the increasing "off" period and decreasing overpotential amplitude.

This effect was quantitatively discussed for the regime of the square-wave pulsating overpotential [23, 31].

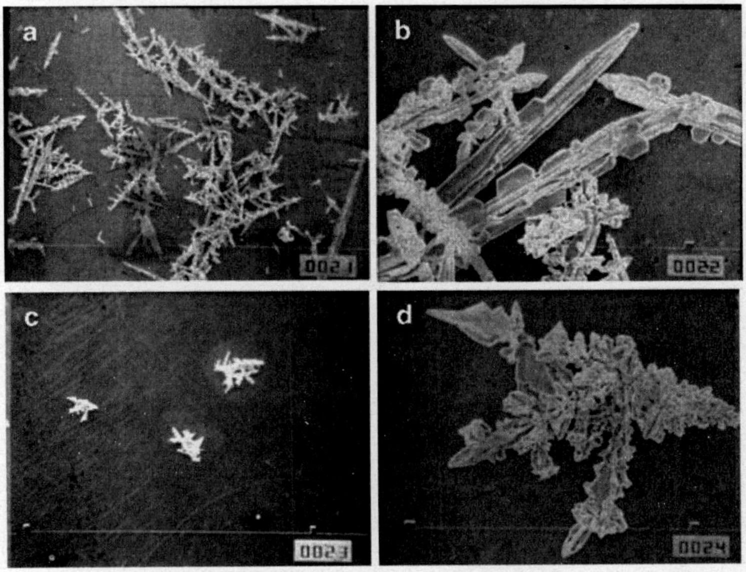

Fig. 2.8 Silver powder particles obtained by square-wave pulsating overpotential. Overpotential amplitude: 300 mV. Pulse duration: 5 ms. (**a**) Pause to pulse ratio: 1, ×100; (**b**) pause to pulse ratio: 1, ×1000; (**c**) pause to pulse ratio: 3, ×100; and (**d**) pause to pulse ratio: 3, ×1000 (Reprinted from [23] with permission from the Serbian Chemical Society.)

Square-wave pulsating overpotential is described by [23, 32]

$$\frac{\partial C}{\partial t} = D\frac{\partial^2 C}{\partial x^2}, \tag{2.2}$$

$$C(x,0) = C_0, \tag{2.3}$$

$$C(\delta,t) = C_0, \tag{2.4}$$

$$\frac{\partial C(0,t)}{\partial x} = \frac{j_0}{nFD}\left[\frac{C(0,t)}{C_0}\exp\left(\frac{2.3\eta}{b_c}\right) - \exp\left(-\frac{2.3\eta}{b_a}\right)\right], \tag{2.5}$$

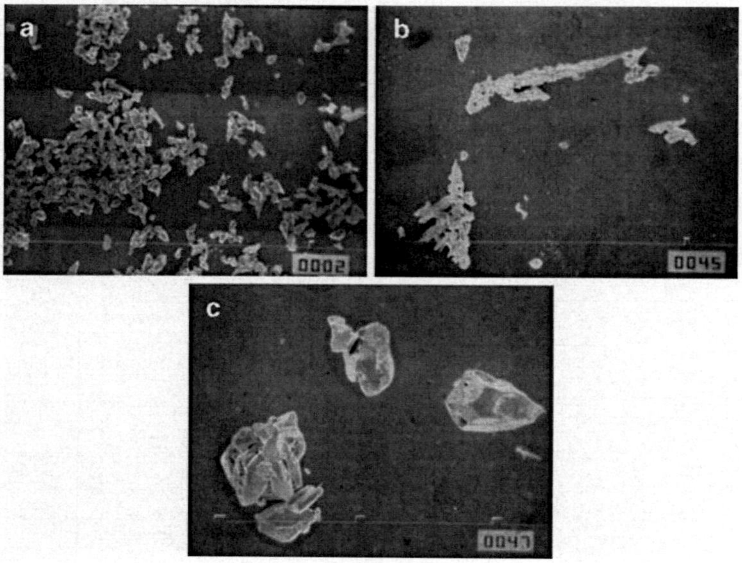

Fig. 2.9 Same as in Fig. 2.8. Overpotential amplitude: 100 mV. Pulse duration: 0.5 s. Pause to pulse ratio: 1: (**a**) ×100, (**b**) ×100; and (**c**) ×500 (Reprinted from [23] with permission from the Serbian Chemical Society.)

where j_0 is the exchange current density, n is the number of transferred electrons, D is the diffusion coefficient, b_c and b_a are the cathodic and anodic Tafel slopes, η is the overpotential, C is the concentration, C_0 is the bulk concentration, t is the time, and x is the coordinate in horizontal direction. η is given by

$$\eta = \begin{cases} \eta_A & \text{for} \quad mT_p < t \leq \left(m + \dfrac{1}{p+1}\right)T_p \\ 0 & \text{for} \quad \left(m + \dfrac{1}{p+1}\right)T_p < t \leq (m+1)T_p \end{cases} \tag{2.6}$$

$$m = 0, 1, 2, \ldots,$$

where η_A is the overpotential amplitude, T_p is the period of pulsation, and p is the pause to pulse ratio.

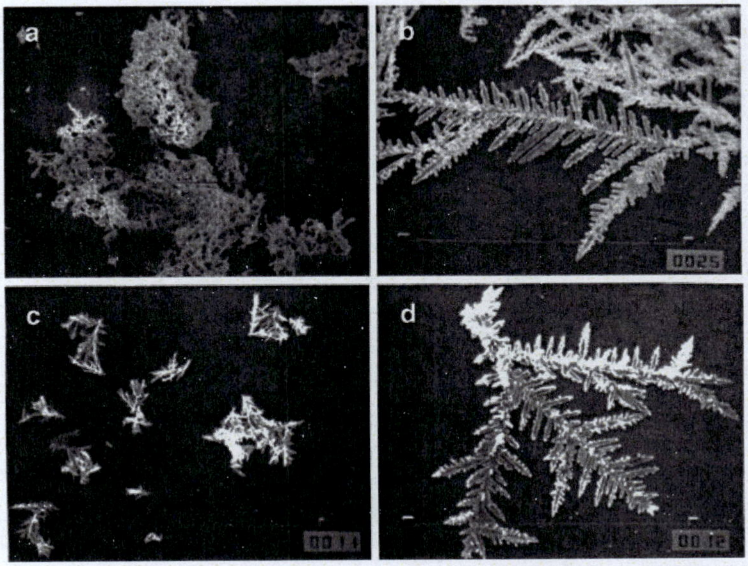

Fig. 2.10 Same as in Fig. 2.8. Overpotential amplitude: 250 mV. Pulse duration: 0.5 s. (**a**) Pause to pulse ratio: 1, ×100; (**b**) pause to pulse ratio: 1, ×1000; (**c**) pause to pulse ratio: 2, ×100; and (**d**) pause to pulse ratio: 2, ×1000 (Reprinted from [23] with permission from the Serbian Chemical Society.)

Assuming that the surface concentration in pulsating overpotential deposition does not vary with time, at sufficiently high frequencies it is easy to show the response of the current density, j, to the input overpotential:

$$j = \begin{cases} j_0\left(1 - \frac{j_{av}}{j_L}\right)\exp\left(\frac{2.3\eta_A}{b_c}\right) - j_0\exp\left(-\frac{2.3\eta_A}{b_a}\right) & \\ \qquad \text{for} \quad mT_p < t \leq \left(m + \frac{1}{p+1}\right)T_p \ \text{(a)} & \\ -j_{av}\frac{j_0}{j_L} \quad \text{for} \quad \left(m + \frac{1}{p+1}\right)T_p < t \leq (m+1)T_p \ \text{(b)} & \end{cases} \tag{2.7}$$

where j_{av} is the average current density and j_L is the limiting diffusion current density.

Equations (2.2)–(2.7) are valid for the flat electrode surfaces or protrusions with sufficiently large tip radii where the surface energy term [25] can be neglected. If it cannot be neglected, then the surface energy term affects the reaction rate [33], and for one electron transfer process, it is valid Eq. (2.8):

$$\frac{\partial C}{\partial x} = \frac{j_0}{nFD}\left[\left(1-\frac{j_{av}}{j_L}\right)\exp\left(-\frac{2\beta\sigma V}{RTr}\right)\exp\left(\frac{2.3\eta}{b_c}\right) - \exp\left(\frac{2(1-\beta)\sigma V}{RTr}\right)\exp\left(-\frac{2.3\eta}{b_a}\right)\right], \quad (2.8)$$

where β is a symmetry factor, T is the temperature and R is the gas constant.

The right-hand side of Eq. (2.5) should be transformed by taking Eq. (2.8) into account. The output current during pauses ($\eta = 0$) becomes

$$j = -j_0 \exp\left(\frac{2(1-\beta)\sigma V}{RTr}\right), \quad (2.9)$$

if $r \to 0$.

It is easy to show that the difference between the current density on the flat surface and at the tip of the dendrites during the "off" period:

$$\Delta j = j_0 - j_0 \exp\left(\frac{2(1-\beta)\sigma V}{RTr}\right), \quad (2.10)$$

if $j_{av} \approx j_L$, which leads to

$$h = h_0 + \frac{Vj_0}{F}\left[1 - \exp\left(\frac{2(1-\beta)\sigma V}{RTr}\right)\right]t \quad (2.11)$$

because of

$$\frac{dh}{dt} = \frac{V\Delta j}{F}, \quad (2.12)$$

where h is the height of protrusion and h_0 is the initial height of protrusion, and taking Eq. (2.10) into account,

$$\frac{\mathrm{d}h}{\mathrm{d}t} = \frac{Vj_0}{F}\left[1 - \exp\left(\frac{2(1-\beta)\sigma V}{RTr}\right)\right]. \tag{2.13}$$

Equation (2.11) represents the change of the height of the protrusion with tip radius r relative to the flat surface or the protrusion with sufficiently large r. In square-wave PO electrodeposition [29, 34], the filaments on the growing grains formed in spongy electrodeposition can be completely dissolved during the pause leading to the formation of compact deposit. In powder electrodeposition by the same regime, the dissolution of branches on the dendrite stalk is also expected.

Hence, the larger the "off" period, the less dendritic particles are obtained. On the other hand, the current density during the "on" period on the tip of dendrites growing inside the diffusion layer is given by:

$$j = j_0 \frac{h}{\delta} \exp\left(\frac{2.3\eta_\mathrm{A}}{b_\mathrm{c}}\right), \tag{2.14}$$

which is a somewhat modified Eq. (2.7) [35], where δ is the thickness of diffusion layer. For the same "on" period the particles will be more dendritic with increasing overpotential amplitude. In the millisecond range the ratio between the overpotential corresponding to bulk diffusion control and the activation overpotential can be reduced to the value corresponding to electrodeposition at lower overpotentials in the constant overpotential regime. Hence, deposits obtained in the PO regimes (at the same η_A and with different p used) are more similar to those obtained in the constant overpotential regime ($p = 0$) at lower overpotentials than the one corresponding to η_A in the PO regimes. The degree of diffusion control decreases with increasing p, even at the limiting diffusion current density, and it can become sufficient to produce the quality of deposits corresponding to mixed, activation, or surface energy control. This will be discussed in more detail in the case of lead electrodeposition.

The above discussion well explains the morphologies of silver deposits obtained in the millisecond range (Fig. 2.8), as well as to some extent in the second range (Figs. 2.9 and 2.10). In the latter case, the anodic current decreases during the "off" period, and this effect is also observed in electrodeposition by the regime of pulsating current (PC). In these cases, the dissolution of disperse metal occurs by the mechanism of galvanic microcells at the higher pause to pulse ratios [36–39]. Obviously, the agglomerate formation cannot be completely prevented in this way except at very low overpotential amplitude.

The second effect of the application of periodically changing regimes of electrolysis on the morphology of powder particles is that the metal adatoms on the surface, which are not in stable positions, dissolve easier than the atoms from the crystal lattice, permitting the formation of ideal crystal planes on powder particles. Powder particles of silver obtained by pulsating and polarity-reversing techniques are almost small monocrystals [1, 24].

2.3 Lead

Electrodeposition of lead belongs to the fast electrochemical processes because it is characterized by a large exchange current density, j_0 [40]. Figure 2.11 shows the polarization curve for lead electrodeposition from solution containing 0.50 M $Pb(NO_3)_2$ in 2.0 M $NaNO_3$. All experiments presented here were performed at the room temperature using cylindrical copper wires as the working electrodes [41]. The reference and counter electrodes were of a pure lead. The polarization curve obtained from this solution consisted of two parts. The characteristic of the first part is linear dependence of the current density on the overpotential. After an overpotential of about 100 mV, the current density increased quickly, and this rapid increase of the current with the overpotential is the characteristic of the second part of the polarization curve.

The linear dependence of the current density on overpotential corresponds to ohmic-controlled electrodeposition [42, 43]. The mechanism of the ohmic-controlled electrodeposition of metals is presented

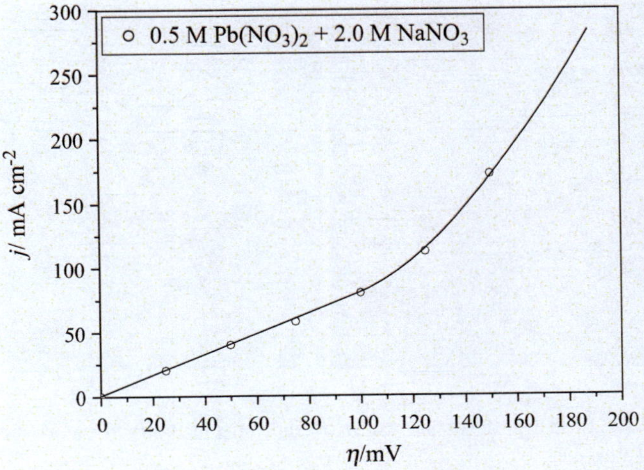

Fig. 2.11 Polarization curve for lead electrodeposition from 0.50 M $Pb(NO_3)_2$ in 2.0 M $NaNO_3$

in Chap. 1. As already mentioned, for sufficiently fast electrode processes ($j_0/j_L \geq 100$), there is no activation or diffusion polarization before the limiting diffusion current density is reached. At current densities lower than the limiting diffusion ones, the measured overpotential is due to the ohmic voltage drop between the electrode and the tip of Lugin capillary [44]. When the limiting diffusion current density is reached, the process of electrochemical deposition is under complete diffusion control. Meanwhile, as seen from Fig. 2.11, instead of the plateau of the limiting current density the inflection point on the polarization curve was observed. This inflection point on the polarization curve corresponds to an overpotential of about 100 mV. The survey of lead surface morphologies obtained at different overpotentials was the most suitable way to analyze this polarization curve and hence the lead electrodeposition system.

Figures 2.12–2.14 show morphologies of lead deposits obtained in the ohmic-controlled deposition (Fig. 2.12), in the transitional zone corresponding to the end of the ohmic-controlled electrodeposition (Fig. 2.13) and to the zone of rapid increase of the current of electrodeposition (Fig. 2.14).

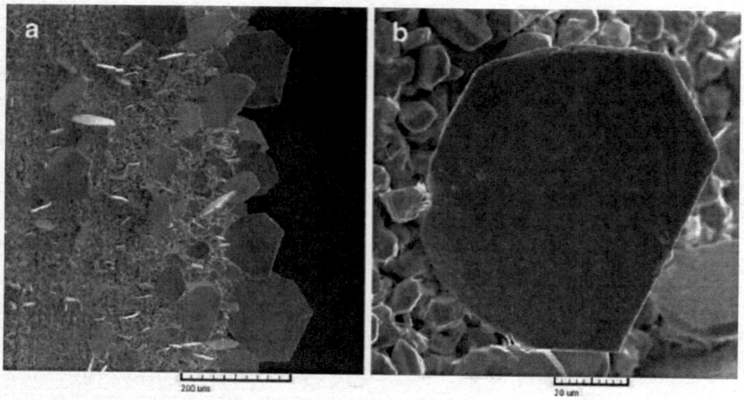

Fig. 2.12 Lead deposits obtained at an overpotential of 50 mV. Time of electrolysis: 180 s

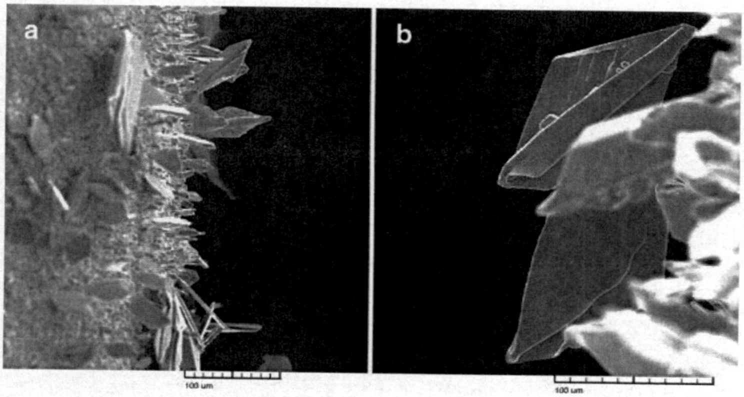

Fig. 2.13 Lead deposits obtained at an overpotential of 100 mV. Time of electrolysis: 60 s

The single lead crystals were obtained in ohmic-controlled electrodeposition at an overpotential of 50 mV (Fig. 2.12). The formation of these single crystals was accompanied by a slight increase in the current during the electrodeposition process. Figure 2.13 shows that the mixture of different regular geometric forms from single crystals to those given in Fig. 2.13b was obtained at an overpotential of 100 mV

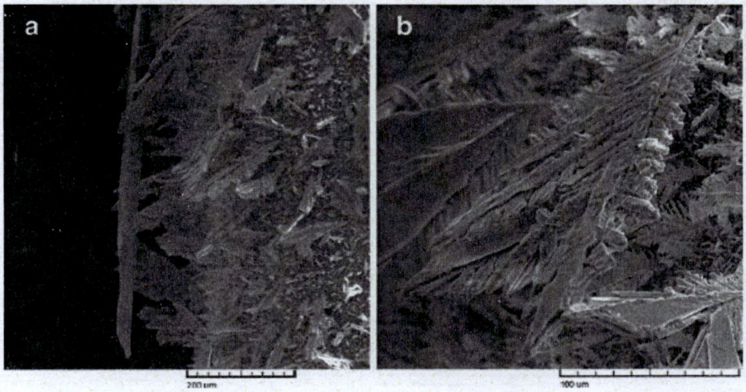

Fig. 2.14 Lead deposits obtained at an overpotential of 150 mV. Time of electrolysis: 20 s

corresponding to the inflection point. The formation of these morphological forms was accompanied by an increase in the current of electrodeposition of about 50% in relation to the initial current of electrodeposition. Finally, the increase of the current during electrodeposition processes at overpotentials higher than 100 mV was very quick and the morphology of electrodeposited lead obtained at an overpotential of 150 mV after double increase of current in relation to the initial electrodeposition current is shown in Fig. 2.14. From Fig. 2.14, it can be clearly seen that the single and two-dimensional (2D) lead dendrites were dominant morphological forms obtained at this overpotential (Fig. 2.14).

To better clarify the formed morphological forms (especially those obtained at an overpotential of 100 mV), these lead deposits were compared with those obtained in the galvanostatic regime. In the galvanostatic regime, lead electrodeposition was performed at current densities of 100 and 160 mA/cm^2. These current densities corresponded to the final currents during electrodeposition in the potentiostatic regime of electrolysis at overpotentials of 100 and 150 mV, respectively. Figures 2.15 and 2.16 show lead deposits obtained at current densities of 100 mA/cm^2 (Fig. 2.15) and 160 mA/cm^2 (Fig. 2.16). The single and 2D (two-dimensional) dendrites were the dominant morphological forms obtained at the

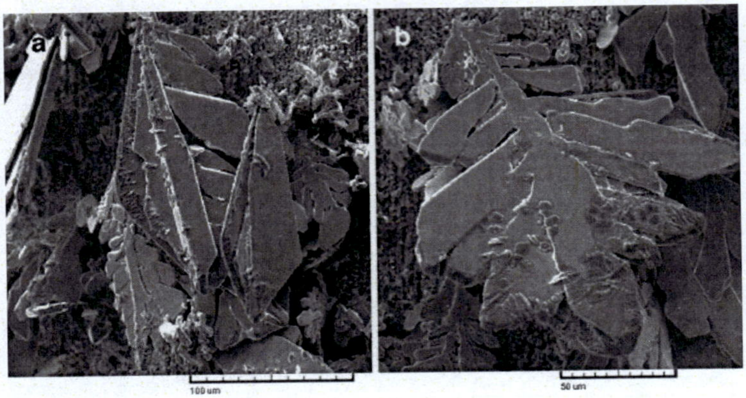

Fig. 2.15 Lead deposits obtained at a current density of 100 mA/cm^2. Time of electrolysis: 60 s

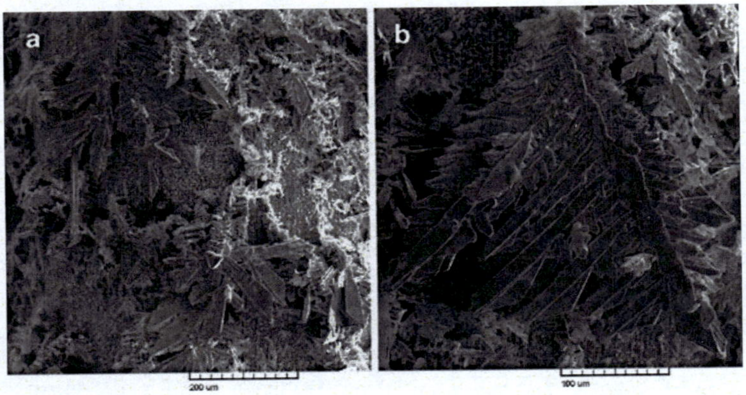

Fig. 2.16 Lead deposits obtained at a current density of 160 mA/cm^2. Time of electrolysis: 20 s

both current densities. The shape of 2D dendrites obtained at a current density of 100 mA/cm^2 clearly indicates that the regular geometric forms obtained at an overpotential of 100 mV (Fig. 2.13b) represents the precursors of dendrites.

It is necessary to note that dendrites obtained in the galvanostatic regime were more branchy structures than those obtained in the potentiostatic regime. One of the reasons for it is the fact that during electrodeposition in the galvanostatic regime at the current density of 160 mA/cm^2, the initial overpotential was about 220 mV, and the overpotential of about 110 mV was attained after electrodeposition with an electrolysis time of 20 s. A similar situation was also observed during lead electrodeposition at 100 mA/cm^2.

The shape of lead dendrites falls under the classical Wranglen's definition of a dendrite. According to Wranglen [45], dendrites consist of stalk and primary and secondary branches. From Figs. 2.14–2.16, these elements of dendrites are clearly visible. From the electrochemical point of view, a dendrite is defined as an electrode surface protrusion that grows under activation control, while electrodeposition to the macroelectrode is predominantly under diffusion control [6, 33, 46, 47]. For very fast electrodeposition processes, the critical overpotential for dendritic growth initiation, η_I, and the critical overpotential for instantaneous dendritic growth, η_c, depend on the metal surface energy and they are of the order of a few millivolts [6]. The initiation of the dendritic growth is followed by the strong increase of the apparent current density because the current density on the tips of formed dendrites is under mixed activation–diffusion or complete activation control. Hence, the strong increase of current with overpotentials higher than 100 mV corresponds to the activation-controlled electrodeposition on the tips of formed dendrites. On the other hand, the inflection point estimated to be at an overpotential of 100 mV corresponds to the transition between ohmic and activation-controlled electrodeposition process.

The two-dimensional forms were only obtained during lead electrodeposition. The two-dimensional nucleation was also a characteristic of silver electrodeposition from nitrate solution (Fig. 2.2) [4]. The polarization curve with an inflection point was also obtained in the case of silver electrodeposition from nitrate solution when the strong increase of current density was accompanied by the formation of dendrites [42, 43]. The common characteristic of these metal electrodeposition processes is affiliation to the same group of metals, i.e., to group of normal metals [Cd, Zn, Pb, Sn, Ag (silver nitrate solutions)] [2] which are characterized by a low melting point, T_m, and high exchange current density, j_0. This indicates a strong relationship between nucleation type and the type of metal.

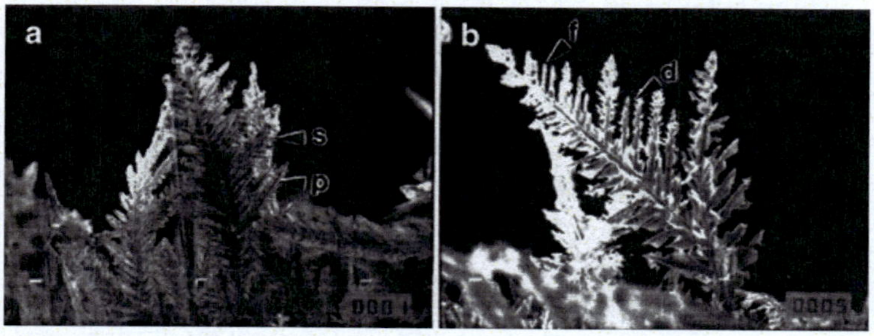

Fig. 2.17 Lead deposit obtained at constant overpotential deposition: (**a**) after 50 s at 75 mV, magnification: ×500 and (**b**) the same as in (**a**) but after 350 s at 0 mV, magnification: ×750 (Reprinted from [48] with permission from Elsevier and [6] with permission from Springer.)

2.3.1 The Effect of the Regime of Pulsating Overpotential on Morphology of Powder Particles

Figure 2.17a shows the lead dendrite obtained in the constant potentiostatic regime at an overpotential of 75 mV with an electrolysis time of 50 s. In all experiments for which results are presented in this section, the processes of electrodeposition were performed from solution containing 0.1 M $Pb(CH_3COO)_2$, 1.5 M $NaCH_3COO$, and 0.15 M CH_3COOH [48]. The other experimental conditions were same as for the previously analyzed solution. It is clearly visible that this dendrite possesses well-developed primary and secondary (indicated by S in Fig. 2.17a) dendrite arms as well as sharp crystallographic morphology. Figure 2.17b shows dendrites formed under the same conditions as those shown in Fig. 2.17a but obtained after a treatment at an overpotential of 0 mV with a time of 350 s. The dissolution of secondary dendrite arms occurs due to the difference in equilibrium potential between points with the different radii of curvature and the degree of dissolution varies from full dissolution (indicated by F in Fig. 2.17b) to the presence of

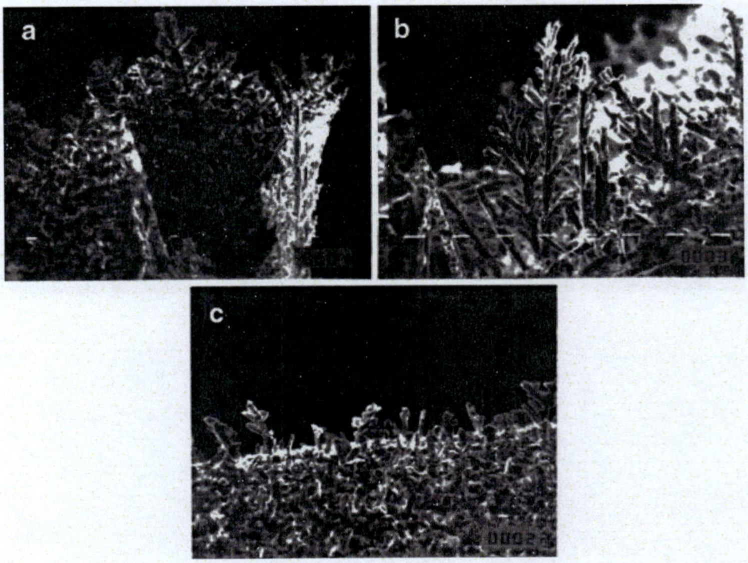

Fig. 2.18 Lead deposit obtained by square-wave pulsating overpotential at an overpotential amplitude of 75 mV with a pulse duration of 5 s and a deposition time of 120 s: (**a**) pause-to-pulse ratio: 1, ×500; (**b**) pause-to-pulse ratio: 3, ×1500, and (**c**) pause-to-pulse ratio 5, ×1000 (Reprinted from [48] with permission from Elsevier.)

secondary arm debris (indicated by D in Fig. 2.17b). Selective dissolution of dendrite arms during the "off" period due to their different tip radii is also seen from Figs. 2.18–2.20. So, increasing the pause-to-pulse ratio p enhanced the formation of more compact and less branched particles.

In addition to dissolution of dendrites, another phenomenon is observed. The sharp crystallographic forms (Figs. 2.18a, 2.19a, 2.20a, c) are absent with increasing p. Also, the sharp crystallographic edges present in the deposited dendrites become partially or well rounded after dissolution.

Of course, the formation of more compact and less branchy powder particles in the PO electrodepositions with the increasing p (at the one and the same η_A) can be ascribed to the fact that

Fig. 2.19 Lead deposit obtained by square-wave pulsating overpotential at an overpotential amplitude of 75 mV with a pulse duration of 0.5 s and a deposition time of 120 s: (**a**) pause-to-pulse ratio: 1, ×500; (**b**) pause-to-pulse ratio: 3, ×1000, and (**c**) pause-to-pulse ratio: 5, ×1000 (Reprinted from [48] with permission from Elsevier and [6] with permission from Springer.)

particles with lower tip radii are dissolved faster than those with larger ones [25]. This selective dissolution is a result of the anodic current density during "off" periods, and although this current density can be neglected in relation to the cathodic one during "on" periods, its effect on the morphology of powder particles is very high. Aside from already observed the effect of the anodic current density on the morphology of silver particles, this effect was also observed during the formation of powder particles of some other metals, such as Sn [35].

This phenomenon can be qualitatively treated and Eqs. (2.2)–(2.7) are also valid for this case. Equations (2.2)–(2.7) are valid for flat electrode surfaces or protrusions with sufficiently large tip radii, where the surface energy term [25] can be neglected. If it cannot be

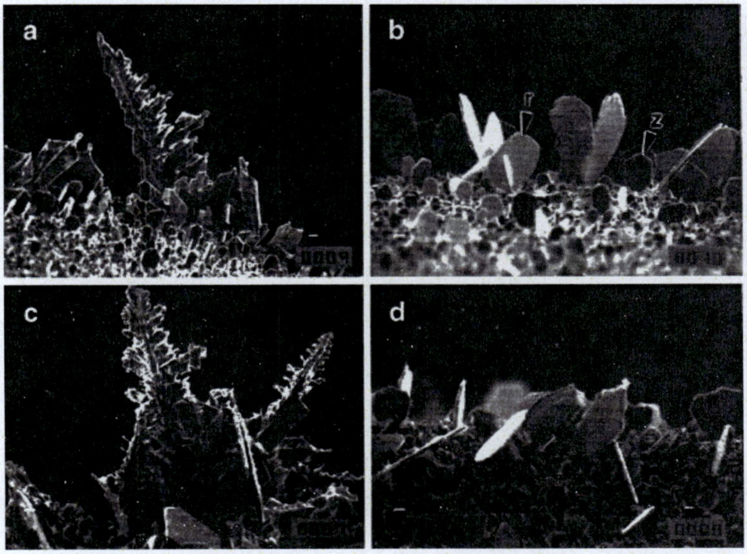

Fig. 2.20 Lead deposit obtained by square-wave pulsating overpotential at an overpotential amplitude of 75 mV with a deposition time of 120 s: (**a**) pulse duration: 0.05 s, pause-to-pulse ratio: 1, ×500; (**b**) pulse duration: 0.05 s, pause-to-pulse ratio: 3, ×1000; (**c**) pulse duration: 0.1 ms, pause-to-pulse ratio: 1, ×500; and (**d**) pulse duration 0.1 ms, pause-to-pulse ratio: 3, ×1000 (Reprinted from [48] with permission from Elsevier.)

neglected, the effect of the surface energy term on the reaction rate [33] for two electron reaction steps is described by Eq. (2.15) [31]:

$$\frac{\partial C(0,t)}{\partial x} = \frac{j_0}{nFD}\left[\left(1-\frac{j_{\mathrm{av}}}{j_{\mathrm{L}}}\right)\exp\left(\frac{2.3\eta}{b_{\mathrm{c}}}\right) - \exp\left(\frac{2\sigma V}{RTr}\right)\exp\left(-\frac{2.3\eta}{b_{\mathrm{a}}}\right)\right]. \quad (2.15)$$

The output current density, j, during pauses ($\eta = 0$) at the tip of the dendrite is presented by Eq. (2.16):

$$j = j_0\left(1-\frac{j_{\mathrm{av}}}{j_{\mathrm{L}}}\right) - j_0\exp\left(\frac{2\sigma V}{RTr}\right). \quad (2.16)$$

The corresponding output current density on the flat surface is given by Eq. (2.7b). The difference between the current density at the tip of the dendrite and on the flat surface during the "off" period is given by Eq. (2.17):

$$\Delta j = j_0 - j_0 \exp\left(\frac{2\sigma V}{RTr}\right), \tag{2.17}$$

if $j_{av} \approx j_L$, which is satisfied in most cases of dendrite growth.

Then, the change of height of surface protrusions with tip radius r relative to the flat surface is given by Eq. (2.18) [34]:

$$\frac{\mathrm{d}h}{\mathrm{d}t} = \frac{V\Delta j}{nF}, \tag{2.18}$$

and finally, for the two electron reaction steps:

$$h = h_0 + \frac{Vj_0}{2F}\left[1 - \exp\left(\frac{2\sigma V}{RTr}\right)\right]t, \tag{2.19}$$

where

$$\left(m + \frac{1}{p+1}\right)T_p < t \leq (m+1)T_p. \tag{2.20}$$

Equation (2.19) represents the height of the dendrite with tip radius r as a function of time, relative to the flat surface or to the protrusion with sufficiently large r. It is obvious that dendrites with very low tip radii can be completely dissolved during the pause. This means that the branching of dendrites can be prevented in square-wave pulsating overpotential deposition. Obviously, the larger p, the greater the degree of dissolution, as follows from Eq. (2.20). The effect is more pronounced if η_A remains constant and p increases, as seen from Figs. 2.18–2.20.

The effect of the "off" period on the micromorphology of metal powder particles can also be seen by the following analysis of lead surface morphology. The lead dendrites obtained in the PO

Fig. 2.21 Lead deposit obtained in constant overpotential deposition at 30 mV with a deposition time of 120 s. Magnification: ×750 (Reprinted from [48] with permission from Elsevier.)

electrodeposition using overpotential amplitude of 75 mV and $p = 1$ at frequencies of 10 and 5,000 Hz (Fig. 2.20a, c) are more similar to those obtained at constant overpotential of 30 mV (Fig. 2.21) than those obtained at 75 mV (Fig. 2.17a), what can be explained as follows.

The average current density in the pulsating overpotential deposition can be obtained from [34]

$$j_{\mathrm{av}} = \frac{j_0}{p+1}\left[\left(1 - \frac{j_{\mathrm{av}}}{j_{\mathrm{L}}}\right)\exp\left(\frac{2.3\eta_{\mathrm{A}}}{b_{\mathrm{c}}}\right) - p\frac{j_{\mathrm{av}}}{j_{\mathrm{L}}}\right], \tag{2.21}$$

if the anodic current density is neglected during overpotential pulses, and the overpotential amplitude can then be obtained in the form

$$\eta_{\mathrm{A}} = \frac{b_{\mathrm{c}}}{2.3}\ln\frac{j_{\mathrm{av}}}{j_0} + \frac{b_{\mathrm{c}}}{2.3}\ln\left(p + 1 + \frac{pj_0}{j_{\mathrm{L}}}\right) + \frac{b_{\mathrm{c}}}{2.3}\ln\left(\frac{1}{1 - j_{\mathrm{av}}/j_{\mathrm{L}}}\right). \tag{2.22}$$

The third term in Eq. (2.22) corresponds to bulk diffusion control. It remains constant for a determined average current density regardless of the pause-to-pulse ratio, whereas the second term, which with the first one represents the activation part of the overpotential, increases with increasing p. In this way the ratio between overpotential corresponding to bulk diffusion control and activation overpotential can be reduced to the value corresponding to the deposition at lower overpotentials in the constant overpotential regime. So, it can be expected that deposits obtained at η_A in pulsating regimes ($p > 0$) are more similar to those obtained at lower overpotentials than at η_A in the constant overpotential regime ($p = 0$).

On the other hand [32], the overpotential and current density on the tips of growing dendrites inside the diffusion layer are related by:

$$j_d = j_0 \frac{h}{\delta} \exp\left(\frac{2.3\eta_A}{b_c}\right) \tag{2.14}$$

and

$$j_{d,av} = j_0 \frac{h}{\delta(p+1)} \exp\left(\frac{2.3\eta_A}{b_c}\right) \tag{2.23}$$

in the constant and pulsating overpotential regimes, respectively. This means that dendritic growth is reduced in pulsating overpotential deposition. This is the second reason why dendrites, or powder particles, obtained in pulsating overpotential deposition are more similar to those obtained at lower overpotentials than to those obtained at overpotential amplitude in constant overpotential deposition.

The effect of the frequency of pulsation on the morphology is also illustrated by Figs. 2.18–2.20. It seems that under deposition at high frequencies (Fig. 2.20) more pronounced anodic dissolution during the pause occurs compared with the deposition at lower frequencies (Figs. 2.18 and 2.19), which leads to a formation of less dendritic deposit.

The difference between the morphologies obtained at lower and higher frequencies can be explained as follows. At sufficiently high frequencies the current response to the input overpotential is given by Eq. (2.7), while at lower frequencies it is given by

$$j = \begin{cases} j_0 \dfrac{C(0,t)}{C_0} \quad \exp\left(\dfrac{2.3\eta_\mathrm{A}}{b_\mathrm{c}}\right) - j_0 \exp\left(-\dfrac{2.3\eta_\mathrm{A}}{b_\mathrm{a}}\right) \\ \qquad \text{for} \quad mT_\mathrm{p} < t \leq \left(m + \dfrac{1}{p+1}\right) T_\mathrm{p} \qquad \text{(a)} . \\ j_0 \left(\dfrac{C(0,t)}{C_0} - 1\right) \text{for} \quad \left(m + \dfrac{1}{p+1}\right) T_\mathrm{p} < t \leq (m+1) T_\mathrm{p} \quad \text{(b)} \end{cases} \tag{2.24}$$

The anodic current during the "off" period is constant and can have significant value at sufficiently high frequencies [Eq. (2.7b)]. At lower frequencies [Eq. (2.24b)], when $C(0,t)$ becomes equal to C_0 the anodic current is zero and the average anodic current during the "off" period can be considerably lower than in the previous case.

Hence, the change of morphology of the deposit can be the indicator of the frequency at which the surface concentration becomes independent of time in pulsating overpotential deposition. This happens at a frequency of 10 Hz as seen from Figs. 2.19 and 2.20.

The formation of silver and lead powdered deposits from nitrate solutions was accompanied by the absence of hydrogen evolution as the second reaction. In the case of silver, the reason for it is the fact that the reversible potential of silver electrode is sufficiently positive to avoid hydrogen evolution reaction. In the case of lead, hydrogen evolution is extremely slow process, and then, hydrogen evolution is negligible at overpotentials at which lead was electrodeposited. The extremely high exchange current densities of these metals permit the formation of dendrites at low overpotentials, and very pronounced effect of selective dissolution during the "off" periods in the pulsating overpotential regime in both the millisecond and second range was observed.

2.4 Cadmium

Cadmium also belongs to the group of normal metals which are characterized by the large exchange current densities. In the case of cadmium, hydrogen evolution is also a slow process and this fact enables the analysis of the formation of cadmium dendrites without the effect of any parallel process.

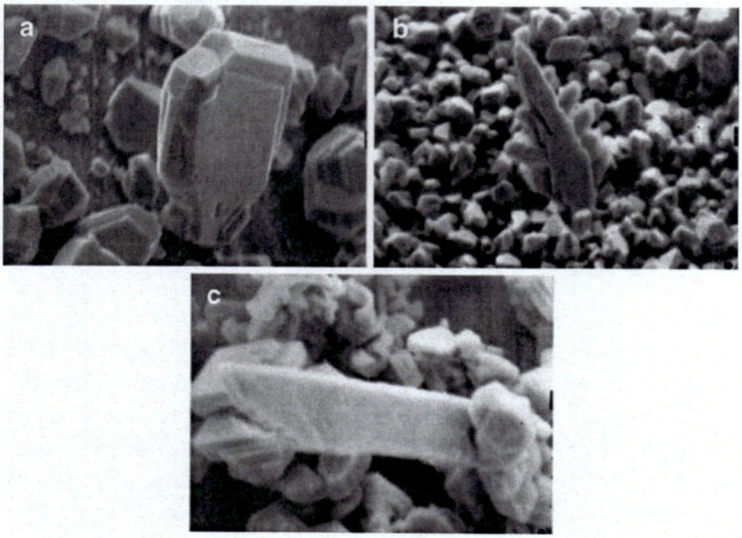

Fig. 2.22 Precursors of cadmium dendrites obtained by electrochemical depositions at overpotentials, η, of: (**a**) $\eta = 50$ mV; deposition time, $t = 2$ min; magnification: ×5000; (**b**) $\eta = 110$ mV; $t = 2$ min; ×3000; and (**c**) $\eta = 130$ mV; $t = 3$ min; ×9000 (Reprinted from [53] and [54] with permission from Elsevier and [6] and [47] with permission from Springer.)

Precursors of cadmium dendrites [47] obtained by the processes of electrochemical deposition from 0.1 M $CdSO_4$ in 0.50 M H_2SO_4 onto cadmium wire electrodes at different overpotentials are shown in Fig. 2.22. It is obvious that further growth of the dendrite precursors shown in Fig. 2.22 leads to the formation of 2D dendrites (Fig. 2.23). Around the tips of dendrite precursors, as well as around the tips of dendrites, spherical or cylindrical diffusion control can occur, which is in good agreement with the requirements of the mathematical model.

There is an induction period before initiation of dendritic growth [25, 33, 49, 50]. During this induction period, dendrite precursors are formed by the growth of suitable nuclei. According to Pangarov and Vitkova [51, 52] the orientation of nuclei is related to the overpotential used. The effect of overpotential of electrodeposition on the shape of cadmium dendrites is illustrated in Fig. 2.23.

Fig. 2.23 Cadmium dendrites obtained at overpotentials, η, of: (**a**) $\eta = 50$ mV; $t = 10$ min; ×200; (**b**) $\eta = 110$ mV; $t = 2$ min; ×400; and (**c**) $\eta = 130$ mV; $t = 3$ min; ×1000 (Reprinted from [53] with permission from Elsevier and [6] and [47] with permission from Springer.)

Most of the data concerning Co, Fe and Ni powders electrodeposition are summarized in Chapter XVIII of the book "Electrodeposition of Powders from Solutions," by Calusaru [55].

2.5 The Iron-Group Metal Powders

According to Calusaru [55] some information on electrodeposition of Co powder is available in patents [56] as well as in reviews [57]. Unfortunately these references are from 1932 and there are no data about the morphology of powder particles, since then there are no published data about electrodeposition of Co powder. Only two micrographs of Co powder, showing that the particles are very fine agglomerates, are presented in the Metallographic Atlas of Powder

Metallurgy [58], but the procedure of their production is not mentioned and it is stated that their typical application is for hard metal production.

Contrary to the limited data for Co and Ni powders electrodeposition, the highest number of 71 references for Fe powder electrodeposition exist in the book of Calusaru [55]. In this introduction only the most important ones are cited.

Because of its application in the manufacturing of porous metalloceramic bearings, of friction materials, parts for machinery, various alloys, in chemical industry, in manufacture of rechargeable batteries, etc., Fe powder is an important industrial product [55]. Significant amount of Fe powder is produced by electrochemical technique and 20% of electrodeposited Fe powders have to be blended with Fe powders produced by other procedures. The main advantage of electrodeposited Fe powder is its volumetric mass (1.5–2.2 g cm^{-2}) and its suitability for pressing, due to dendritic particle shape.

Generally speaking, in most of the electrolytes for Fe powder electrodeposition fragile deposits were obtained and transformed into powders by removal from the cathode surface and subsequent grinding [59–70]. Very few high purity Fe powders were obtained by electrodeposition. Usually these powders contain certain amount of oxides, mainly due to powder oxidation during the washing and drying procedure.

Three types of electrolytes were used for Fe powder electrodeposition (1) electrolytes based on sulfate salts [59–63]; (2) electrolytes based on chloride salts [65–69]; and (3) alkaline electrolytes: reduction of $Fe(OH)_2$ suspension in alkaline media [71]; formation of $Fe(OH)_3$, and its reduction from the suspension of Fe_3O_4 in alkaline solution [70].

Because of the possibility for $Fe(OH)_2$ or $Fe(OH)_3$ formation the suggested pH values for the electrolytes are pH < 3.

An interesting method for producing very fine Fe powder by the electrolysis in a two-layer electrolytic bath using a hydrocarbon solvent from an oil refining fraction as an upper organic layer has also been suggested [72]. It was also shown that in the presence of chelating agents, fine, nondendritic Fe powders could be obtained [73].

Two types of electrolytes for Ni powder electrodeposition were investigated: acid electrolytes [74–82] and ammoniacal electrolytes [74, 77, 83]. For acid electrolytes it is characteristic that the increase of current density and decrease of nickel ion concentration in solution cause a lowering of powder fragility. The powder is free from oxides and basic salts and dry powder could be stored in a dry place indefinitely without oxidation or structure change [74–82]. Characteristic of ammoniacal electrolytes is that the increase of ammonia concentration cause disperse and pure (without hydroxide impurities) deposit formation consisting of dark nickel particles of 4–10 μm and of larger particles of about 400 μm [74, 77, 83].

The influence of pulsating and reversing current regime on the Ni powder electrodeposition was investigated in the paper of Pavlović et al. [84]. It is found that the increase of frequency of pulsating current induces a decrease in particle size, while in the case of reversing current regimes the size of powder particles increase with increasing average current density.

2.5.1 Characterization of the Polarization Curves for Co, Fe, Ni Powders Electrodeposition

It is well known that the electrodeposition of the iron group metals (Co, Fe, Ni) occurs with simultaneous hydrogen evolution [85–89]. In such a case it is practically impossible to determine the diffusion limiting current density for their electrodeposition from the polarization curves (contrary to the case of the electrodeposition of Ag, Cu, Pd, etc) and define the beginning of powder formation expressed with the increase of current density over the diffusion limiting current density, corresponding to the process of simultaneous hydrogen evolution. To obtain correct polarization curves for the electrodeposition of these systems it is necessary to apply IR drop compensation technique. In all our cases the current interrupt technique was used, while polarization curves were recorded onto glassy carbon electrode.

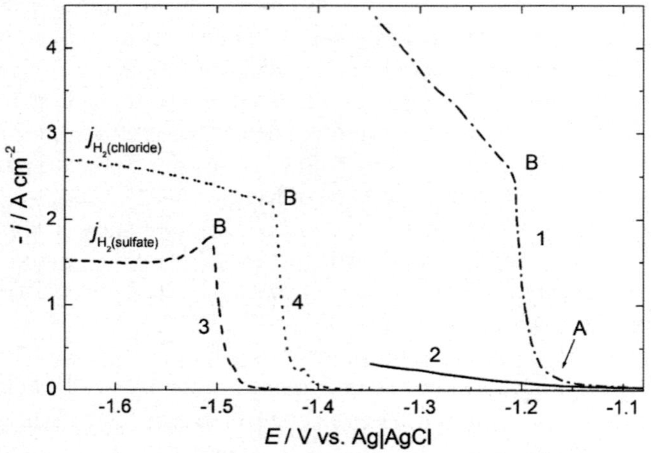

Fig. 2.24 Typical polarization curves for the electrodeposition of Co powder from ammonium sulfate–ammonium hydroxide supporting electrolyte with (1) and without (2) IR drop correction. Polarization curves for hydrogen evolution in different supporting electrolytes are also presented in the figure (*curves* 3 and 4) (Reprinted from [90, 91] with the permission of Elsevier.)

Polarization curves were recorded in a three-compartment standard electrochemical cell at the room temperature. The sweep rate was 1 mV s^{-1} and current interrupt was applied at each 0.5 s. The platinum foil counter electrode and the reference—saturated silver/silver chloride, Ag|AgCl—electrode were placed in separate compartments. The working electrode was glassy carbon disc ($d = 0.3$ cm) placed parallel to the counter electrode.

Typical polarization curves with (1) and without IR drop correction (2) are presented in Fig. 2.24. While recording polarization curves with and without IR drop correction for all investigated alloy powders electrodeposition, almost identical difference between curves with (1) and without IR drop correction (2) for all supporting electrolytes, as well as for all investigated systems is obtained. In Fig. 2.24 is shown an example for Co electrodeposition from ammonium sulfate containing supporting electrolyte [90, 91].

As can be seen after IR drop correction (dash-dot line—1) significantly different current response is obtained than the one measured without IR drop correction (solid line—2), being characterized by a sudden increase of current density at the commencement of the electrodeposition process. As far as the electrodeposition started, hydrogen evolution also started and the processes of all metal electrodeposition were accompanied by hydrogen evolution in the whole range of investigated potentials. As a consequence, extremely high current densities were recorded (since at the pH 9.2–9.5 hydrogen is evolving from water molecules) and accordingly correction for IR drop caused significant change in the shape of the polarization curves. At the same time it is important to note that because of intensive hydrogen evolution already at the commencement of the process of the electrodeposition, Co powder formation took place at current densities as low as 0.3 A cm^{-2} (low in comparison with the maximum current density of 3–6 A cm^{-2}). The characteristic for all curves is that after sudden increase of current (in this case at about -1.17 V—point A) additional inflection point appeared on the polarization curves (B) at the potential of about -1.21 V with further change of current density showing linear increase with the potential.

2.5.1.1 Polarization Curves for Hydrogen Evolution onto Glassy Carbon Electrode

As shown in Fig. 2.24 in a pure supporting electrolytes (1 M $(NH_4)_2SO_4$ + 0.7 NH_4OH or 1 M NH_4Cl + 0.7 NH_4OH) polarization curves for hydrogen evolution, corrected for IR drop, are different than those for metal or alloy electrodeposition, and their shapes and positions on the potential scale depend on the anions present in a supporting electrolyte. In both electrolytes the increase of current density corresponding to the hydrogen evolution onto glassy carbon electrode (3 and 4) is seen to take place at more negative potentials than in the case of metal or alloy electrodeposition. It is interesting to note that the overpotential for hydrogen evolution in sulfate containing supporting electrolyte is for about 0.1 V higher than that in chloride containing supporting electrolyte. Such a behavior is a clear indication that the processes of Co electrodeposition catalyze the reaction of hydrogen evolution in both supporting electrolytes.

Concerning the shape of both polarization curves for hydrogen evolution it appears that such shape is a consequence of the so called electrode effect [92–97]. This phenomenon is connected with the hydrogen bubble formation at high current densities, so that at the inflection point B the formation of bubbles becomes the rate limiting step of the electrochemical process [96, 97]. Accordingly, the shape of the polarization curves for iron group metal and their alloy electrodeposition is also defined by the process of bubble formation since the current density for hydrogen evolution in all cases amounts to 70–85% of total current density.

2.5.1.2 Current Efficiency Determination

Taking into account that simultaneous hydrogen evolution occurs in all cases, it was necessary to determine the current density for hydrogen evolution (j_H), subtract it from the measured (corrected for IR drop) current density values (j_{tot}) given in Fig. 2.25a to obtain current densities for powders electrodeposition (in this case Co powder, j_{Co}). Hence, several values of current density on the polarization curve for Co powder electrodeposition were chosen and the volume of evolved hydrogen was determined in the burette. The current for hydrogen evolution (○) was obtained using the equation [98]

$$I_{H_2} = \frac{nFV}{tV_n} = \frac{nF}{V_n} \times \frac{V}{tT} = 2.3522 \times 10^3 \times \frac{V}{tT}, \qquad (2.25)$$

where V—experimentally determined volume of the evolved hydrogen at P_{at} and $T = 298$ K corrected to the normal conditions (P^{Θ} and $T = 273$ K); t—time of hydrogen evolution under constant current; V_n—volume of 1 mol of hydrogen at normal conditions (22.4 dm^3 mol^{-1}); n—number of exchanged electrons, and F—Faraday's constant. Using this procedure polarization curve corresponding to Co powders electrodeposition (△) was obtained.

The current efficiency (○) for the electrodeposition process was obtained from the relation

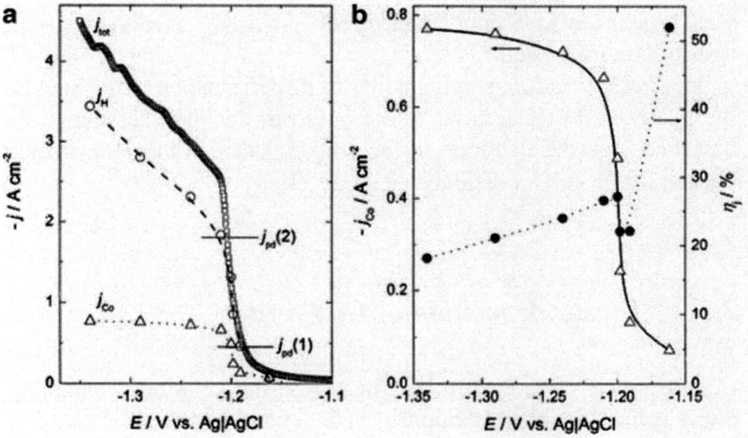

Fig. 2.25 (**a**) Polarization curve for the electrodeposition of Co powder after IR drop correction (j_{tot}), polarization curve for hydrogen evolution (j_H), and polarization curve for Co powder electrodeposition (j_{Co}) after subtraction of the current density for hydrogen evolution. (**b**) Polarization curve for Co powder electrodeposition (j_{Co}) and corresponding current efficiency curve (η_j) (Reprinted from [90, 91] with the permission of Elsevier.)

$$\eta_j(\%) = \frac{j_{Co}}{j_{tot}} \times 100 = \frac{j_{tot} - j_H}{j_{tot}} \times 100. \tag{2.26}$$

Corresponding values are presented in Fig. 2.25b.

As can be seen already at the beginning of the deposition process significant amount of hydrogen is evolving, causing the value of the current efficiency of about 52%. At more negative potentials this value decreases to about 20–25%, with the sharp decrease taking place in the region of the sharp increase in current density.

All powder samples for XRD analysis and morphology investigations were electrodeposited at the room temperature in the cylindrical glass cell of the total volume of 1 dm^3 with cone-shaped bottom of the cell to collect powder particles in it. Working electrode was a glassy carbon rod of the diameter of 5 mm, with the total surface area of 7.5 cm^2 immersed in the solution and placed in the middle of the cell. Cylindrical Pt–Ti mesh placed close to the cell walls was used as a counter electrode providing excellent current distribution in the cell.

In all cases certain, small amount of disperse deposit remained on the glassy carbon electrode, while only powder particles that were detached from the electrode surface and collected at the cone-shaped bottom of the cell were analyzed.

2.5.2 *Electrodeposition of Co Powder*

Cobalt powder was electrodeposited from the solutions containing either sulfate or chloride anions.

2.5.2.1 Electrodeposition of Co Powder from Sulfate Electrolyte

The composition of the electrolyte was 1M $(NH_4)_2SO_4$ + 0.7 M NH_4OH + 0.1 M $CoSO_4$, pH 9.5. Polarization curves were recorded on the glassy carbon disc by the procedure explained in Sect. 2.5.1. The powders for SEM and XRD analyses were electrodeposited at two constant current densities (see Fig. 2.25a): $j_{pd}(1) = -0.50$ A cm^{-2} and $j_{pd}(2) = -1.85$ A cm^{-2}.

The polarization curve and the corresponding current efficiency potential dependence are presented in Fig. 2.25b. The process of Co electrodeposition commences at about -1.10 V with the second inflection point being placed at -1.20 V. The curve for Co powder electrodeposition (j_{Co}) is characterized with well-defined current density plateau in the potential range between -1.20 V and -1.35 V, indicating that at potentials more negative than -1.20 V diffusion limiting current density for Co powder electrodeposition is reached in Fig. 2.25a. The current efficiency (η_j), obtained using equation (2.26), is seen to decrease sharply from about 52% to about 20% in the potential range of sharp increase of current density on the polarization curve j_{Co} vs. E in Fig. 2.25b. At potentials more negative than -1.20 V, η_j vs. E dependence slightly decreases from about 27% to about 20%.

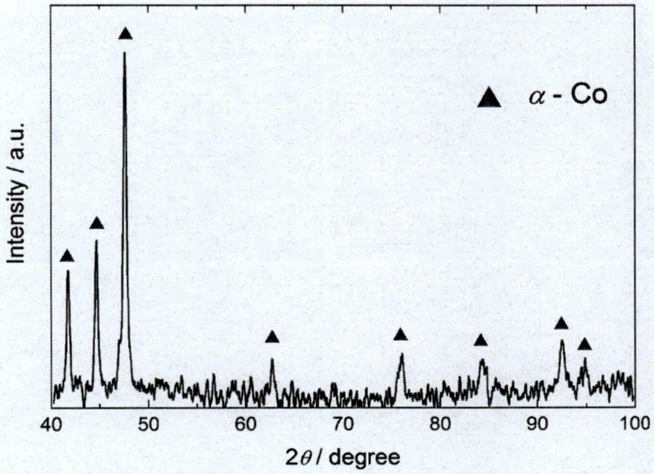

Fig. 2.26 X-ray diffraction pattern of electrodeposited cobalt powders (Reprinted from [99] with the permission of Springer-Verlag.)

X-ray diffraction pattern for Co powder samples electrodeposited either with $j_{pd}(1)$ or $j_{pd}(2)$, from sulfate or chloride electrolytes, is shown in Fig. 2.26. As can be seen the powder consists only of the hexagonal close-packed α-cobalt phase with the lattice parameters of $a = 2.5007$ Å and $c = 4.0563$ Å. No hydroxide or oxide impurities were detected [99].

In the case of Co powder electrodeposition at lower current density ($j_{pd}(1) = -0.50$ A cm^{-2}), generally two types of agglomerates are detected [99]:

1. Dendrite particles varying in the size from about 5 μm to about 50 μm, as shown in Fig. 2.27a.
2. Different types of agglomerates varying in the size from about 100 μm to about 500 μm, as it is shown in Fig. 2.27b–d. These agglomerates can further be divided into three groups:

   (a) Compact agglomerates of the size of about 200 μm to about 500 μm, composed of smaller agglomerates of the size of about 20 μm to about 50 μm, Fig. 2.27b. These agglomerates are characterized with the presence of deep cylindrical

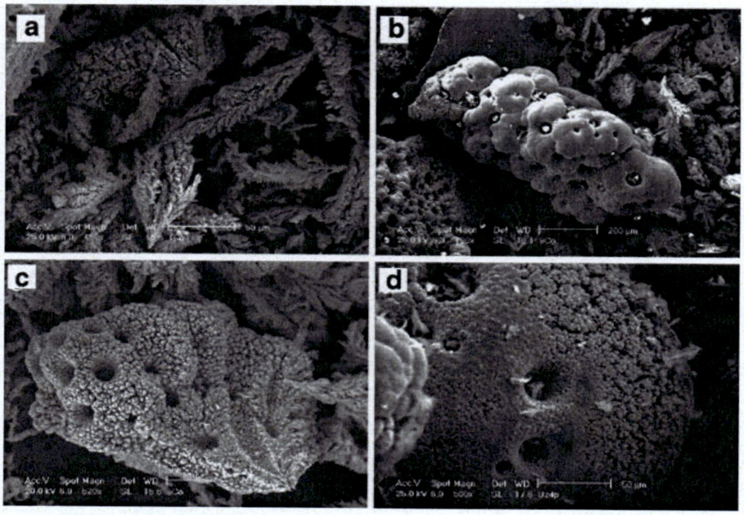

Fig. 2.27 Typical particles detected in the electrodeposited cobalt powder (Reprinted from [99] with the permission of Springer-Verlag.)

cavities and fern-like dendrites formed on the bottom of most of the cavities.

(b) Spongy-like agglomerates of different shapes varying in size from about 100 μm to about 200 μm with the fern-like dendrites also formed on the bottom of most of the cavities, Fig. 2.27c.

(c) Balls of the size of about 200 μm containing deep cavities with the fern-like dendrites formed on the bottom of most of the cavities and more or less dense cauliflower structure on the surface of these balls, as shown in Fig. 2.27d.

In Figs. 2.28a–d and 2.29a, b cross sections of some of the agglomerates detected in the Co powder deposit are shown.

As can be seen they are in good agreement with the SEM results: Fig. 2.28b ball-like particles; Fig. 2.28c spongy particles; Fig. 2.28d compact agglomerates. Although for Co powder different types of agglomerates are present, it is quite obvious that all Co powder

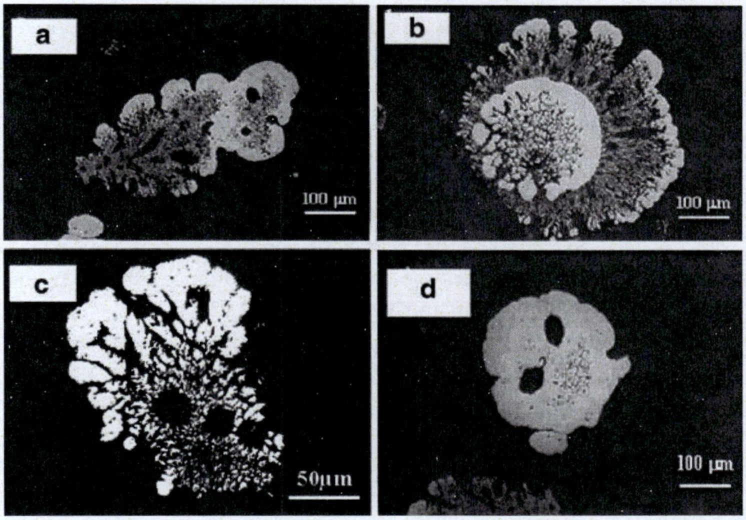

Fig. 2.28 Cross sections of some of the agglomerates detected in the Co powder deposit (Reprinted from [99] with the permission of Springer-Verlag.)

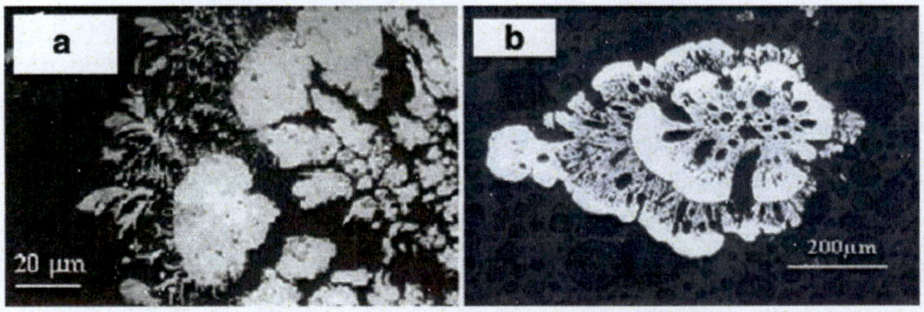

Fig. 2.29 Cross sections of some of the agglomerates detected in the Co powder deposit (Reprinted from [99] with the permission of Springer-Verlag.)

agglomerates are characterized with the presence of large number of cavities. A common characteristic of these cavities for all types of agglomerates (Fig. 2.27) is the presence of small fern-like dendrites on the bottom of most of the cavities.

2.5.2.2 Mechanism of Co Powder Electrodeposition

Considering cross sections of some of the agglomerates detected in the Co powder deposit [99], Figs. 2.28 and 2.29, it appears that they are all different, not only because different types of agglomerates have been detected but also because some pictures represent cross sections of the agglomerate parallel to the line of its growth (Fig. 2.28a), while some pictures represent cross sections of the agglomerate normal to the line of its growth (Fig. 2.28b–d). It has already been explained in the literature [100] that in the case of dendrite particles, depending on their length, it is possible that the morphology of the dendrite changes from disperse one to a compact one, as shown in Fig. 2.28a. This agglomerate seems to grow from the left to the right side.

With the time of growth the disperse agglomerate is branching in different directions and at the tip of each branch spherical diffusion is taking over the planar one, providing conditions for the growth of compact deposit as a consequence of the decrease of the local current density on the tip of each branch. After some time, these branches form compact deposit all over the agglomerate surface and the same agglomerate further grows as a compact one (right side of the particle), until it falls off from the electrode surface. In Fig. 2.28b it is most likely that a compact part of the agglomerate represents the picture of the cross section normal to the line of its growth. Depending on the way of growth, the moment of detaching agglomerate from the electrode surface and the position at the electrode surface where some agglomerate started to grow, different shapes of agglomerates are obtained (Fig. 2.28a–d), but all of them show the same characteristic of transforming disperse into compact deposit with increase in their size.

A special case is the formation of the balls of the size from about 200 μm containing deep cavities with the fern-like dendrites formed on the bottom of cavities and more or less dense cauliflower structure on the surface of these balls, Fig. 2.27d. The cross section of such agglomerate is shown in Fig. 2.28b. For some reasons this agglomerate started to grow as a ball, again starting from disperse (in the middle) and finishing with compact deposit at the surface for a reason already explained. After some time of the growth of compact deposit the conditions for planar diffusion were restored causing the formation of dendrites all over the ball surface. It should be noted that at the same time there was a local increase of overpotential due to reduction of the active surface area as a result of compact deposit formation. Dendrites were growing normal to the ball surface, and again after some time spherical diffusion took over the control of the planar one and dendrites started transforming into compact deposit. Since dendrites were not dense, they were not able to form compact deposit all over the agglomerate surface when the agglomerates detached from the electrode surface.

A cross section of one of the agglomerates parallel to the line of its growth is shown in Fig. 2.29b. As can be seen, in accordance with the previous statement, after the first front of compact deposit has been formed, disperse deposit started to grow and for the same reasons its growth finished as a compact deposit at the moment when this agglomerate detached from the electrode surface.

It is necessary to note that almost all morphological forms presented in this chapter can be explained by the above discussion, being dependent on the stage of the agglomerates, i.e., the moment when they detached from the electrode surface. It is most likely that the form presented in Fig. 2.30 should be considered as the initial stage of the growth of a second generation of dendrites, clearly detected on a cross section of ball-like particles presented in Fig. 2.29a.

Quite unique feature of all agglomerates detected in Co powder deposit is the presence of deep cavities on their surfaces and the fern-like dendrites on their bottom for most of the cavities. This is illustrated in Fig. 2.27b for compact agglomerates and in Fig. 2.27d for ball-like agglomerates. The most interesting one is the cavity detected in the ball-like agglomerates. More detailed micrograph for such cavity is shown in Fig. 2.31.

It is most likely that these cavities are the consequence of hydrogen bubble formation, preventing deposition inside the cavity. Once the bubble is liberated, the conditions for the growth of fern-like dendrites are fulfilled at the bottom of the cavity due to current distribution and restored planar diffusion. Since this agglomerate is not dense, crystals of different size can be seen in the cavity, with less

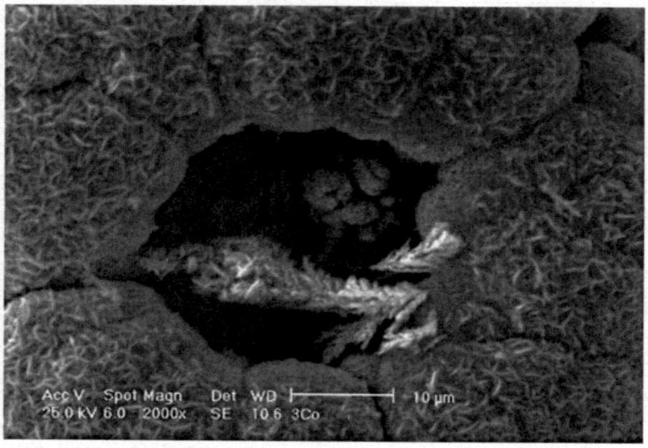

Fig. 2.30 SEM micrograph of the surface of compact agglomerate with the initial stage of the growth of a second generation of dendrites (Reprinted from [99] with the permission of Springer-Verlag.)

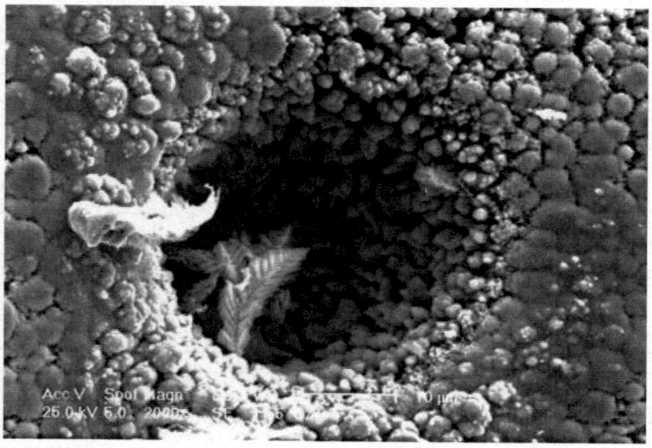

Fig. 2.31 SEM micrograph of the cavity detected on the ball-like agglomerate (Reprinted from [99] with the permission of Springer-Verlag.)

dense ones placed at the bottom and more dense ones placed close to the top of the cavity. We believe that such distribution of crystals is the consequence of current distribution over the hydrogen bubble, while at the moment of bubble liberation fern-like dendrite starts growing at the bottom of cavity in the same way as the dendrite precursors in the diffusion layer of the macroelectrode [101]. At that moment spherical diffusion is restored at the edges of cavity and compact deposit is obtained all around the cavity.

Concerning spongy deposits, it should be emphasized here that in comparison with other morphologies, traditionally called spongy, presented in the literature [47, 102] so far, the morphology of particles shown in Fig. 2.27c represents the real spongy deposit and that such similarity with the sponge has never been presented so far.

The morphology of Co powder electrodeposited at higher current density ($j_{pd}(2) = -1.85$ A cm^{-2}) is much more uniform in comparison with the morphology of Co powder electrodeposited at a lower current density.

Independently of the substrate (Co powders were electrodeposited onto GC, Co, and Ni electrodes) all agglomerates possessed spongy-like shapes, some of them being bigger (about 600 μm, Fig. 2.32a), while most of the agglomerates were of the order of 100–200 μm, Fig. 2.32b, c. The best sample for the formation of such agglomerates is presented in Fig. 2.32c. They start growing as dendrites (marked with arrow ←1) and after some time, for the reasons explained earlier, their surfaces become compact and characterized with the presence of two types of cavities (as a consequence of hydrogen bubbles formation on the agglomerate surfaces): cone-shaped ones (marked with arrow 2→) and cylindrical ones (marked with arrow ←3, identical to those obtained for Co powders electrodeposited at lower current density, $j_{pd}(1) = -0.50$ A cm^{-2}). It is interesting to note that fern-like dendrites were not formed on the bottom of cylindrical cavities, indicating that the agglomerates were detached from the electrode surface before the liberation of hydrogen bubbles formed on their surfaces. Hence, the electrolyte did not enter the cavities providing conditions for planar diffusion inside the cavity and the growth of fern-like dendrites. This is reasonable to expect since the time for agglomerates detaching should be much shorter for more intensive hydrogen evolution at $j_{pd}(2)$.

Considering Fig. 2.32 it can be concluded that most of the agglomerate surfaces are covered with very flat nodular endings. By the investigation of these surfaces at higher magnification it is discovered

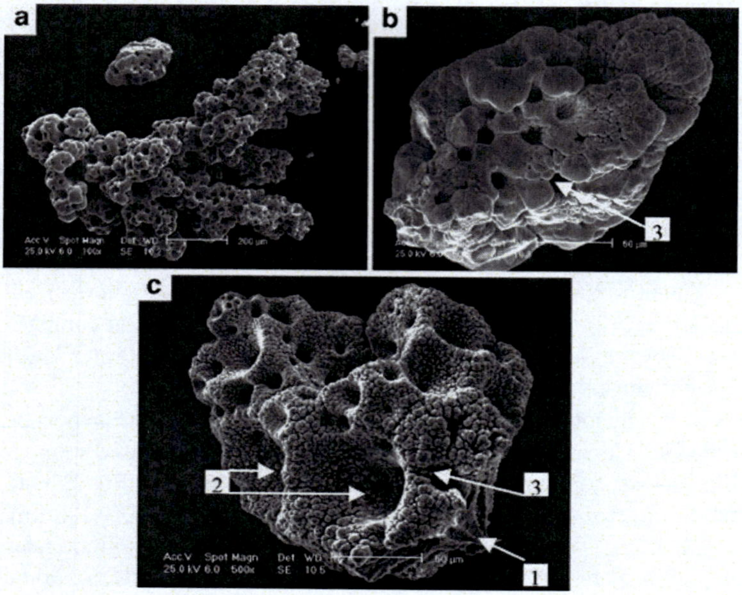

Fig. 2.32 SEM micrographs of the agglomerates electrodeposited at high current density

that these flat surfaces are actually composed of very thin (about 0.1 μm) and about 1 μm long crystals of Co, as shown in Fig. 2.33.

2.5.2.3 Electrodeposition of Co Powder from Chloride Electrolyte

The composition of the electrolyte was 1M NH_4Cl + 0.7 M NH_4OH + 0.1 M $CoCl_2$, pH 9.4. Polarization curves were recorded on the glassy carbon (GC) and Co discs. The powders for SEM and XRD analyses were electrodeposited at a constant current density of -2.00 A cm^{-2} onto GC electrode and -0.77 A cm^{-2} onto Co electrode.

The polarization curve and the corresponding current efficiency potential dependence recorded onto GC electrode are presented in Fig. 2.34. The shape of the polarization curve in Fig. 2.34a is

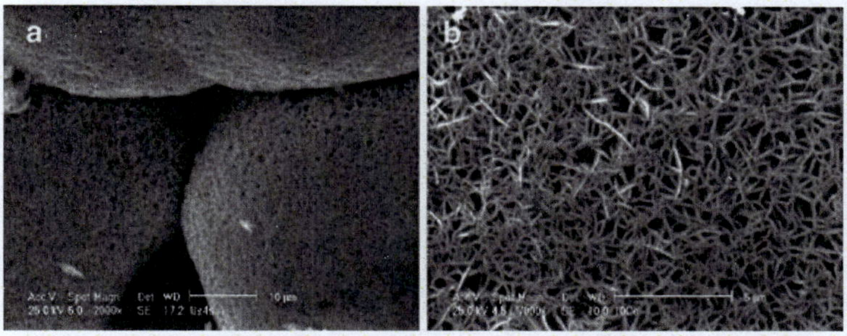

Fig. 2.33 Structure of the flat nodular endings detected on the surfaces of agglomerates electrodeposited at high current density. The same structure was detected on compact agglomerates electrodeposited at low current density

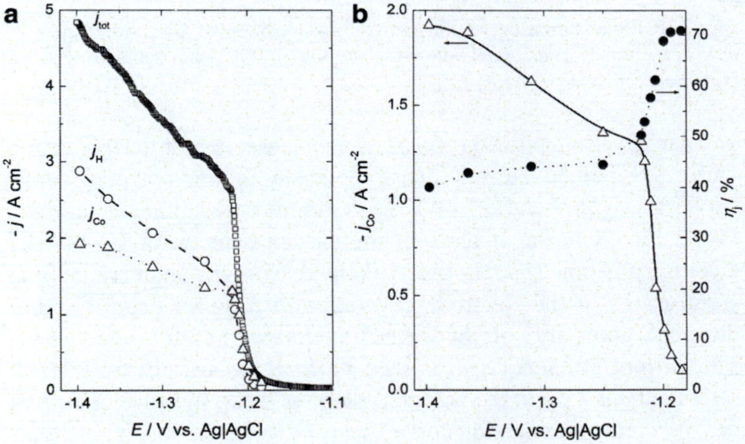

Fig. 2.34 (**a**) Polarization curve for the electrodeposition of Co powder after IR drop correction (j_{tot}), polarization curve for hydrogen evolution (j_H) and polarization curve for Co powder electrodeposition (j_{Co}) after subtraction of the current density for hydrogen evolution. (**b**) Polarization curve for Co powder electrodeposition (j_{Co}) and corresponding current efficiency curve (η_j) (Reprinted from [90, 91] with the permission of Elsevier.)

identical to the one recorded for Co powders electrodeposition from sulfate electrolyte (Fig. 2.25a). As can be seen in Fig. 2.34 the process of Co electrodeposition commences at about −1.10 V with the second inflection point being placed at −1.21 V. The curve for Co

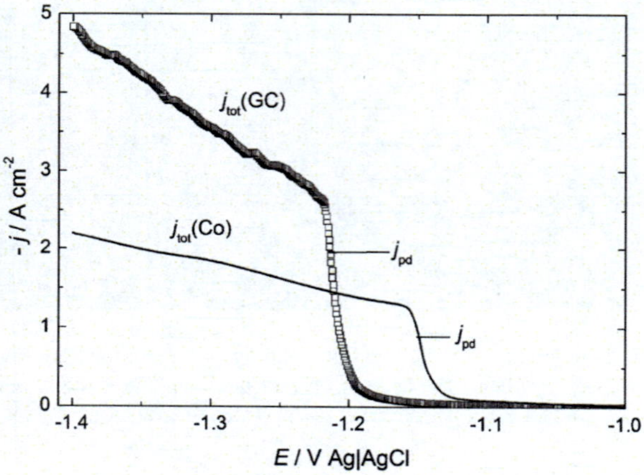

Fig. 2.35 Polarization curves for the electrodeposition of Co powder after IR drop correction (j_{tot}) recorded onto GC electrode (j_{tot}(GC)) and onto Co electrode (j_{tot}(Co))

powder electrodeposition (j_{Co}) in this case is not characterized with well-defined current density plateau in the potential range between −1.21 V and −1.40 V as was in the case sulfate electrolyte, Fig. 2.25a. A linear increase of the current density for Co powder electrodeposition (j_{Co}) is most likely due to the increase of real surface area of the electrode covered with compact deposit and/or different intensity of hydrogen evolution in this electrolyte. The current efficiency (η_j) is seen to decrease sharply from about 70% to about 45% in the potential range of sharp increase of current density on the polarization curve j_{Co} vs. E in Fig. 2.25b. At potentials more negative than −1.20 V, η_j is practically independent of potential value.

Comparing polarization curves (j_{tot}) recorded onto GC and Co electrodes (Fig. 2.35) one can see that the electrodeposition process onto Co electrode commences at more positive potentials, while the current density of the second inflection point, as well as the current density at potentials more negative than that point, is smaller. Such behavior is reasonable to expect since the overvoltage for Co electrodeposition should be higher for GC than for Co electrodes, while at the same time the increase of the real surface area onto Co

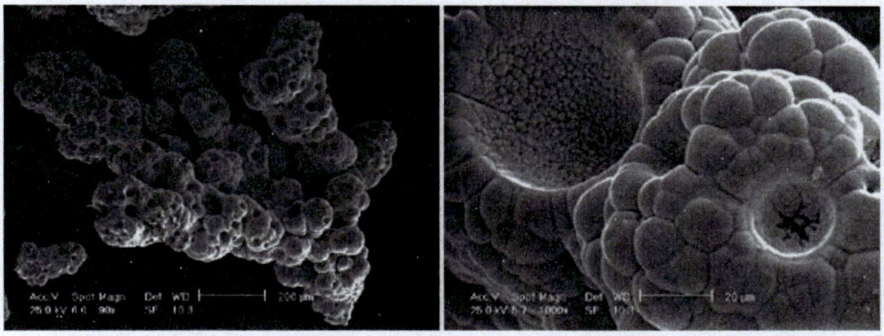

Fig. 2.36 SEM micrographs of the agglomerates electrodeposited at j_{pd} marked in Fig. 2.11 for GC and Co electrode (Reprinted from [91] with the permission of Elsevier.)

electrode should be smaller than that on the GC electrode since the epitaxial growth of Co onto Co electrode should be expected at the beginning of the electrodeposition process.

The morphology of Co powder agglomerates electrodeposited at j_{pd} (marked in Fig. 2.35), independently of the substrate (GC or Co), is presented in Fig. 2.36. It is practically the same as that for Co powder electrodeposited at $j_{pd}(2)$ from sulfate electrolyte [90], except that all nodular endings possess flat surface [91] which is actually composed of very thin (about 0.1 μm) and about 1 μm long crystals of Co, as shown in Fig. 2.33.

2.5.3 Electrodeposition of Fe Powder

Iron powder was electrodeposited from the solution containing 1 M NH_4Cl + 0.2 M $Na_3C_6H_5O_7$ + 0.1 M $FeCl_2$, pH 4.0. Polarization curves were recorded on the glassy carbon disc by the procedure explained in Sect. 2.5.1. The powders for SEM and XRD analyses were electrodeposited at the current density j_{pd} (marked in Fig. 2.37) by the procedure explained in Sect. 2.5.1.

The polarization curve for Fe powder electrodeposition is shown in Fig. 2.37. The current efficiency for Fe powder electrodeposition is very small (<2%) and the polarization curves for Fe powder

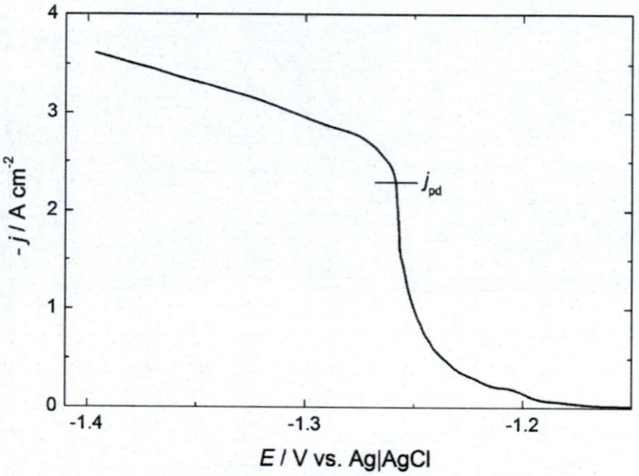

Fig. 2.37 Polarization curve for Fe powder electrodeposition

electrodeposition (j_{tot}) and for hydrogen evolution (j_H) practically overlap. Accordingly, η_j vs. E is not presented [103].

The diffractogram of electrodeposited Fe powder is presented in Fig. 2.38. The dimensions of crystallites were about 20 nm. Because of very small dimensions of crystallites only phases with the highest intensity were determined with high certainty, and these were peaks of the α—Fe phase (▲).

Agglomerates of the order of 100–200 μm with big cone-shaped cavities in Fig. 2.39a and small number of deep cylindrical cavities Fig. 2.39b are characteristic for this powder, as shown in Fig. 2.39 [103].

The surface of nodular endings on the agglomerates is not as flat as in the case of Co powder agglomerates, and the formation of new crystals on compact surface can be detected (Fig. 2.39b).

2.5.4 Electrodeposition of Ni Powder

Electrodeposition of Ni powder was investigated from three different solutions: 1 M $(NH_4)_2SO_4$ + 0.7 M NH_4OH + 0.1 M $NiSO_4$—sulfate solution; 1 M NH_4Cl + 0.7 M NH_4OH + 0.1 M $NiCl_2$—chloride

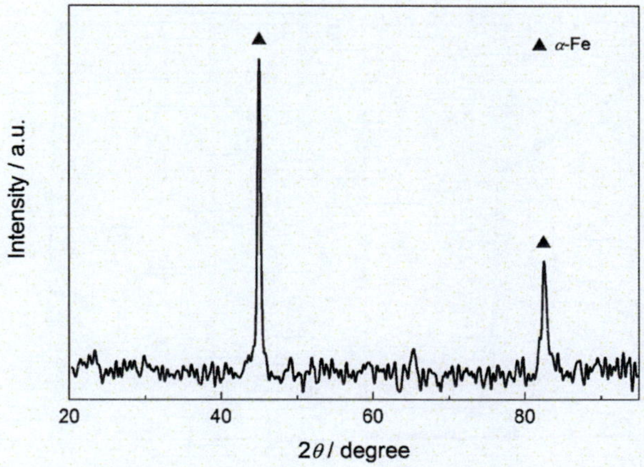

Fig. 2.38 X-ray diffraction pattern of electrodeposited Fe powder

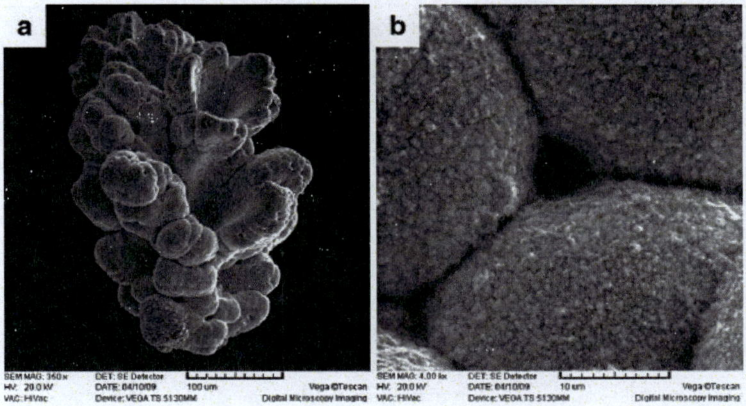

Fig. 2.39 (**a**) SEM micrograph of the agglomerates obtained in electrodeposited Fe powder. (**b**) Deep cylindrical cavity surrounded with spherical compact surface covered with new crystals (Reprinted from [103] with the permission of Elsevier.)

solution; and 1M NH_4Cl + 0.2 M $Na_3C_6H_5O_7$ + 0.1 M $NiCl_2$—citrate solution.

The polarization curve and the corresponding current efficiency potential dependence recorded onto GC electrode from sulfate solution are shown in Fig. 2.40.

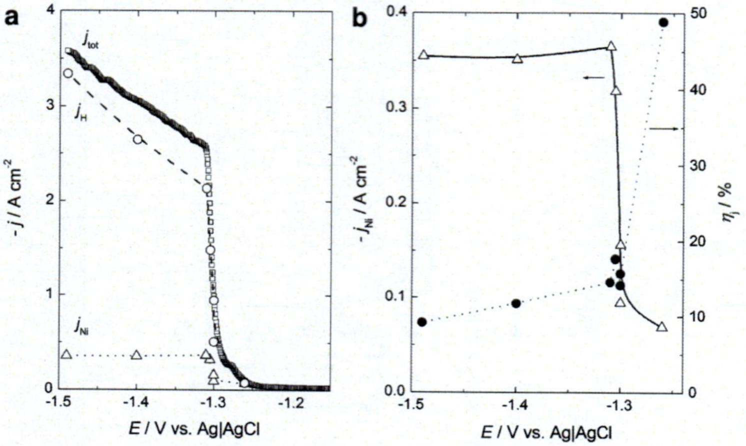

Fig. 2.40 (**a**) Polarization curve for the electrodeposition of Ni powder from sulfate solution after IR drop correction (j_{tot}), polarization curve for hydrogen evolution (j_H) and polarization curve for Ni powder electrodeposition (j_{Ni}) after subtraction of the current density for hydrogen evolution. (**b**) Polarization curve for Ni powder electrodeposition (j_{Ni}) and corresponding current efficiency curve (η_j) (Reprinted from [90] with the permission of Elsevier.)

The process of Ni electrodeposition commences at about −1.25 V with the second inflection point at −1.30 V. The curve for Ni powder electrodeposition (j_{Ni}) is characterized with well-defined current density plateau in the potential range between −1.30 V and −1.40 V. The current efficiency (η_j) is seen to decrease sharply from about 50% to about 15% in the potential range of sharp increase of current density on the polarization curve j_{Ni} vs. E in Fig. 2.40b. At potentials more negative than −1.30 V η_j slightly decreases to about 10% at −1.40 V.

The polarization curve and the corresponding current efficiency potential dependence recorded onto GC electrode from chloride solution [91] are shown in Fig. 2.41. As can be seen in Fig. 2.41a the electrodeposition of Ni in this solution commences at slightly more negative potential of about −1.28 V, while j_H is lower than that recorded in sulfate solution and accordingly the current efficiency (η_j) is higher, sharply decreasing from about 90% to about 45% in the potential range of sharp increase of current density on the

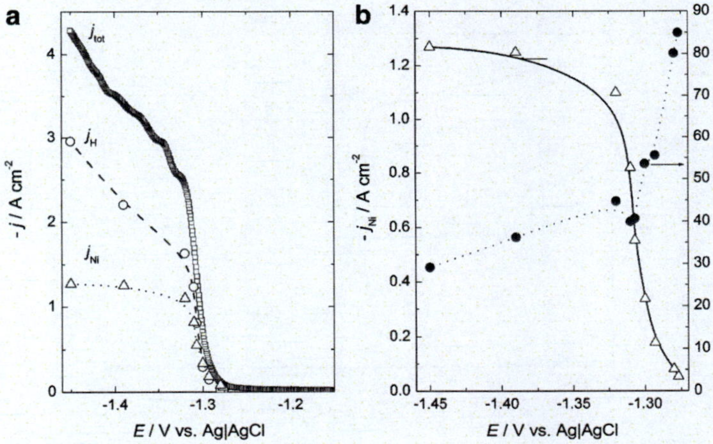

Fig. 2.41 (**a**) Polarization curve for the electrodeposition of Ni powder from chloride solution after IR drop correction (j_{tot}), polarization curve for hydrogen evolution (j_H), and polarization curve for Ni powder electrodeposition (j_{Ni}) after subtraction of the current density for hydrogen evolution. (**b**) Polarization curve for Ni powder electrodeposition (j_{Ni}) and corresponding current efficiency curve (η_j) (Reprinted from [91] with the permission of Elsevier.)

polarization curve j_{Ni} vs. E in Fig. 2.41b. At potentials more negative than −1.30 V η_j slightly decreases to about 30% at −1.40 V.

The polarization curve recorded onto GC electrode from citrate solution is compared with those from sulfate and chloride solutions in Fig. 2.42. Similar polarization curves from all solutions are obtained, except that the current density after the second inflection point for citrate solution is half of that for sulfate and chloride solutions. The (η_j) vs. E dependence recorded in citrate solution is similar to those for sulfate and chloride solutions, being practically constant (20%) in the potential range of current density plateau on the polarization curve j_{Ni} vs. E.

Samples for SEM, EDS, and XRD analyses were electrodeposited at the j_{pd} values marked in the figure, $j_{pd}(sul) = -0.5$ A cm^{-2}, $j_{pd}(chl) = -2.0$ A cm^{-2}, $j_{pd}(cit) = -1.4$ A cm^{-2}.

X-ray diffraction pattern of nickel powder sample is shown in Fig. 2.43. The same diffractograms are obtained for Ni powders electrodeposited from all three solutions. As can be seen the powder

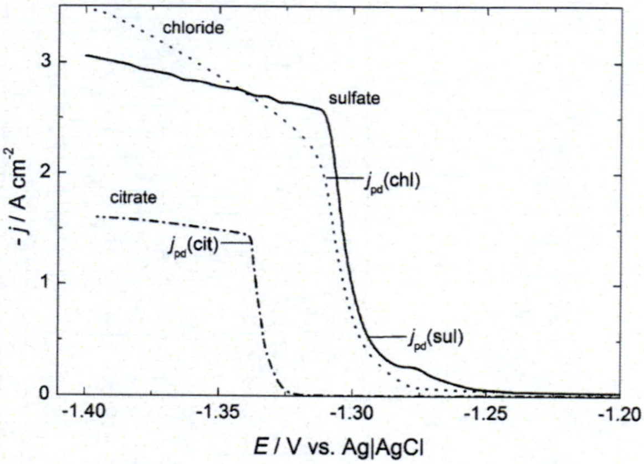

Fig. 2.42 Polarization curves for the electrodeposition of Ni powder from citrate, sulfate and chloride solutions after IR drop correction; corresponding current densities for powder electrodeposition, j_{pd}(cit), j_{pd}(sul), and j_{pd}(chl) are marked in the figure

consists only of the face-centered cubic nickel phase (β-Ni) with the lattice parameter of $a = 3.5231$ Å.

In Fig. 2.44a, b are shown SEM micrographs of Ni powder electrodeposited from sulfate solution at $j_{pd}(\text{sul}) = -0.5$ A cm^{-2}. As can be seen typical cauliflower-type agglomerates are obtained with the size of agglomerates varying from about 5 μm to about 50 μm. No hydroxide or oxide impurities were detected, which is in accordance with the literature data [74, 77, 83]. Ni powder agglomerates are very similar to those of copper [102] and since the mechanism of their growth has already been explained, it is not necessary to discuss the mechanism of their growth.

SEM micrographs of Ni powder electrodeposited from chloride solution at $j_{pd}(\text{chl}) = -2.0$ A cm^{-2} are shown in Fig. 2.45a, b. Typical spongy-like agglomerates are obtained with the size of agglomerates varying from about 200 μm to about 600 μm. As in the case of Co powder electrodeposited at high current densities from chloride and sulfate solution, cone-shaped (Fig. 2.45a) and cylindrical cavities (Fig. 2.45b) are obtained on the surface of Ni agglomerates. No fern-like dendrites were formed on the bottom of cylindrical cavities, indicating that the agglomerates were detached

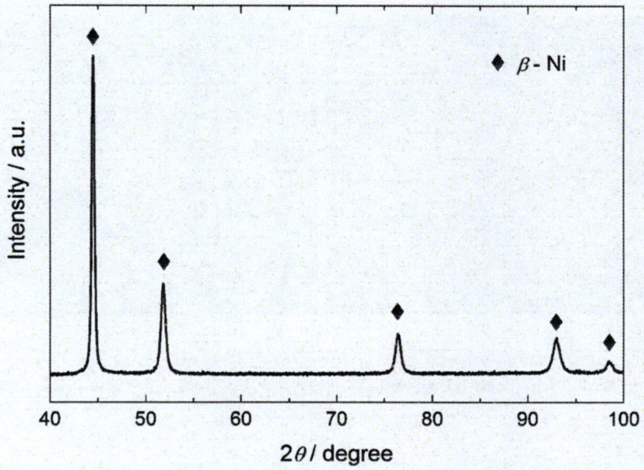

Fig. 2.43 X-ray diffraction pattern of electrodeposited Ni powder (Reprinted from [99] with the permission of Springer-Verlag.)

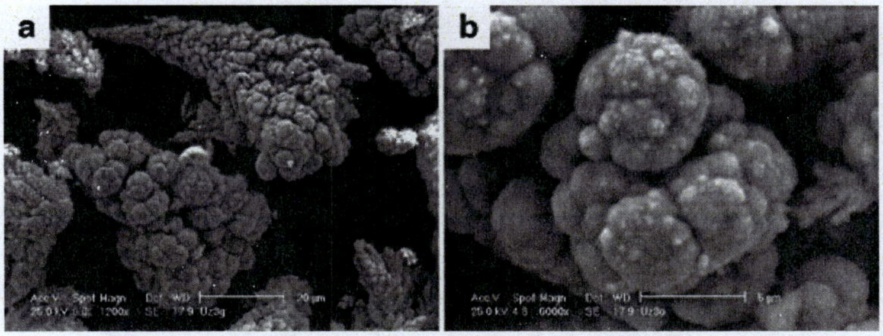

Fig. 2.44 SEM micrographs of the agglomerates electrodeposited at j_{pd}(sul) marked in Fig. 2.42 for GC electrode (Reprinted from [90] with the permission of Elsevier.)

from the electrode surface before the liberation of hydrogen bubbles formed on their surfaces, as explained in Sect. 2.5.2.2.

From the citrate solution completely different morphology of electrodeposited Ni powder is obtained [103], as shown in Fig. 2.46. Typical Ni powder is characterized by the presence of flake-like particles of the maximum size of about 50 μm covered with nodules of mainly flat surfaces. The formation of a second zone of

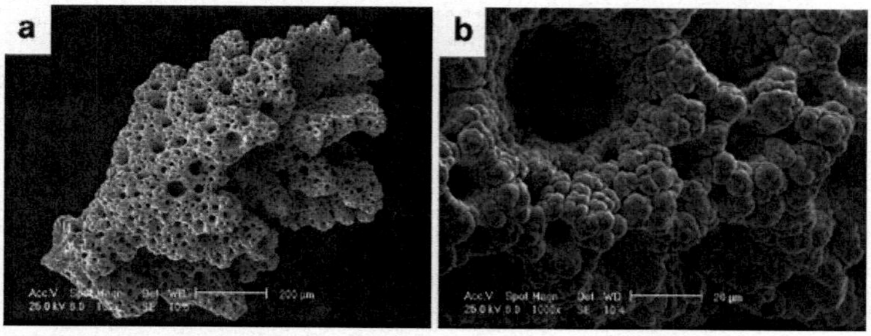

Fig. 2.45 SEM micrographs of the agglomerates electrodeposited at j_{pd}(chl) marked in Fig. 2.42 for GC electrode (Reprinted from [91] with the permission of Elsevier.)

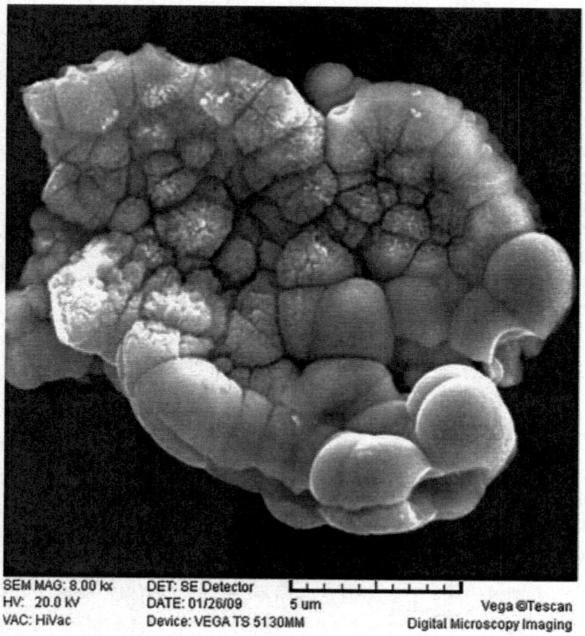

Fig. 2.46 Typical Ni powder particle electrodeposited at j_{pd}(cit) in citrate solution (Reprinted from [103] with the permission of Elsevier.)

dendrites (beginning with the formation of small crystals in the upper part of the particle presented in Fig. 2.46), typical for powder deposition [99], could be detected on some particles.

2.6 Conclusions

Powders of both academically and technologically important metals, such as silver, lead, cadmium, cobalt, nickel, and iron, were produced by the electrolytic processes. The constant regimes of electrolysis and the regime of pulsating overpotential (PO) were used for the electrochemical synthesis of powders of these metals. Morphologies of such obtained powders were characterized by the scanning electron microscopy and optical microscopy, as well as by X-ray diffraction techniques.

The processes of electrochemical deposition of silver from nitrate solutions, lead, and cadmium are characterized by high exchange current densities, and single and two-dimensional (2D) dendrites were largely formed by electrodepositions of these metals. On the other hand, very branchy dendrites constructed from corncob-like elements (similar to copper dendrites) were obtained by the electrodeposition of silver from ammonium solution which is characterized by the medium exchange current density. Also, the application of the PO regime strongly affected the morphology of silver and lead powder particles and mathematical model was presented to explain the change of morphology of particles in relation to the one obtained in the constant potentiostatic mode.

Ni powder is characterized by cauliflower-like type particles in ammoniacal electrolytes, while in citrate electrolytes the flakes of the maximum size of about 50 μm covered with nodules of mainly flat surfaces were detected. On some particles the formation of a second zone of dendrites (beginning with the formation of small crystals) could be detected. The Fe powder particles of the size of about 200 μm also contain nodules of flat surfaces, but they are characterized by the presence of cone-shaped cavities. In the case of Co powder, generally two types of particles were detected

(1) dendrite particles and (2) different types of agglomerates being characterized by the presence of cavities and fern-like dendrites formed on the bottom of most of them. These agglomerates can further be divided into three groups (a) compact agglomerates; (b) spongy-like agglomerates of different shapes, and (c) balls of the size of about 200 μm with more or less dense cauliflower structure on their surface. The growth mechanism of all agglomerates detected in the electrodeposited Co powder has been explained. The type of the agglomerate depends on its growth stage, i.e., on the moment when it becomes detached from the electrode surface.

Acknowledgments This work was financially supported by the Ministry of Education and Science of the Republic of Serbia through the Projects Nos. 1806/2002, 142032/2006, 172046/2011 and 172054/2011.

The authors are indebted to V.M. Maksimović from the Institute of Nuclear Sciences—Vinča, Belgrade, Serbia, for the XRD and SEM analyses of some investigated systems, as well as to Dr. G. Branković and E.R. Ivanović for SEM analysis of lead deposits.

The authors also wish to express their gratitude to M.G. Pavlović, Institute of Electrochemistry ICTM, Belgrade, Serbia, for useful discussions in the case of Co and Ni powders electrodeposition.

References

1. Pavlović MG, Popov KI (2005) Electrochem Encycl. http://electrochem.cwru.edu/ed/encycl/
2. Winand R (1998) Electrochim Acta 43:2925
3. Price PB, Vermilyea DA (1958) J Chem Phys 28:720
4. Djokić SS, Nikolić ND, Živković PM, Popov KI, Djokić NS (2011) ECS Trans 33:7
5. Popov KI, Krstajić NV, Popov SR (1984) Surf Technol 22:245
6. Popov KI, Djokić SS, Grgur BN (2002) Fundamental aspects of electrometallurgy. Kluwer Academic/Plenum, New York
7. Nikolić ND, Popov KI, Pavlović LjJ, Pavlović MG (2006) J Electroanal Chem 588:88
8. Nikolić ND, Popov KI, Pavlović LjJ, Pavlović MG (2006) Surf Coat Technol 201:560
9. Nikolić ND, Popov KI, Pavlović LjJ, Pavlović MG (2007) Sensors 7:1
10. Nikolić ND, Branković G, Pavlović MG, Popov KI (2008) J Electroanal Chem 621:13

11. Nikolić ND, Popov KI (2010) Hydrogen co-deposition effects on the structure of electrodeposited copper. In: Djokić SS (ed) Electrodeposition: theory and practice, vol 48, Modern aspects of electrochemistry. Springer, New York, pp 1–70
12. Nikolić ND, Pavlović LjJ, Pavlović MG, Popov KI (2007) Electrochim Acta 52:8096
13. Nikolić ND, Pavlović LjJ, Branković G, Pavlović MG, Popov KI (2008) J Serb Chem Soc 73:753
14. Nikolić ND, Pavlović LjJ, Pavlović MG, Popov KI (2008) Powder Technol 185:195
15. Nikolić ND, Branković G, Pavlović MG, Popov KI (2009) Electrochem Commun 11:421
16. Nikolić ND, Branković G, Maksimović VM, Pavlović MG, Popov KI (2010) J Solid State Electrochem 14:331
17. Nikolić ND, Branković G, Maksimović VM, Pavlović MG, Popov KI (2009) J Electroanal Chem 635:111
18. Nikolić ND, Branković G, Popov KI (2011) Mater Chem Phys 125:587
19. Nikolić ND, Branković G (2010) Electrochem Commun 12:740
20. Nikolić ND, Branković G, Maksimović V (2012) J Solid State Electrochem 16:321
21. Popov KI, Vojnović M, Rikovski G (1968) Hemijska Industrija 8:1392 (in Serbian)
22. Trasatti S (1972) J Electroanal Chem 39:163
23. Popov KI, Pavlović MG, Stojilković ER, Radmilović V (1996) J Serb Chem Soc 61:47
24. Pavlović MG, Maksimović MD, Popov KI, Kršul MB (1978) J Appl Electrochem 8:61
25. Barton JL, Bockris JO'M (1962) Proc R Soc A268:485
26. Bek RYu, Kudryavtsev NT (1961) Zh Prikl Khim 34:2013
27. Bek RYu, Kudryavtsev NT (1961) Zh Prikl Khim 34:2020
28. Arouete S, Blurton KF, Oswin HG (1969) J Electrochem Soc 116:166
29. Popov KI, Keča DN, Anđelić MD (1978) J Appl Electrochem 8:19
30. Popov KI, Anđelić MD, Keča DN (1978) Glasnik Hem Društva Beograd 43:67
31. Popov KI, Pavlović MG, Remović GŽ (1991) J Appl Electrochem 21:743
32. Popov KI, Maksimović MD, Zečević SK, Stojić MR (1986) Surf Coat Technol 27:117
33. Despić AR, Popov KI (1972) Transport controlled deposition and dissolution of metals. In: Conway BE, Bockris JO'M (eds) Modern aspects of electrochemistry, vol 7. Plenum, New York, pp 199–313
34. Popov KI, Maksimović MD (1989) Theory of the effect of electrodeposition at periodically changing rate on the morphology of metal deposition. In: Conway BE, Bockris JO'M, White RE (eds) Modern aspects of electrochemistry, vol 19. Plenum press, New York, pp 193–250
35. Popov KI, Pavlović MG, Jovićević JN (1989) Hydrometallurgy 23:127

36. Romanov VV (1963) Zh Prikl Khim 36:1050
37. Romanov VV (1961) Zh Prikl Khim 34:2692
38. Romanov VV (1963) Zh Prikl Khim 36:1057
39. Popov KI, Maksimović MD, Simičić NV, Krstajić NV (1984) Surf Technol 22:159
40. Vijh AK, Randin JP (1977) Surf Technol 5:257
41. Nikolić ND, Lačnjevac U, Branković G. J Solid State Electrochem doi: 10.1007/s10008-011-1626-y
42. Popov KI, Živković PM, Krstić SB, Nikolić ND (2009) Electrochim Acta 54:2924
43. Popov KI, Živković PM, Nikolić ND (2010) The effect of morphology of activated electrodes on their electrochemical activity. In: Djokić SS (ed) Electrodeposition: theory and practice, vol 48, Modern aspects of electrochemistry. Springer, New York, pp 163–213
44. Bockris JO'M, Reddy AKN, Gamboa-Aldeco M (2000) Modern electrochemistry: fundamentals of electrodics, vol 2A, 2nd edn. Kluwer Academic/Plenum, New York, p 1107
45. Wranglen G (1960) Electrochim Acta 2:130
46. Diggle JW, Despić AR, Bockris JO'M (1969) J Electrochem Soc 116:1503
47. Popov KI, Krstajić NV, Čekerevac MI (1996) The mechanism of formation of coarse and disperse electrodeposits. In: White RE, Conway BE, Bockris JO'M (eds) Modern aspects of electrochemistry, vol 30. Plenum, New York, pp 261–312
48. Popov KI, Stojilković ER, Radmilović V, Pavlović MG (1997) Powder Technol 93:55
49. Popov KI, Maksimović MD, Trnjančev JD, Pavlović MG (1981) J Appl Electrochem 11:239
50. Popov KI, Pavlović MG, Maksimović MD (1982) J Appl Electrochem 12:525
51. Pangarov NA (1964) Electrochim Acta 9:21
52. Pangarov NA, Vitkova SD (1966) Electrochim Acta 11:1733
53. Popov KI, Čekerevac MI (1989) Surf Coat Technol 37:435
54. Čekerevac MI, Popov KI (1989) Surf Coat Technol 37:441
55. Calusaru A (1979) Electrodeposition of powders from solutions. Elsevier, New York, pp 363–389
56. Mathers, Turner (1932) British Patent No. 403281
57. Mathers (1932) Met Ind (NY) 30:321, 396, 368
58. Huppomann WJ, Dalal K (1986) Metallographic atlas of powder metallurgy. Werlag Schmid GmbH, pp 21–63
59. Kudryavtsev N, Petrova A (1932) Novosti tehniki, N.K.T.R 171
60. Hardy C, Mantell C (1937) French Patent No. 814500
61. Hardy C, Mantell C (1938) US Patent No. 2157699
62. Kudryavtsev N, Tereshkocitch E (1948) Zhur Prikl Khim 12:1298
63. Borok B, Olhov I (1948) Metallurgy of powders, pp 34–35
64. Kalaida T, Rozenzweig S (1936) Collect Works Chem Cells 1:48
65. Balshin M (1935) NIIMASH 12:5

66. Mantell C (1939) British Patent No. 503306
67. Casey H (1949) US Patent No. 2481079
68. Wranglen G (1950) Acta Polytechnica Electr Eng Ser 2:69
69. Sanvordenkar K, Tendolkar G (1954) J Indian Chem Soc Ind News Ed 17:13
70. Rozenzweig S (1936) Sov avt 46284
71. Gardam G (1951) Powder Metall 6:75; (1947) Spec Rep 38:3
72. Zhelibo EP, Aryupina KA, Natanson EM (1973) Powder Metall 122:14
73. Chu CM, Wan CC (1992) J Mater Sci 27:6700
74. Fedorova O (1938) Zh Obsch Khim 8:1711
75. Hardy C, Mantell C (1937) French Patent No. 815500
76. Loshkarev M, Gernostaleva O, Kriukova A (1946) Zhur Prikl Khim 19:739
77. Levin A (1946) Zhur Prikl Khim 19:779
78. Hiruma K (1949) J Electrochem Soc Jpn 17:160
79. Drozdov B (1955) Zhur Prikl Khim 1:45
80. Nicol A (1946) CR 222:1043
81. Wranglen G (1950) Acta Polytechnica Electr Eng Ser 2:69
82. Kuroda M, Yto G, Shimizu Y (1953) Rep Sci Res (Japan) 29:429
83. Mantell C (1941) US Patent No. 2233103
84. Pavlović MG, Hadžismajlović DŽ, Popov KI (1991) Chem Ind 45:39
85. Santos JS, Matos R, Trivinho-Strixino F, Pereira EC (2007) Electrochim Acta 53:644
86. Gabe DR (1997) J Appl Electrochem 27:908
87. Palomar-Pardave M, Scharifker BR, Arce EM, Romero-Romo M (2005) Electrochim Acta 50:4736
88. Rojas M, Fan CL, Miao HJ, Piron DL (1992) J Appl Electrochem 22:1135
89. Diaz SL, Calderon JA, Barcia OE, Mattos OR (2008) Electrochim Acta 53:7426
90. Jović VD, Jović BM, Pavlović MG (2006) Electrochim Acta 51:5468
91. Jović VD, Jović BM, Maksimović V, Pavlović MG (2007) Electrochim Acta 52:4254
92. Vogt H, Balzer RJ (2005) Electrochim Acta 50:2073
93. Vogt H (1997) Electrochim Acta 42:2695
94. Vogt H (1999) J Appl Electrochem 29:137
95. Wüthrich R, Fascio V, Bleuler H (2004) Electrochim Acta 49:4005
96. Wüthrich R, Bleuler H (2004) Electrochim Acta 49:1547
97. Wüthrich R, Commninellis Ch, Bleuler H (2005) Electrochim Acta 50:5242
98. Lowenheim FA (1974) Modern electroplating, 3rd edn. Wiley-Interscience, New York
99. Jović VD, Maksimović V, Pavlović MG, Popov KI (2006) J Solid State Electrochem 10:373
100. Popov KI, Krstić SB, Obradović MC, Pavlović MG, Pavlović LjJ, Ivanović ER (2004) J Serb Chem Soc 69:43
101. Murashova IB, Pomosov AV (1989) Itogi nauki i tekhniki, Seria Elektrokhimiya, vol 30. VINITI, Moskva, p 90
102. Popov KI, Krstajić NV (1983) J Appl Electrochem 13:775
103. Lačnjevac U, Jović BM, Jović VD (2009) Electrochim Acta 55:535

Chapter 3
Electrodeposition of Copper Powders and Their Properties

Nebojša D. Nikolić and Konstantin I. Popov

3.1 Introduction

A powder is a finely divided solid, smaller than 1,000 μm in its maximum dimension. A particle is defined as the smallest unit of a powder. The particles of a powder may assume various forms and sizes, whereas powders, an association of such particles, exhibit, more or less, the same characteristics as if they were formed under identical conditions and if the manipulation of the deposits after removal from the electrode was the same [1, 2]. The size of particles of many metal powders can vary in a quite wide range from a few nanometers to several hundreds of micrometers. The most important properties of a metal powder are the specific surface, the apparent

N.D. Nikolić (✉)
ICTM-Institute of Electrochemistry, University of Belgrade, Njegoševa 12, P.O.B. 473,11001 Belgrade, Serbia
e-mail: nnikolic@tmf.bg.ac.rs

K.I. Popov
ICTM-Institute of Electrochemistry, University of Belgrade, Njegoševa 12, P.O.B. 473,11001 Belgrade, Serbia

Faculty of Technology and Metallurgy, University of Belgrade, Karnegijeva 4, P.O.B. 3503,11001 Belgrade, Serbia
e-mail: kosta@tmf.bg.ac.rs

Stojan S. Djokić (ed.), *Electrochemical Production of Metal Powders*, Modern Aspects of Electrochemistry 54, DOI 10.1007/978-1-4614-2380-5_3,

density, the flowability, and the particle grain size and distribution. These properties, called decisive properties, characterize the behavior of a metal powder.

Different methods for the production of metal powders including mechanical commuting, chemical reaction, electrolysis, and liquid metal atomization are used in practice [1]. Powders of about 60 metals can successfully be produced by electrolysis. The majority of metallic powders are obtained by molten-salts electrolysis. However, due to technological advantages and various industrial applications most of the practically useful powders, e.g., copper, iron, and nickel, are produced from aqueous solutions [3].

Electrodeposited metallic powders are mainly produced in a dendritic form. The dendrites can spontaneously fall off or can be removed from the electrode by tapping or other similar techniques. Also, the powders are obtained as flakes or needles, fibrous or spongy forms, etc., depending on the conditions of electrodeposition and on the nature of the metal.

The formation of powders by electrolysis is an economical processing method with a low capital investment and operational cost. The main advantages of this method in relation to other methods of metal powder production are high purity of the produced powder, which can be easily pressed and sintered, and low oxygen content [1, 2]. It is environmentally a friendly way of powder production which enables working in a closed circuit [4]. Metal powders can be formed by both potentiostatic and galvanostatic regimes of electrolysis [2, 5, 6]. In addition, metal powders can be produced under periodically changing regimes of electrolysis, such as pulsating overpotential (PO), pulsating current (PC), and reversing current (RC) [5, 6].

3.1.1 Formation of Metal Powders by Electrolysis: Comparison of Potentiostatic and Galvanostatic Regimes

All metals which can be electrodeposited exhibit a tendency to appear in the form of powders at current densities larger than a certain critical value, j_{cr}. This value is equal to the limiting diffusion current density in galvanostatic deposition, as shown by Hirakoso [7, 8] and Ibl [9, 10]. At the same time Kudra et al. [11, 12] observed that the product of the current density used and the square root of the time of powder formation t_i is a constant quantity. The time for powder formation at current densities equal to j_{cr} and larger can be observed visually as the electrode is seen to turn suddenly from a lustrous to a black appearance. During this induction period a compact deposit is formed. Ibl and Schadegg [13] showed that at sufficiently high deposition times, powdered deposits can be obtained at all overpotentials which correspond to the limiting diffusion current plateau. It is known that the limiting diffusion current plateau covers a wide range of overpotentials, because of a large change of overpotential for extremely small changes of current density. Therefore, as pointed out by Calusaru [3], the formation of electrolytic powder cannot be localized at a certain point on the current density versus polarization curve by using only current density measurements. Calusaru et al. [3, 14, 15] showed that there are three ranges of overpotential which can be determined from studies of deposit structure. Similar facts were reported by Russev [16] and Theis et al. [17]. According to Calusaru [3], there are regions of overpotential in which compact, rough, and really powdered deposits are obtained. Popov et al. [18] showed that in potentiostatic deposition two critical values of overpotential can be determined: the critical overpotential for dendrite growth initiation, η_i, and the critical overpotential for powder formation, η_c. Simultaneously, it was shown that dendritic deposits are obtained at all overpotentials between η_i and η_c after a sufficiently long induction period, which is in agreement with the findings of Ibl and Schadegg [13]. Also, it was shown [3] that during potentiostatic deposition with sufficiently long deposition times, dendritic and powdered deposits can be obtained at current densities lower than the limiting diffusion value. This is not possible in the case of galvanostatic deposition.

According to Popov et al. [18], the minimum overpotential at which dendritic growth, η_i, is possible at a macroelectrode is given by

$$\eta_i = \frac{b_c}{2.3} \ln \frac{4 j_L}{j_0} + \Delta\eta \tag{3.1}$$

and the minimum overpotential at which instantaneous dendritic growth, η_c, is possible is given by

$$\eta_c = \frac{b_c}{2.3} \ln \frac{j_L \delta}{j_0 h_0} + \Delta\eta, \tag{3.2}$$

where b_c is the cathodic Tafel slope, j_L is the limiting diffusion current density, j_0 is the exchange current density, δ is the diffusion layer thickness, and h_0 is the initial protrusion height.

$\Delta\eta$ in Eqs. (3.1) and (3.2) represents the difference in the reversible potential of the tip of the dendrite and a planar surface [19], and it is presented by Eq. (3.3):

$$\Delta\eta = \frac{2\sigma V}{nFR}, \tag{3.3}$$

where nF is the number of Faradays per mole of consumed ions, R is the molar gas constant, V is the molar volume of deposited metal, and σ is the interfacial energy between metal and solution. $\Delta\eta$ is the order of a few millivolts and it can be neglected in the consideration of both η_i and η_c overpotentials. Also, the following forms of Eqs. (3.1) and (3.2) are often found in the literature: [5]

$$\eta_i = \frac{b_c}{2.3} \ln \frac{j_L}{j_0} \tag{3.4}$$

and

$$\eta_c = \frac{b_c}{2.3} \ln \frac{j_L \delta}{j_0 h_0}. \tag{3.5}$$

The relationship between overpotential and current density in mixed controlled metal electrodeposition is given by

$$\eta = \frac{b_c}{2.3} \ln \frac{j}{j_0} \frac{1}{(1 - j/j_L)}, \tag{3.6}$$

where j is the current density of electrodeposition.

Current densities j_{i} and j_{cr} which correspond to η_{i} and η_{c} can be obtained by eliminating η from (3.1), (3.2), and (3.6) as[1]

$$j_{\mathrm{i}} = 0.8 j_{\mathrm{L}} \tag{3.7}$$

and

$$j_{\mathrm{cr}} = \frac{j_{\mathrm{L}}(\delta/h_0)}{1 + (\delta/h_0)} \tag{3.8}$$

or

$$j_{\mathrm{i}} = j_{\mathrm{L}} \tag{3.9}$$

for $\delta/h_0 \gg 1$. Hence, it can be concluded that dendritic growth is not possible at [19]

$$j < j_{\mathrm{i}} \tag{3.10}$$

but growth is possible after an induction time at

$$j_{\mathrm{i}} \leq j < j_{\mathrm{L}} \tag{3.11}$$

and instantaneous growth is possible at

$$j \geq j_{\mathrm{L}} \tag{3.12}$$

in potentiostatic electrodeposition. The fact that dendritic growth in potentiostatic deposition is possible at $\eta < \eta_{\mathrm{c}}$, and hence $j < j_{\mathrm{L}}$, (regardless $j \sim j_{\mathrm{L}}$), was explained by the effect of nondendritic surface roughness amplification during the induction time of dendritic growth [18].

This effect in galvanostatic electrodeposition will be in the opposite direction. It was shown by Maksimović et al. [20, 21] that the

[1] The elimination η from Eqs. (3.4) to (3.6), $j_{\mathrm{i}} = 0.5\ j_{\mathrm{L}}$, and this dependence can also be found in the literature.

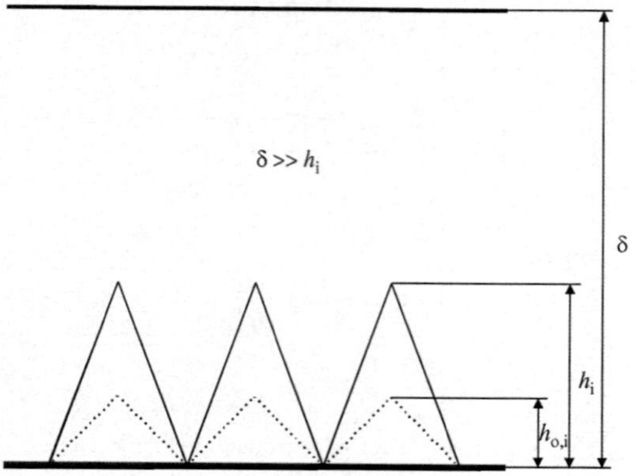

Fig. 3.1 Schematic representation of a rough electrode surface area

surface roughness amplification of a protrusion with an initial height, h_0, in galvanostatic deposition (for $t \ll \tau$) obeys the same relation as in the case of potentiostatic electrodeposition [22–27], i.e.,

$$h = h_0 \exp(t/\tau), \qquad (3.13)$$

where $\tau = (\delta^2/V_m DC_0)$, if the condition $\delta \gg h$ is satisfied. In Eq. (3.13), D is diffusion coefficient and C_o is the bulk concentration. It is easy to show that for the electrode surface presented in Fig. 3.1, the real electrode surface area will increase with time according to

$$S = S_0 \exp(t/\tau) \qquad (3.14a)$$

since

$$S = k \sum_{i=1}^{N} h_i \qquad (3.14b)$$

and

$$S_0 = k \sum_{i=1}^{N} h_{0,i}. \qquad (3.14c)$$

Obviously, the real current density will decrease according to

$$j = j^0 \exp(-t/\tau), \tag{3.15}$$

where j^0 is initial current density, while the overpotential will decrease according to

$$\eta = \frac{b_c}{2.3} \ln \frac{j^0 \exp(-t/\tau) j_L}{j_0 [j_L - j^0 \exp(-t/\tau)]}, \tag{3.16}$$

where Eq. (3.16) is obtained by substitution of j from Eq. (3.15) into Eq. (3.6).

The critical overpotential for instantaneous dendritic growth is given by Eq. (3.5) for protrusions with an initial height h_0. In potentiostatic deposition, an overpotential lower than η_c can belong to the limiting diffusion range. Nondendritic surface roughness amplification in the limiting diffusion current density range does not depend on overpotential, leading to an increase of height of the protrusion. Substitution of h from Eq. (3.13) in Eq. (3.5) shows the change of critical overpotential of instantaneous dendritic growth with time caused by nondendritic surface roughness amplification as

$$\eta_{c,t} = \frac{b_c}{2.3} \ln \frac{j_L \delta}{j_0 h} = \eta_c - \frac{b_c}{2.3} \frac{t}{\tau}. \tag{3.17}$$

Hence, the overpotential of deposition remains constant and the critical overpotential of instantaneous dendritic growth decreases, and at $t = t_i$ these values become equal and dendritic growth starts. In galvanostatic conditions, nondendritic amplification causes a decrease in the critical overpotential for dendritic growth according to Eq. (3.17), but at the same time the overpotential of deposition decreases according to Eq. (3.16). The time t_i in which these two overpotentials become equal can be obtained by the elimination of η from Eqs. (3.16) and (3.17) as

$$t_i = -2.3\tau \log \frac{j_L}{j^0} \tag{3.18}$$

if $\delta \gg h_0$. Hence, $t_i = 0$ for $j^0 = j_L$ and instantaneous dendritic growth is possible; at $j^0 < j_L$, for $t_i < 0$ dendritic growth is not possible. In this way, the induction period for the dendritic growth becomes equal to the transition time.

The above discussion is valid for galvanostatic powder electrodeposition. Due to the increase in real surface area of the electrode, the working current density should be many times larger than limiting diffusion one [28]. In this way, the decrease in the current density below the limiting diffusion one can be avoided. Hence, hydrogen codeposition in galvanostatic powder electrodeposition is inevitable. The situation is somewhat different under the potentiostatic conditions. In this regime, dendrites of deposited metal also appear in the limiting diffusion current density range. For some metals, however, hydrogen codeposition is not a necessary factor influencing the powder formation [5, 29]. In the absence of vigorous hydrogen evolution, powder particles are well-developed dendrites or parts of them. At overpotentials larger than that of vigorous hydrogen evolution, the conditions of the powder formation become similar to those in galvanostatic deposition.

There are at least two basic consequences of the vigorous hydrogen evolution during the metal powder particle formation. First, the hydrodynamic regime in the vicinity of the electrode can be changed due to stirring of the solution by evolved hydrogen, thus resulting in the increase of the limiting diffusion current density [30]. This causes the formation of morphological forms appearing before dendrite growth initiation at specified overpotential. Second, the formation of hydrogen bubbles strongly influences the current density distribution over electrode surface and hence the powder particle formation [31, 32].

3.2 Copper Powdered Deposits

3.2.1 Basic Facts

The most often employed electrolytes for the electrodeposition of copper are those based on aqueous solutions of cupric sulfate ($CuSO_4$) and sulfuric acid (H_2SO_4) [33]. There is an ionic equilibrium of a

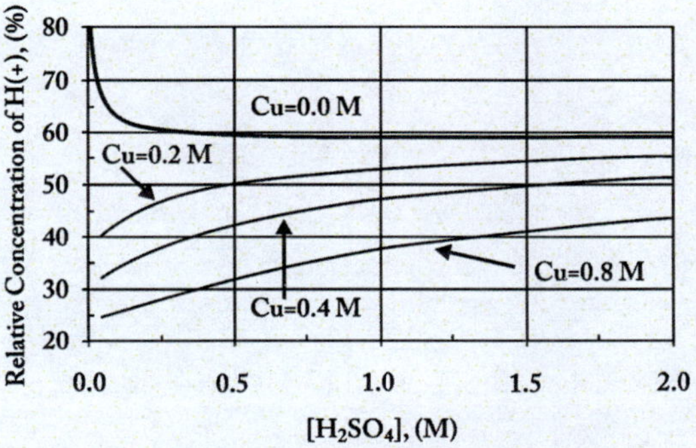

Fig. 3.2 Relative concentration of hydrogen ions as a function of sulfuric acid and total copper concentrations, at 25°C ($C_{RH+} = [H^+]/[H_T]$) (Reprinted from [36] with permission from Elsevier and [38] with permission from Springer.)

lot of species, such as bisulfate ions (HSO_4^-), cupric ions (Cu^{2+}), aqueous cupric sulfate ($CuSO_{4(aq)}$), hydrogen ions (H^+), and sulfate ions (SO_4^{2-}), in the $CuSO_4$–H_2SO_4–H_2O system [34–36]. Using Pitzer's model [37], the ionic equilibrium of these species over a wide range of concentrations and temperatures was calculated [36], and the dependence of the relative concentration of hydrogen ions (H^+) on H_2SO_4 concentration for different copper concentration is shown in Fig. 3.2. From Fig. 3.2, it can be seen that increasing the copper concentration produces a sharp decrease in the hydrogen ion concentration, while increasing the concentration of sulfuric acid produces an increase in the hydrogen ion concentration [36].

Irregular or powdered copper deposits are formed by electrodeposition from acid sulfate solutions at current densities and overpotentials corresponding to the plateau of the limiting diffusion current density and at higher ones. At these overpotentials and current densities, parallel to copper electrodeposition processes, hydrogen evolution reaction occurs [30]. Hence, it is very clear that the formation of powdered copper deposits is closely related with hydrogen evolution reaction as the second reaction in copper electrochemical deposition processes at high overpotentials and current densities [38].

Due to the ionic equilibrium of the species in the $CuSO_4$–H_2SO_4–H_2O system (Fig. 3.2), the quantities of evolved hydrogen and hence morphologies of powdered copper deposits depend strongly on the

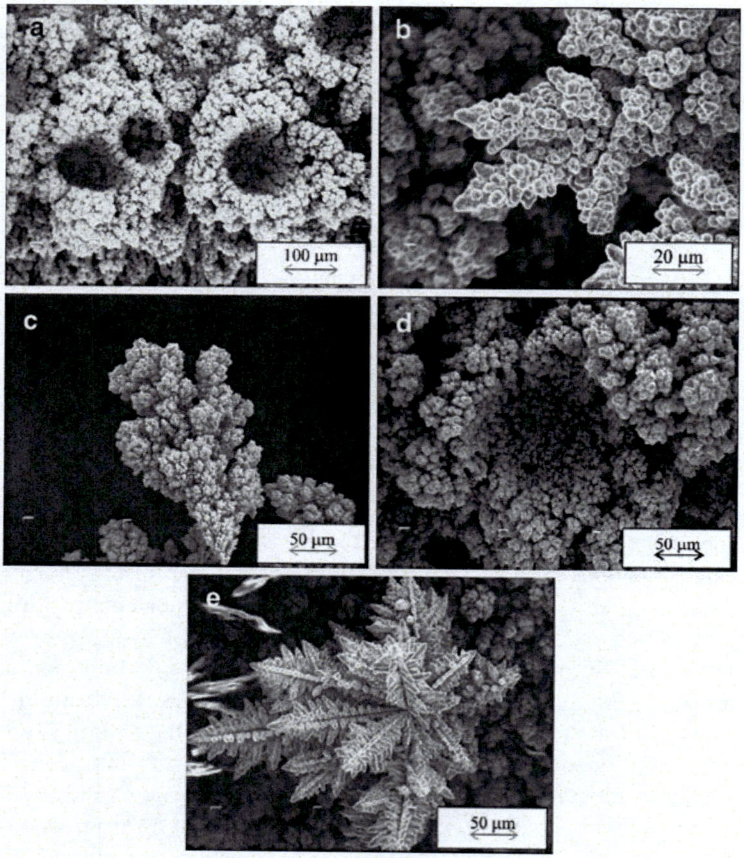

Fig. 3.3 Copper deposits obtained at an overpotential of 800 mV from (**a**) 0.075 M $CuSO_4$ in 0.50 M H_2SO_4; (**b**)–(**d**) 0.30 M $CuSO_4$ in 0.50 M H_2SO_4; and (**e**) 0.60 M $CuSO_4$ in 0.50 M H_2SO_4 (Reprinted from [39] with permission from Elsevier and [38] with permission from Springer.)

used concentrations of $CuSO_4$ and H_2SO_4 [39, 40]. Figure 3.3 shows copper deposits obtained by potentiostatic electrodepositions at an overpotential of 800 mV from copper solutions containing 0.075, 0.30, and 0.60 M $CuSO_4$ in 0.50 M H_2SO_4. For all these solutions, an overpotential of 800 mV was about 50 mV outside the plateau of the limiting diffusion current density [31, 39]. The quantities of hydrogen generated during electrodeposition processes at 800 mV from these solutions corresponded to the average current efficiencies of hydrogen evolution, $\eta_{I,av}(H_2)$, of 42.2% for 0.075 M $CuSO_4$ in 0.50 M H_2SO_4 (the ratio of $CuSO_4/H_2SO_4 = 0.15$), 3.5% for 0.30 M $CuSO_4$ in 0.50 M H_2SO_4 (the ratio of $CuSO_4/H_2SO_4 = 0.60$), and 0.66% for 0.60 M $CuSO_4$ in 0.50 M H_2SO_4 (the ratio of $CuSO_4/H_2SO_4 = 1.2$) [39].

The honeycomb-like structure was formed by electrodeposition from 0.075 M $CuSO_4$ in 0.50 M H_2SO_4 (Fig. 3.3a). A mixture of dendritic forms (Fig. 3.3b), degenerate dendrites (Fig. 3.3c), and holes formed due to the attached hydrogen bubbles (Fig. 3.3d) was obtained by electrodeposition from 0.30 M $CuSO_4$ in 0.50 M H_2SO_4. Finally, very branchy dendrites were formed by electrodeposition from 0.60 M $CuSO_4$ in 0.50 M H_2SO_4 (Fig. 3.3e). It is very clear that the change of morphology of electrodeposited copper with the increasing Cu(II) ions concentration is a consequence of the decrease of effectiveness of solution stirring by evolved hydrogen caused by the decrease of the relative concentration of H^+ ions with the increase of Cu concentration (Fig. 3.2). The critical quantity of evolved hydrogen which causes the effective solution stirring and leads to the change of hydrodynamic conditions in the near-electrode layer was estimated to correspond to $\eta_{I,av}(H_2)$ of 10.0% [31]. For the H_2SO_4 concentration of 0.50 M, the maximal concentration of Cu(II) ions enabling the formation of the honeycomb-like structures is estimated to correspond to 0.15 M $CuSO_4$ [31].

This is a vertical analysis of the ionic equilibrium of species in the $CuSO_4$–H_2SO_4–H_2O system. Horizontal analysis of this equilibrium is made keeping the Cu(II) ions constant and varying H_2SO_4 concentrations. Morphologies of electrodeposited copper obtained at an overpotential of 800 mV from 0.15 M $CuSO_4$ in 0.125, 0.25, and 1.0 M H_2SO_4 are shown in Fig. 3.4. For these solutions, an overpotential of 800 mV was about 50 mV outside the plateaus of the limiting diffusion current density [41]. The average current efficiencies of hydrogen evolution, $\eta_{I,av}(H_2)$, were 4.83% (for the copper solution containing 0.15 M $CuSO_4$ in 0.125 M H_2SO_4; the $CuSO_4/H_2SO_4$ ratio = 1.2), 9.05% (for the copper solution containing 0.15 M $CuSO_4$ in 0.25 M H_2SO_4; the $CuSO_4/H_2SO_4$ ratio = 0.60), and 23.3% (for the copper solution containing

Fig. 3.4 Copper deposits obtained at an overpotential of 800 mV from: (**a**)–(**c**) 0.15 M $CuSO_4$ in 0.125 M H_2SO_4; (**d**) 0.15 M $CuSO_4$ in 0.25 M H_2SO_4; and (**e**) 0.15 M $CuSO_4$ in 1.0 M H_2SO_4 (Reprinted from [39] with permission from Elsevier and [38] with permission from Springer.)

0.15 M $CuSO_4$ in 1.0 M H_2SO_4; the $CuSO_4/H_2SO_4$ ratio = 0.15) [39]. Please note that the same $CuSO_4/H_2SO_4$ ratios were analyzed in both cases.

A channel structure (Fig. 3.4a), degenerate dendrites (Fig. 3.4b), and cauliflower-like forms (Fig. 3.4c) were formed by copper electrodeposition from 0.15 M $CuSO_4$ in 0.125 M H_2SO_4. Holes originating from attached hydrogen bubbles were formed by electrodeposition from 0.15 M $CuSO_4$ in 0.25 M H_2SO_4 (Fig. 3.4d). Degenerate dendrites and cauliflower-like forms, similar to those shown in Fig. 3.4b, c, were also formed by electrodeposition from this solution. Finally, the honeycomb-like structure, constructed from holes formed due to attached hydrogen bubbles and cauliflower-like agglomerates of copper grains between them, was formed by the electrodeposition from 0.15 M $CuSO_4$ in 1.0 M H_2SO_4 (Fig. 3.4e).

The observed morphologies of copper deposits were in a good agreement with the prediction of the ionic equilibrium of the species in the $CuSO_4$–H_2SO_4–H_2O system. The addition of excess H_2SO_4 to the electroplating solution increases the H^+ ion concentration, which is confirmed by the higher values of the average current efficiencies of hydrogen evolution from the solutions with higher concentrations of H_2SO_4 and by the change of copper morphology from cauliflower-like forms and degenerate dendrites to the honeycomb-like structure. For a constant H_2SO_4 concentration, the H^+ ion concentration decreases with increasing copper concentration, which is manifested by the smaller quantity of evolved hydrogen from solutions with higher $CuSO_4$ concentrations and by the change of morphology of electrodeposited copper from the honeycomb-like structure to dendrites [39].

The effect of $CuSO_4$ and H_2SO_4 concentrations on both hydrogen evolution rate and morphology of electrodeposited copper was also observed during electrochemical deposition processes at overpotentials of 650 and 1,000 mV [31, 41, 42]. During electrodeposition processes at 650 mV, the branching of dendrites decreased with the increasing $CuSO_4$ concentration [31], while the number of formed dendritic forms as well as their branching increased with the decrease of H_2SO_4 concentration [41]. Although holes originating due to detachment of hydrogen bubbles were obtained by electrodepositions at 1,000 mV from all analyzed solutions, the number, shape, and size of holes strongly depended on the $CuSO_4$ and H_2SO_4 concentrations used [42].

It is necessary to note that effects on morphology of powdered deposits similar to those obtained by the use of more solutions of different $CuSO_4$ and H_2SO_4 concentrations can be attained by the use of the only one solution if electrodeposition processes were performed at a periodically changing rate.

3.2.2 The Application of Periodically Changing Regimes of Electrolysis in the Formation of Powdered Deposits

Electrodeposition at a periodically changing rate is based on the periodic repetition of current or overpotential waves [27, 43]. The application of periodically changing regimes of electrolysis, such as pulsating overpotential (PO), pulsating (PC), and reversing current (RC), in metal electrodeposition processes is of great academic and practical significance [5, 43]. The most important regime from the theoretical point of view is obviously pulsating overpotential. On the other hand, the most important regime from practical point of view is the reversing current.

Deposition at a periodically changing rate offers a number possibilities for changing the deposition conditions at one and the same deposition rate [43]. Deposits with desired composition, structure, porosity, and hydrogen content, the enhanced throwing power especially in the holes, and the reduction of the use of additives are some of advantages of the use of pulse regimes of electrolysis. From the point of view of powder formation, powder particles with different grain size and morphology can be obtained by varying the wave of periodically changing current or overpotential [6].

In the hydrogen codeposition range, the effect of the PO regime on microstructural characteristics of the honeycomb-like structures was similar to those attained by the application of additives in electrodeposition processes [44–46]. Some of conveniences of the application of these regimes of electrolysis on the formation of various disperse morphological forms of copper can be presented as follows: in this section, the presented copper morphologies were obtained by electrodepositions from 0.15 M $CuSO_4$ in 0.50 M H_2SO_4 by the regime of pulsating current (PC). Electrodepositions were performed at the room temperature using cylindrical copper wires as working electrodes. The current density amplitude of 0.20 A cm^{-2} was used [47].

The first set of experiments was done applying square-waves PC with a constant pause duration, t_p, of 10 ms, and deposition pulses, t_c, of 1, 2, and 50 ms (pause to pulse ratios, p, where $p = t_p/t_c$ were 10, 5, and 0.2, respectively). The average current efficiencies for hydrogen evolution reaction, $\eta_{I,av}(H_2)$, determined using the experimental procedure adapted to the PC regime were 5.5, 10.3, and 27.2% for pause to pulse ratios of 10, 5, and 0.2, respectively [47].

Morphologies of copper deposits obtained with deposition pulses of 1, 2, and 50 ms and a pause duration of 10 ms are shown in Fig. 3.5. Holes formed by attached hydrogen bubbles, very branchy dendrites, and small agglomerates of copper grains are formed when the applied deposition pulse was 1 ms (Fig. 3.5a, b). The mixture of holes and degenerate dendrites was formed with a deposition pulse of 2 ms (Fig. 3.5c, d). Honeycomb-like copper structure constructed of holes and cauliflower-like agglomerates of copper grains formed around them was obtained with a deposition pulse of 50 ms (Fig. 3.5e, f).

In the second set of experiments, the constant deposition pulse of 10 ms and pause durations of 100, 50, and 2 ms were analyzed (pause to pulse ratios: 10, 5, and 0.2, respectively). In this case, the average current efficiencies for hydrogen evolution reaction, $\eta_{I,av}(H_2)$, were 5.6, 12.4, and 26.8% for duration of pause of 100, 50, and 2 ms, respectively [47]. Morphologies of copper deposits obtained with a deposition pulse of 10 ms and pause durations of 100, 50, and 2 ms are shown in Fig. 3.6. Holes which the walls were constructed of dendrites and dendritic particles formed between holes were primarily formed by square-wave PC with a pause of 100 ms (Fig. 3.6a, b). These copper dendrites were considerably smaller and formed over the whole electrode surface in relation to those formed by the PC regime of the same pause to pulse ratio ($p = 10$) but with a deposition pulse of 1 ms and a pause of 10 ms (Fig. 3.5b). The shortening of pause duration from 100 to 50 ms led to the increase of number of holes formed by attached hydrogen bubbles (Fig. 3.6c), and the change of morphology of electrodeposited copper from dendritic particles to agglomerates consisted of copper grains and rare small dendrites on them (Fig. 3.6d). Finally, the typical honeycomb-like structure is formed with a pause of 2 ms (Fig. 3.6e, f). The increase of the number of holes, as well as the change of morphology of electrodeposited copper from dendrites to agglomerates of copper grains, is a result of the increasing quantity of evolved hydrogen and the increased effectiveness of solution stirring by evolved hydrogen with the shortening of pause duration.

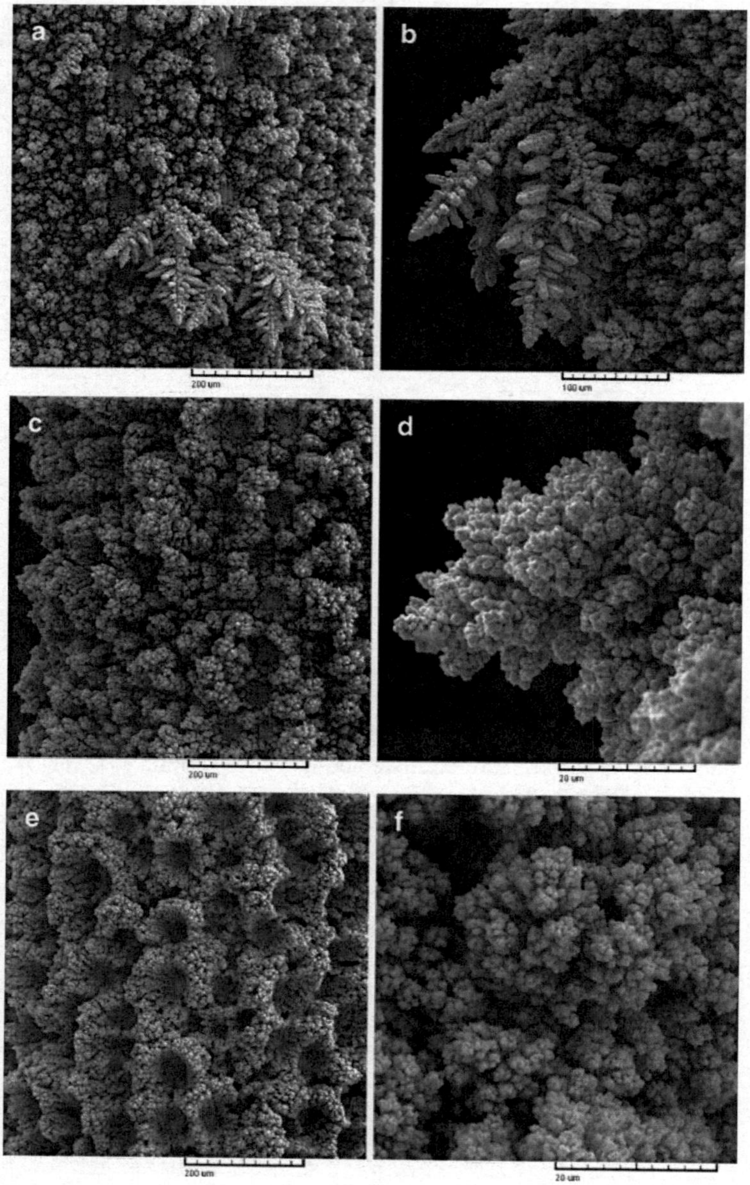

Fig. 3.5 Copper deposits obtained by the PC regimes with pause duration of 10 ms and deposition pulses of (**a**) and (**b**) 1 ms; (**c**) and (**d**) 2 ms; and (**e**) and (**f**) 50 ms (Reprinted from [47] with permission from Elsevier.)

Fig. 3.6 Copper deposits obtained by the PC regimes with a deposition pulse of 10 ms and pause durations of (**a**) and (**b**) 100 ms; (**c**) and (**d**) 50 ms; and (**e**) and (**f**) 2 ms (Reprinted from [47] with permission from Elsevier.)

The selected parameters of square-wave PC enabled a comparison of the obtained morphologies of electrodeposited copper with those obtained by electrodepositions in the hydrogen codeposition range at the constant overpotential from six solutions of different concentrations of $CuSO_4$ and H_2SO_4 (see Figs. 3.3 and 3.4) [39]. It is clear that the effect of increasing concentration of Cu(II) ions was equivalent to the effect of the decreasing deposition pulses, while the constant pause duration was equivalent to H_2SO_4 concentration used. In a similar way, the effect of different H_2SO_4 concentrations (for the constant $CuSO_4$ concentration) can be correlated with the effect of different pause durations. The effect of increasing H_2SO_4 concentration was equivalent to the effect of decreasing the pause duration, while the constant $CuSO_4$ concentration is equivalent to the constant deposition pulse used [47]. Of course, the constant overpotential used corresponds to the amplitude current density used [47].

Hence, effects attained by the choice of appropriate parameters of square-wave PC were equivalent to those obtained by electrodeposition at the constant overpotential in the hydrogen codeposition range from solutions of different concentrations of $CuSO_4$ and H_2SO_4. In this way, the ionic equilibrium in the $CuSO_4$–H_2SO_4–H_2O system, determined by the dependence of the relative concentration of H^+ ions on H_2SO_4 concentration for different copper concentration, was successfully simulated. Also, the substitution of more different electrodeposition solutions by the one solution was achieved. This is of potential high technological significance because it enables saving of chemicals for the preparation of electrodeposition baths, as well as saving of place in the plating plants due to the reduced number of needed electrochemical cells.

The presented disperse or irregular morphological forms of copper may be useful in the production of powders [6, 32], while the honeycomb-like deposit type is due to an open porous structure with the extremely high surface area ideally suited to be used as electrodes in many electrochemical devices, such as fuel cells, batteries, and sensors [48].

It is easy to use the pulsating overpotential in laboratory-scale cells, but difficulties arise on a practical scale because of the demands for higher power and a high-speed potentiostat. It is easier to obtain pulsating or reversing currents on a practical scale, and, because of this, the deposition of metal powders with the desired properties obtained by pulsating and reversing currents is of greater practical importance than that obtained by the pulsating overpotential.

3.3 Copper Powder Particles

3.3.1 Basic Facts

Pure copper powder is used in the electrical and the electronics industries because of its excellent electrical and thermal conductivities [49]. Alloyed with tin, zinc, nickel, and other elements, copper in powder form is used in structural parts and friction materials. Brasses, bronzes, and other copper alloys produced by powder metallurgy methods have the physical and mechanical properties of their cast or wrought counterparts. Copper is also used as an alloying element in iron powder components to enhance the mechanical properties and control dimensional changes during sintering, the addition being made either by mixing or by infiltration.

Probably the best known application of copper powder is the self-lubricating bearing which was the first major application and still accounts for about 70% of the granular copper powder used. This application takes advantage of the ability to produce a component with controlled interconnected and surface-connected porosity. The production of metallic filters also takes advantage of this ability.

In addition to the above applications of granular copper powder, a large quantity of copper and copper alloy powder is used in flake form, i.e., as a powder whose thickness is small in relation to its other dimensions. Such powders are used, for example, in antifouling paints, decorative and protective coatings, and printing inks.

Copper and copper alloy powders are also used in such nonstructural applications as brazing, cold soldering, and mechanical plating, as well as for medals and medallions, metal–plastic decorative products, and a variety of chemical and medical purposes.

Due to the ionic equilibrium of species in the $CuSO_4$–H_2SO_4–H_2O system, the formation of copper powder particles by electrolytic technique is closely related to the quantity of hydrogen generated at the cathode surface during electrolysis. The shape and size of electrolitically formed powder particles strongly depend on concentrations of deposition ion and supporting electrolyte (that are $CuSO_4$ and H_2SO_4 concentrations for acid sulfate solutions), temperature of electrolysis, the applied current density or overpotential of electrodeposition, as well as of the time of removal of powder from the electrode surface, the type of working electrode used, rotation speed of cathode, and electrolyte circulation rate [4, 6, 50–53].

On the other hand, due to the fact that morphology of a deposit is determined by overpotential of electrodeposition, the structure of powder particles will not depend on electrolysis time in the potentiostatic conditions of electrodeposition. In the galvanostatic regime of electrolysis, overpotential of electrodeposition changes with electrolysis time, and for that reason, this regime of electrolysis is not suitable for basic experiments required for a theoretical consideration.

3.3.2 Correlation Between Morphology of Powder Particles Obtained by the Different Regimes of Electrolysis and the Quantity of Evolved Hydrogen

3.3.2.1 Morphologies of Powdered Deposits and Powder Particles Obtained by the Constant Potentiostatic Regime

In the dependence of the quantity of evolved hydrogen, the two types of powdered deposits are formed [32, 54]. The typical powdered deposits electrodeposited from 0.075 M $CuSO_4$ in 0.50 M H_2SO_4 at an overpotential of 650 mV (plateau of the limiting diffusion current density) and at an overpotential of 1,000 mV (about 250 mV above the plateau) are shown in Fig. 3.7a, b, respectively.

Very branchy dendrites and cauliflower-like agglomerates of copper grains were formed during electrodeposition at an overpotential of 650 mV (Fig. 3.7a). The amount of hydrogen evolved

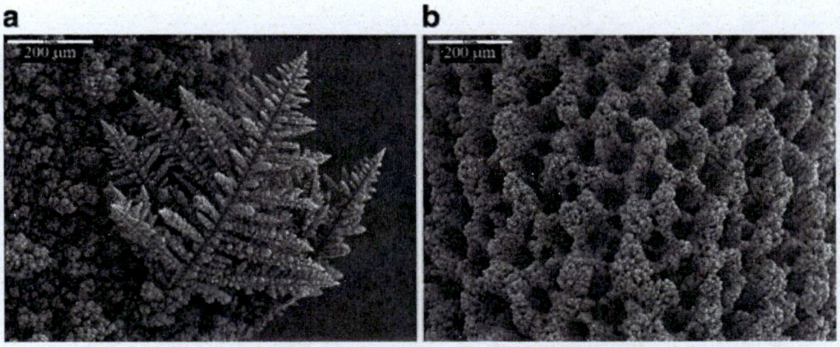

Fig. 3.7 Macrostructures of copper powdered deposits electrodeposited at overpotentials of: (**a**) 650 mV and (**b**) 1,000 mV (Reprinted from [54] with permission from Electrochemical Society.)

for the formation of this deposit type corresponds to an average current efficiency of hydrogen evolution of 7.5% [31], and it was below the critical value for the change of hydrodynamic conditions in the near-electrode layer [31]. The electrodeposition, where the formation of highly branched dendrites takes place, is rather diffusion than electron transfer controlled process [5].

The second type of powdered deposits is formed by electrodeposition at an overpotential of 1,000 mV (Fig. 3.7b). As expected, it is a typical honeycomb-like structure composed of holes formed by attached hydrogen bubbles and cauliflower-like agglomerates of copper grains formed around them. An average current efficiency of hydrogen evolution of 68.7% for this case was reported [42], and it was above the critical value for the change of hydrodynamic conditions in the near-electrode layer. The amount of evolved hydrogen was enough to cause effective solution stirring in the near-electrode layer leading to a decrease of the cathode diffusion layer thickness and to an increase of the limiting diffusion current density [30].

The dendritic particle obtained by tapping of the copper deposit presented in Fig. 3.7a is shown in Fig. 3.8a. The dendritic character of this particle is made of the corncob-like elements as shown in Fig. 3.8b. The ultrasonic treatment of copper dendrites showed that the corncob-like forms were the basic elements of which copper dendrites are composed [32]. A further analysis of the corncob-like

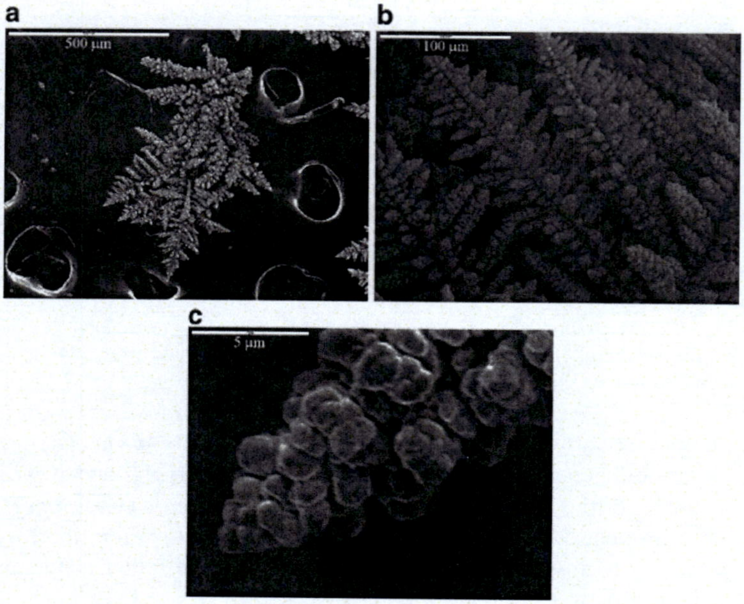

Fig. 3.8 (**a**) Dendritic particle obtained by tapping of the copper deposit electrodeposited at an overpotential of 650 mV, (**b**) corncob-like elements of which dendrites are composed; and (**c**) microstructure of the corncob-like element (Reprinted from [54] with permission from Electrochemical Society.)

elements at the microlevel found that they are composed of small agglomerates of copper grains (Fig. 3.8c).

In the dependence of concentration of Cu(II) ions, and hence the quantity of evolved hydrogen, the corncob-like elements can be grouped the different forms of dendritic particles from tree-like to those formed as flowers (Fig. 3.9a, b) or alternatively can be formed individually at the electrode surface (Fig. 3.9c) [31].

A particle obtained by tapping of the copper deposit electrodeposited at an overpotential of 1,000 mV is shown in Fig. 3.10a. Channel structure generated through the interior of the particle by the simultaneous copper nucleation and strong hydrogen evolution *in situ* can easily be seen from Fig. 3.10a. This type of powder consists of an aggregate of small cauliflower-like particles (Fig. 3.10b). Top view of the powder shown in Fig. 3.10a clearly revealed its

Fig. 3.9 Copper dendrites electrodeposited at an overpotential of 650 mV from (**a**) 0.075 M, (**b**) 0.30 M; and (**c**) 0.60 M $CuSO_4$ in 0.50 M H_2SO_4 (Reprinted from [32] with permission from Elsevier and [38] with permission from Springer.)

cauliflower-like character (Fig. 3.10c). An analysis of the cauliflower-like forms at the microlevel showed that they were composed of small agglomerates of copper grains (Fig. 3.10d). When this powder was ultrasonically treated the results showed that the basic element of these particles has the shape of a degenerate dendrite [32].

Anyway, the macrostructure of the formed powdered deposits was very different (Fig. 3.7). On the other hand, the similarity of the deposits at the microlevel is evident (Figs. 3.8c and 3.10d). The obvious difference is in the size of the individual copper grains of which the particles are comprised. The smaller size of the individual copper grains produced at an overpotential of 1,000 mV in comparison with the size of those deposited at 650 mV is attributable to the higher nucleation rate at an increased overpotential.

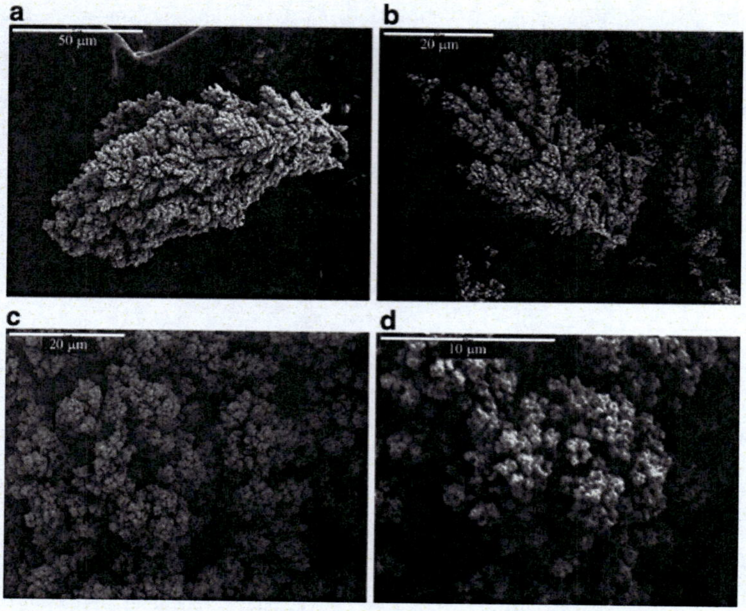

Fig. 3.10 (**a**) Powder particle obtained by tapping of the copper deposit electrodeposited at an overpotential of 1,000 mV, (**b**) and (**c**) cauliflower-like character of this type of the powder particle, and (**d**) microstructure of the cauliflower-like particle (Reprinted from [54] with permission from Electrochemical Society.)

The special case was the formation of powder by electrolysis from 0.30 M $CuSO_4$ in 0.50 M H_2SO_4 at an overpotential of 1,000 mV. This solution was denoted as transitional one between solutions with higher and lower concentrations of Cu(II) ions (in 0.50 M H_2SO_4) [42]. This deposit (Fig. 3.11a) contains characteristic of both types of powdered deposits: holes formed by attached hydrogen bubbles (Fig. 3.11a) and very branchy dendrites formed around them (Fig. 3.11b, c) [55]. The amount of evolved hydrogen spent for the formation of this deposit corresponded to the average current efficiency of hydrogen evolution of 16.0% [42]. The typical dendritic particle obtained by the tapping of this deposit is shown in Fig. 3.11d.

Fig. 3.11 Morphology of the copper deposit obtained by electrodeposition from 0.30 M $CuSO_4$ in 0.50 M H_2SO_4 at an overpotential of 1,000 mV: (**a**) top view, (**b**) and (**c**) magnified parts from (**a**); and (**d**) dendritic particle obtained by the tapping of the powdered deposit shown in (**a**)

3.3.2.2 Morphologies of Powdered Deposits and Powder Particles Obtained by the Regime of Pulsating Overpotential

Figure 3.12a, b shows the honeycomb-like structures electrodeposited from 0.15 M $CuSO_4$ in 0.50 M H_2SO_4 by the regimes of pulsating overpotential (PO) with the overpotential amplitude of 1,000 mV, a pause duration of 10 ms, and deposition pulses of 3 ms (Fig. 3.12a) and

Fig. 3.12 Powdered deposits electrodeposited by the pulsating overpotential (PO) regime with a deposition pulse of (**a**) 3 ms, (**b**) 20 ms, (**c**) magnified part from (**a**), and (**d**) magnified part from (**b**). Pause duration: 10 ms. Overpotential amplitude: 1,000 mV. Solution: 0.15 M $CuSO_4$ in 0.50 M H_2SO_4

20 ms (Fig. 3.12b). Electrodepositions were performed at the room temperature using cylindrical copper wires as working electrodes. The amount of hydrogen generated during the formation of these powdered deposits corresponded to the average current efficiencies of hydrogen evolution of 16.4% for a deposition pulse of 3 ms and 28.1% for a deposition pulse of 20 ms [56]. The increased details from Fig. 3.12a, b which reveal the surface morphology around holes are shown in Fig. 3.12c, d, respectively. Very branchy dendrites are formed

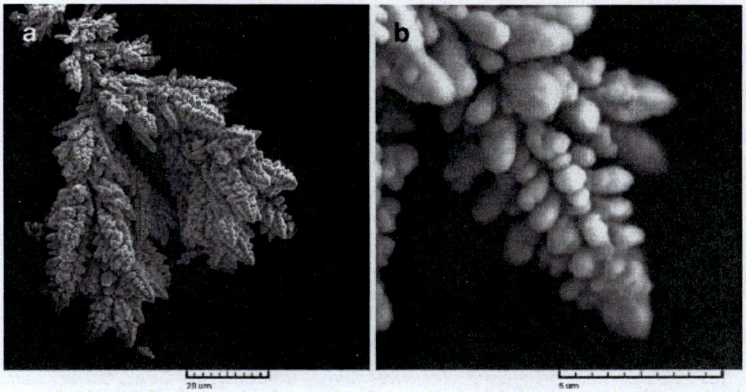

Fig. 3.13 Dendritic particle obtained by tapping of the powdered deposit electrodeposited by the pulsating overpotential (PO) regime with a deposition pulse of 3 ms (**a**) top view and (**b**) magnified part from (**a**). Pause duration: 10 ms. Overpotential amplitude: 1,000 mV. Solution: 0.15 M $CuSO_4$ in 0.50 M H_2SO_4

with a deposition pulse of 3 ms (Fig. 3.12c), while cauliflower-like agglomerates of copper grains were formed with a deposition pulse of 20 ms (Fig. 3.12d).

Dendritic particle obtained by tapping of the powdered deposit obtained with a deposition pulse of 3 ms is shown in Fig. 3.13a. Copper dendrites are constructed from corncob-like elements. Microanalysis of corncob-like elements revealed that they are composed of small agglomerates of copper grains (Fig. 3.13b).

The SEM micrograph of the particle obtained by tapping of the powdered deposit obtained with a deposition pulse of 20 ms is shown in Fig. 3.14a. A channel structure formed by *in situ* by the simultaneous processes of copper nucleation and strong hydrogen evolution can be easily noticed in this structure. Analysis of this particle at the higher magnification (Fig. 3.14b) showed that the particle is constructed from aggregates of small cauliflower-like agglomerates of copper grains.

Hence, the applied parameters of square-waves PO strongly affected hydrogen evolution reaction and hence morphology of copper powder particles. At the first sight, the effect of the shortening of deposition pulse duration was equivalent to the decrease of the

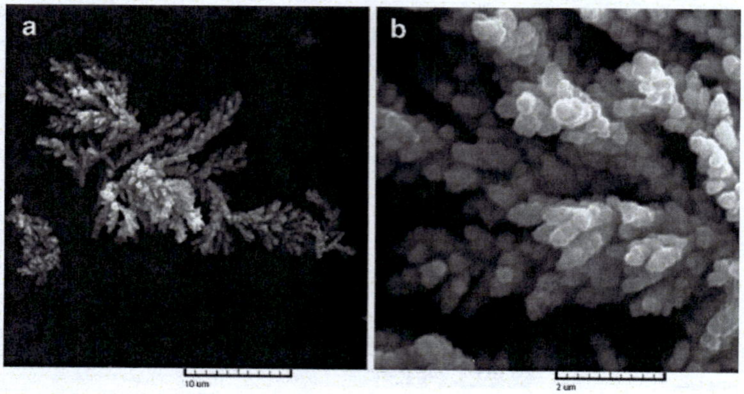

Fig. 3.14 Cauliflower-like particle obtained by tapping of the powdered deposit electrodeposited by the pulsating overpotential (PO) regime with a deposition pulse of 20 ms (**a**) top view and (**b**) magnified part from (**a**). Pause duration: 10 ms. Overpotential amplitude: 1,000 mV. Solution: 0.15 M $CuSO_4$ in 0.50 M H_2SO_4

overpotential of electrodeposition in the potentiostatic regime of electrolysis [32, 54]. It was shown [32, 54] that very branchy dendrites were electrodeposited from the same solution at overpotentials of 650 mV (Fig. 3.7a) [54] or 700 mV [32, 38], while aggregates of small cauliflower-like particles were obtained at an overpotential of 1,000 mV [32, 38].

3.3.2.3 Comparison of Morphologies of Powdered Deposits and Powder Particles Obtained by the Constant Potentiostatic Regime and by the Regime of Pulsating Overpotential

Considering the strong effect of parameters of the PO regime on both hydrogen evolution reaction and copper electrodeposition rate (Figs. 3.12–3.14), equivalence between morphology of copper powder particles obtained by the regime of PO and those obtained by electrodeposition in the potentiostatic regime from solutions of different concentrations of $CuSO_4$ and H_2SO_4 can be made and presented in the following way.

From Figs. 3.11 and 3.12, it is very clear that macromorphologies of deposits electrodeposited from 0.30 M $CuSO_4$ in 0.50 M H_2SO_4 at 1,000 mV and by the PO regime with a deposition pulse of 3 ms were very similar to each other; in both cases, holes formed by attached hydrogen bubbles surrounded by very branchy dendrites were formed.

It is necessary to note that both the powdered deposits were formed with approximately the same average current efficiency of hydrogen evolution ($\eta_{I,av}(H_2)$ of 16.0% for the deposit obtained in the potentiostatic deposition and $\eta_{I,av}(H_2)$ of 16.4% for the deposit obtained by square-wave PO with a deposition pulse of 3 ms).

In the PO regimes, the average current efficiency for hydrogen evolution reaction increases with the prolongation of deposition pulse duration approaching to the one obtained at the constant overpotential [56]. Then, copper electrodeposition from 0.15 M $CuSO_4$ in 0.50 M H_2SO_4 by the PO regime with a deposition pulse of 20 ms ($\eta_{I,av}(H_2) = 28.1\%$) should compare with the one from the same solution at an overpotential of 1,000 mV ($\eta_{I,av}(H_2) = 30.0\%$). Meanwhile, having in view the fact that macrostructures of the honeycomb-like deposits obtained with the average current efficencies of hydrogen evolution of 30.0% and with higher ones were consisted of holes formed by attached hydrogen bubbles and cauliflower-like agglomerates of copper grains around them [38], powder particles formed with a deposition pulse of 20 ms can be compared with those formed by electrodeposition from 0.075 M $CuSO_4$ in 0.50 H_2SO_4 at an overpotential of 1,000 mV (Figs. 3.7b and 3.10). In both cases, particles composed of small cauliflower-like agglomerates of copper grains were formed.

Comparing morphologies of powder particles (as well as the powdered deposits) obtained by the PO regimes with the different length of deposition pulse with those obtained by electrodeposition in the potentiostatic regime at 1,000 mV from solutions of different concentrations of $CuSO_4$ and H_2SO_4, it is clear that effect of the increase of deposition pulse duration on both hydrogen evolution reaction and copper electrodeposition rate was equivalent to the one observed by the decrease of $CuSO_4$ concentration (for the constant H_2SO_4 concentration). The effect of the constant pause duration ($t_p = 10$ ms) was equivalent to the constant H_2SO_4 concentration.

Of course, the overpotential amplitude in the PO regimes corresponded to the overpotential of electrodeposition in the constant regimes of electrolysis.

3.3.2.4 General Discussion

In the dependence of solution composition, regime of electrolysis and preparing of the working electrode, the copper surface morphologies can be grouped into the three hydrogen codeposition range:

(I) The range of the average current efficiencies of hydrogen evolution, $\eta_{I,av}(H_2)$, between 0 and 10.0%—the dominant presence of branchy dendrites, independently formed cauliflower-like forms, and the possible formation of individual holes of detached hydrogen bubbles (dish-like hole) [30, 31, 38, 41].

(II) The range of the average current efficiencies of hydrogen evolution, $\eta_{I,av}(H_2)$, between 10.0 and 20.0%—the mixture of dish-like holes and holes constructing the honeycomb-like structure with independently formed copper grains agglomerates among them [42], as well as the mixtures of holes and cauliflower-like forms [30, 38], and holes and branchy dendrites (Figs. 3.11a and 3.12a).

(III) With $\eta_{I,av}(H_2)$ above 20.0%—the mixture of holes and cauliflower-like copper grains agglomerates formed around holes without the appearing of dendrites [30, 31, 38, 41, 42].

In general, the number of holes increased with intensification of hydrogen evolution reaction while the shape of holes changed from dish-like holes to holes constructing the honeycomb-like structure. The honeycomb-like structures are formed with $\eta_{I,av}(H_2)$ larger than 10.0% [30, 38]. The quantity of the evolved hydrogen corresponding to the average current efficiency of hydrogen evolution, $\eta_{I,av}(H_2)$, of 10.0% was just denoted as the critical quantity leading to the formation of the honeycomb-like structures [31]. This critical quantity of generated hydrogen is mostly spent for the creating of holes in the honeycomb-like type of structure and it does not cause a stirring of the solution, and hence it does not affect hydrodynamic conditions in the near-electrode layer. On the other hand, with the increasing quantity of evolved hydrogen, the change of surface morphology from dendrites to cauliflower-like agglomerates of copper grains formed around holes was observed.

The analysis of the copper surface morphologies classified in all three hydrogen codeposition ranges indicates that the overall quantity of evolved hydrogen can be divided into two parts. The one part of evolved hydrogen is spent for the creating of holes and their increase with electrolysis time, while the other part of evolved hydrogen (or the rest of the overall quantity of evolved hydrogen) determines morphology of deposits. This can be explained as follows.

In the initial stage of electrodeposition, both hydrogen evolution and copper nucleation occur at the most active energy sites of the electrode surface [38, 57]. The formed hydrogen bubbles isolate substrate and current lines are concentrated around them forming rings composed of agglomerates of copper grains. Simultaneously, small agglomerates of copper grains are formed at the electrode surface of copper nucleus formed in the initial stage of electrodeposition among hydrogen bubbles. In the growth process, due to the effect of current density distribution, further copper nucleation and hydrogen evolution primarily occur at top of both groups of copper grains agglomerates. Some of new, freshly formed hydrogen bubbles will coalesce with hydrogen bubbles formed in the initial stage of electrodeposition leading to their growth with electrolysis time. When the critical size of these hydrogen bubbles to detach from the electrode surface is attained, they will detach forming holes at the electrode surface. This quantity of evolved hydrogen does not contribute to stirring of solution and to the change of hydrodynamic conditions in the near-electrode layer.

Meanwhile, some of new, freshly formed hydrogen bubbles at the top of agglomerates around initially formed hydrogen bubbles will not coalesce with them, because they are situated between freshly formed copper nuclei which represent barrier to find a path and to coalesce with initially formed hydrogen bubbles. Also, they cannot develop in large hydrogen bubbles for the same reasons. The similar situation occurs at top of those copper grains agglomerates which are formed of initially formed copper nuclei. These hydrogen bubbles will detach from the electrode surface very fast forming "current of hydrogen" which cause stirring of the solution and strongly affect hydrodynamic conditions in the near-electrode layer. Morphology of copper deposits is determined by this quantity of evolved hydrogen.

The fact that morphology of electrodeposited copper is determined by the difference between the overall quantity of evolved hydrogen and those spent for the creating of holes can be confirmed by the following consideration.

As already mentioned, the following disperse copper structures were formed in the range of $\eta_{I,av}(H_2)$ between 10.0 and 20.0% (group (II)): the mixture of dish-like holes and holes constructing the honeycomb-like structure [42], as well as the honeycomb-like structures constructed from holes surrounded by relatively compact cauliflower-like agglomerates of copper grains ($\eta_{I,av}(H_2) = 10.8\%$) [30] or branchy dendrites (Figs. 3.11a and 3.12a). Making the difference between the overall quantity of evolved hydrogen and the critical quantity of evolved hydrogen spent for the formation of the honeycomb-like structures ($\eta_{I,av}(H_2)$ of 10.0%), the excellent agreement between the morphological forms obtained in this hydrogen codeposition range and those obtained in the hydrogen codeposition range between 0 and 10.0% was observed. Very compact copper grains cauliflower-like agglomerates, branchy dendrites, and individual dish-like holes are obtained with a quantity of evolved hydrogen in the range of the average current efficiencies of hydrogen evolution between 0 and 10.0%. The diffusion layer of the macroelectrode was not disturbed by the formation of these morphological forms indicating that these quantities of evolved hydrogen are insufficient to cause the solution stirring and to the change of hydrodynamic conditions in the near-electrode layer.

Meanwhile, dendrites are not formed by electrodeposition processes accompanied by $\eta_{I,av}(H_2)$ larger than 20.0%. Honeycomb-like structures consisted of holes and very disperse cauliflower-like agglomerates of copper grains were formed by these electrodeposition processes [38]. The difference between the overall quantity of evolved hydrogen and the critical value for the creating of holes in the honeycomb-like ones gives the effective quantity of evolved hydrogen larger than 10.0%. This quantity of evolved hydrogen was sufficient to cause effective solution stirring leading to the change of hydrodynamic conditions in the near-electrode layer.

3.4 Analysis of Decisive Properties of Powders

As stated in Introduction, some properties, called the decisive properties, characterize the behavior of metal powder. The most important of them are the specific surface, the apparent density, the flowability, the particle grain size, and the particle size distribution [1]. These properties were analyzed by Popov et al. [58–73] which showed that some of them can be mutually related, as well as that the specific surface of copper powder can be related to the overpotential of electrodeposition [58, 59].

3.4.1 Correlation Between Specific Surface and Overpotential of Electrodeposition

The specific surface of a powder and a powdered deposit is determined as the surface per unit of the mass of powder.

It is well known [74, 75] that the surface coarseness during potentiostatic electrodeposition in the mixed activation–diffusion control increases with the increase of the current density of electrodeposition. The surface coarseness also increases during potentiostatic electrodeposition at the limiting diffusion current density with an increase of overpotential [18, 76], resulting in the formation of dendrites. In metal electrodeposition in the limiting diffusion current density range, the real current density remains constant regardless of the overpotential used. Simultaneously, the larger overpotential of electrodeposition is used, the more disperse deposit characterized by the increased specific surface is formed.

The last effect can be qualitatively discussed as follows.

The overpotential of electrodeposition, η, in the region of mixed activation–diffusion control is given by Eq. (3.19):

$$\eta = \frac{b_c}{2.3}\ln\frac{j}{j_0} + \frac{b_c}{2.3}\ln\frac{1}{1-\frac{j}{j_L}}. \tag{3.19}$$

The activation part of electrodeposition overpotential required for the charge transfer, η_{act}, is given by Eq. (3.20):

$$\eta_{\mathrm{act}} = \frac{b_{\mathrm{c}}}{2.3} \ln \frac{j}{j_0} \tag{3.20}$$

and the rest of the overpotential, η_{diff}, given by Eq. (3.21) is due to mass transfer limitations:

$$\eta_{\mathrm{diff}} = \frac{b_{\mathrm{c}}}{2.3} \ln \frac{1}{1 - \frac{j}{j_{\mathrm{L}}}}. \tag{3.21}$$

Hence, if $j \rightarrow j_{\mathrm{L}}$,

$$\eta_{\mathrm{act}} = \frac{b_{\mathrm{c}}}{2.3} \ln \frac{j_{\mathrm{L}}}{j_0}, \tag{3.22}$$

being equal to the critical overpotential for the initiation of dendritic growth [5, 27]. Simultaneously, for $j \rightarrow j_{\mathrm{L}}$

$$\eta_{\mathrm{diff}} \rightarrow \infty. \tag{3.23}$$

It is obvious that a very small increase of current density in the limiting diffusion current density range causes a large increase in deposition overpotential. Hence, the charge transfer overpotential and the ohmic drop in the solution remain the same for all overpotentials belonging to the limiting diffusion current density plateau, regardless of electrodeposition overpotential. This is due to the fact that both the charge transfer overpotential and the ohmic drop only depend on the current density. On the other hand, an increase of the deposition overpotential in the limiting diffusion current density range causes a strong increase of the dispersity and hence an increase of the specific surface area of metal deposits. Hence, it seems reasonable to assume that the difference in overpotential of electrodeposition can be related to the increase of the specific surface of the deposited metal by the relation [58, 59]

$$\gamma(S_2 - S_1) = (\eta_2 - \eta_1) \int_0^t I\mathrm{d}t, \tag{3.24}$$

where I is a current of electrodeposition, t is a time of electrodeposition, $\int_0^t I\mathrm{d}t$ is the quantity of passed electricity, η_2 and η_1 are overpotentials of electrodeposition belonging to the limiting diffusion current density range, S_2 and S_1 are the surface area of disperse deposits at the overpotentials η_2 and η_1, respectively, and γ is the surface energy of disperse deposit formation.

Equation (3.24) can be rewritten in the form

$$\gamma = \frac{(\eta_2 - \eta_1) \int_0^t I\mathrm{d}t}{(S_2 - S_1)} \tag{3.25}$$

for the direct determination of the energy of disperse solid copper surface formation in sulfate solutions.

The energy of disperse solid copper surface formation, γ, calculated by Eq. (3.25) includes all energetic loss during electrodeposition in the range of the limiting diffusion current densities.

For the estimation of the surface energy of disperse deposits formation, γ, according to Eq. (3.25), it is necessary to determine surface area of deposits, S_2 and S_1, at overpotentials η_2 and η_1 belonging to the limiting diffusion current density range. Obviously, the surface area of a deposit S_1 corresponds to an overpotential η_1 at the beginning of the limiting diffusion current density plateau, while the surface area of a deposit S_2 corresponds to an overpotential η_2 at the end of the limiting diffusion current density plateau. During the depositions, I–t dependences were recorded and the quantity of electricity, $\int_0^t I\mathrm{d}t$ was determined by a graphical integration.

Due to very high surface areas of powdered (or disperse) deposits, the determination of real surface area of this deposit type can represent serious problem [71]. The determination of the real surface area of disperse metal deposits by some common methods, such as the use of STM and AFM techniques (using the option surface area diff., in the accompanying software package), is not possible, because these techniques are suitable for the determination of the real surface area of only compact and relatively smooth surface area [77, 78].

For that reason, the new and relatively easy way for the estimation of the real surface area of disperse deposits was proposed by Popov et al. [58, 59, 71]. For copper solution containing 0.15 M $CuSO_4$ in 0.50 M H_2SO_4, this estimation can be presented in the following way: working electrode (copper or platinum) of surface area $S_{0,i}$ is covered by a thin copper film by electrodeposition at an overpotential of 300 mV during 2 min. After a relaxation of the diffusion layer for 15 min, current at an overpotential of 50 mV, I_0, is recorded, being proportional to the original electrode surface area $S_{0,i}$. The overpotential is then adjusted to the desired value, and electrodeposition is carried out. After the determined quantity of electricity had been reached, the overpotential is decreased to 50 mV, and after relaxation of the diffusion layer for 15 min, the current I_η, corresponding to the surface area S_f generated during electrodeposition, is determined. The surface area of the deposit is then calculated using Eq. (3.26):

$$S_f = S_{0,i}\frac{I_\eta}{I_0} - S_{0,i} = S_{0,i}\left(\frac{I_\eta}{I_0} - 1\right). \tag{3.26}$$

It is well known [6] that dendrites are not formed at overpotentials of electrodeposition lower than a critical overpotential for dendritic growth initiation, η_i, and that powdered deposits characterized by a very large surface area are formed at overpotentials higher than some critical value η_c.

For $\eta_2 \geq \eta_c$, $\eta_1 \geq \eta_i$ and $S_2 >> S_1$, Eq. (3.25) can be rewritten in the form

$$S_2 = \frac{(\eta_2 - \eta_1)\int_0^t I\mathrm{d}t}{\gamma}. \tag{3.27}$$

On the other hand, the quantity of electrodeposited metal m is given by

$$m = \frac{M}{nF}\int_0^t I\mathrm{d}t, \tag{3.28}$$

assuming the current efficiency for metal electrodeposition $\eta_I(M)$ to be 1, where M is the atomic mass of deposited metal. From Eqs. (3.27) and (3.28), the specific powder (or powdered deposit) surface S_{sp} is

$$S_{sp} = \frac{S_2}{m} = \frac{(\eta_2 - \eta_1)nF}{\gamma M}. \tag{3.29}$$

If the current efficiency for metal electrodeposition, $\eta_I(M)$, is lower than 1, Eq. (3.28) becomes

$$m = \eta_I(M)\frac{M}{nF}\int_0^t I\mathrm{d}t \tag{3.30}$$

and Eq. (3.29) becomes:

$$S_{sp} = \frac{(\eta_2 - \eta_1)nF}{\eta_I(M)\gamma M}. \tag{3.31}$$

Equations (3.30) and (3.31) are valid in the hydrogen codeposition range at overpotentials lower than the critical one for the change of the growth of dendrites. The situation is dramatically different in galvanostatic electrodeposition of powder. In this case, due to the increase of the surface coarseness, the low increase of the limiting diffusion current density caused by the increase of the surface area of a deposit leads to the strong decrease of overpotential of the electrodeposition in the limiting diffusion current density range even with the short deposition times. The internal structure of powder particles does not change with time during potentiostatic electrodeposition, and it only depends on the presence or the absence of hydrogen codeposition. In the galvanostatic case, the structure of particles becomes a more compact with the increasing time of electrodeposition and it can be changed from dendrites at the beginning of the electrodeposition process to the compact one with the longer electrodeposition time, as shown in Fig. 3.15a (*Powder A*) [60, 66]. In the hydrogen codeposition range, the overpotential of electrodeposition is determined by hydrogen reduction and at sufficiently large initial

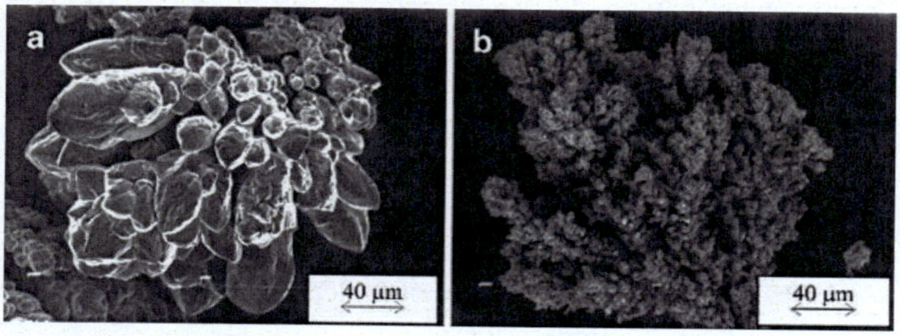

Fig. 3.15 SEM photomicrographs of copper powder particles obtained in constant current deposition. $c(Cu^{2+}) = 15$ g/dm^3, $c(H_2SO_4) = 140$ g/dm^3, $Q = 0.11$ dm^3/min^1, $t = (50 \pm 2)$°C, fraction (149–177) μm: (**a**) $j = 1{,}800$ A/m^2, $\tau_r = 1.5$ h, apparent density 1.122 g/cm^3 and (**b**) $j = 3{,}600$ A m^2, $\tau_r = 15$ min, apparent density 0.524 g/cm^3 (Reprinted from [60] with permission from NOVA publishers, [66] with permission from the Serbian Chemical Society and copied by permission from the "Electrochemistry Encyclopedia" (http://electrochem.cwru.edu/ed/encycl/) on 04/25/2007. The original material is subject to periodical changes and updates.)

current densities, $j \gg j_L$, as well as with enough short electrodeposition times, the formation of copper powder in a galvanostatic deposition becomes similar to the one in potentiostatic electrodeposition, as shown in Fig. 3.15b (*Powder B*).

The described method for the determination of the specific surface of electrodeposited copper is applicable if some kind of a Faradaic cage is not formed on the surface of deposit, i.e., when the formed structure is "open" to the bulk of electrolyte solution in potentiostatic deposition.

According to Calusaru [3], the specific surface of copper powder is 500–3,000 cm^2 g^{-1} depending on the electrodeposition conditions. The critical overpotential for the dendritic growth initiation in copper sulfate acid solution is about 0.30 V, and the critical overpotential for the instantaneous dendritic growth initiation, which is equal to the critical overpotential of copper powder formation, is about 0.7 V [18]. Assuming that electrodeposition is carried out at an overpotential of the instantaneous dendritic growth initiation, we can show the following. Using the difference in the actual deposition overpotential and the overpotential corresponding to the beginning of the plateau of the limiting diffusion density as 0.40 V, γ determined in the presented way as 2.7 J cm^{-2} [58, 59], and η_I(Cu) = 1, the minimum specific powder surface, S_{sp}, of 500 cm^2 g^{-1} is calculated, which is in good agreement with the findings of Calusaru [3].

If the electrodeposition overpotential and the overpotential of dendritic growth initiation from the examined electrolyte are known, it is obvious that the specific surface of copper powder can be calculated by Eq. (3.31) using this value of γ and the corresponding value of the current efficiency for copper electrodeposition. The upper limit of the value of the copper powder specific surface can be estimated as follows. Assuming that electrodeposition is carried out at 1.0 V with the current efficiency for copper electrodeposition of 0.5, $S_{sp} \cong$ 2,800 cm^2 g^{-1} is obtained using Eq. (3.31) which is in accordance with the data of Calusaru. In this way, one of the most important characteristics of copper powder is related to overpotential of the electrodeposition and hence to the electrodeposition conditions.

Also, the specific surface of powdered deposits can be related to morphology of deposits obtained at overpotentials belonging to the plateau of the limiting diffusion current density, as well as at the higher ones. This is presented by the analysis of copper electrodeposition processes from 0.15 M $CuSO_4$ in 0.50 M H_2SO_4 at overpotentials of 550 and 700 mV (the plateau of the limiting diffusion current density), as well as at overpotentials of 800 and 1,000 mV which are about 50 and 250 mV outside the plateau of the limiting diffusion current density [30, 38].

According to Eqs. (3.26), (3.30), and (3.31), the specific surface of the electrodeposited copper, $S_{sp,m}$, can be given by Eq. (3.32), where $\eta_I(M) = \eta_{I,av}(Cu)$:

$$S_{sp,m} = \frac{S_{0,i}\left(\frac{I_\eta}{I_0} - 1\right)}{\frac{\eta_{I,av}(Cu)\int_0^t I dt}{nF} M} = \frac{nFS_{0,i}}{\eta_{I,av}(Cu)M} \frac{1}{\int_0^t I dt} \left(\frac{I_\eta}{I_0} - 1\right). \tag{3.32}$$

In this case, the number of electrons involved in the electrodeposition reaction, n, is 2, the atomic mass of copper, M, is 63.55 g mol^{-1}, and Faraday constant, F, is 96,485 As mol^{-1}. In this investigation, the original electrode surface before electrodeposition, $S_{0,i}$, was 0.50 cm^2.

Table 3.1 The average current efficiency for copper electrodeposition, $\eta_{I,av}(Cu)$ (%) obtained at different overpotentials

Overpotential, η (mV)	550	700	800	1,000
The average current efficiency for copper electrodeposition, $\eta_{I,av}(Cu)$ (%)	100	98.03	89.2	70.0

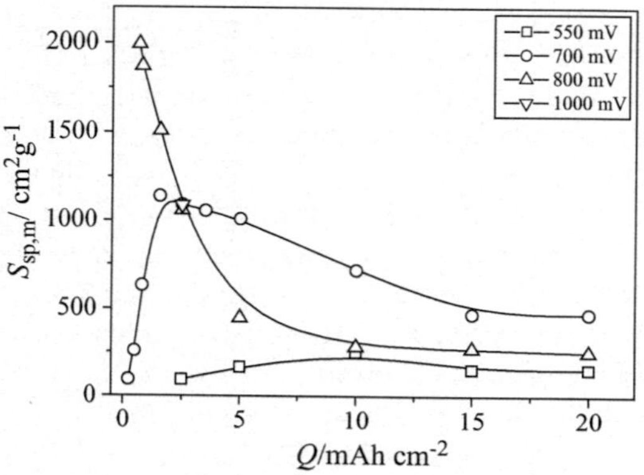

Fig. 3.16 The dependence of the specific surface of the electrodeposited copper on the quantity of the electricity, Q

Hydrogen evolution at an overpotential of 550 mV was not observed. The average current efficiencies of hydrogen evolution, $\eta_{I,av}(H_2)$ in %, at overpotentials of 700, 800, and 1,000 mV were 1.97, 10.8, and 30.0%, respectively [30, 38].

Since $\eta_{I,av}(Cu) + \eta_{I,av}(H_2) = 1$, the average current efficiencies for copper electrodeposition can be simply calculated and the obtained values are summarized in Table 3.1.

Current I_η and I_0 are obtained in an already described way.

Figure 3.16 shows the dependence of the specific surface of the electrodeposited copper on the quantity of the electricity. Two groups of the dependence of the specific surface of the electrodeposited copper on the quantity of the electricity are observed by the analysis of this figure.

In the first group are inserted the dependences of $S_{sp,m}$ on the quantity of the electricity Q obtained at overpotentials of 550 and 700 mV.

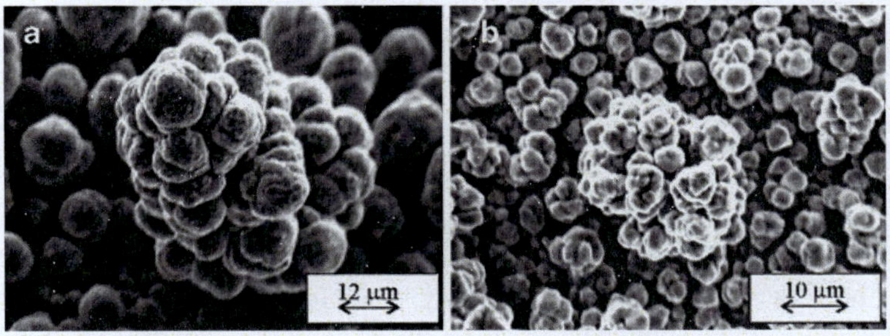

Fig. 3.17 Cauliflower-like forms electrodeposited at an overpotential of (**a**) 550 mV, quantity of the electricity: 10 mAh cm^{-2} and (**b**) 700 mV, quantity of the electricity: 2.5 mAh cm^{-2}

Fig. 3.18 Copper deposits obtained at an overpotential of (**a**) 550 mV; the quantity of the electricity: 20 mAh cm^{-2}, and (**b**) 700 mV; the quantity of the electricity: 10 mAh cm^{-2} (Reprinted from [79] with permission from Elsevier.)

At these overpotentials, there is no hydrogen evolution (550 mV) or it is very small (700 mV). The characteristic of this group is the existence of maximum of the specific surface at the determined quantity of the electricity. The increase of the overpotential led to the decrease of the quantity of the electricity needed to reach this maximum. The cauliflower-like forms, as those shown in Fig. 3.17, were obtained with a quantity of electricity of 10 and 2.5 mAh cm^{-2} at overpotentials of 550 and 700 mV, respectively.

The mixture of cauliflower-like forms and individual dendrites is formed at an overpotential of 550 mV with a quantity of electricity of 20 mAh cm^{-2} (Fig. 3.18a) [79], while very branchy copper dendrites (Fig. 3.18b) are formed at an overpotential of 700 mV with a quantity of larger electricity of 2.5 mAh cm^{-2} [71]. Copper dendrites were consisted of corncob-like elements, and the whole electrode surfaces were covered with them after electrodeposition with quantities of the electricity larger (approximately 5.0 mAh cm^{-2}).

Anyway, the values of $S_{sp,m}$ obtained at an overpotential of 700 mV before initiation of dendritic growth, i.e., with the low quantities of passed electricity, are similar to those obtained at 550 mV with the larger quantities of electricity. The further increase of $S_{sp,m}$ with the increasing quantity of the electricity before dendritic growth initiation is due to the decrease of grain size of a deposit. After formation of both the precursors of dendrites and dendrites, $S_{sp,m}$ strongly increases and remains approximately constant up to the quantity of electricity at which the overlap of dendrites begins. With larger quantities of electricity, the $S_{sp,m}$ value decreases up to constant value which probably corresponds to the established front of dendrites growing to the bulk of the solution.

The second group of the dependences of $S_{sp,m}$ on Q is obtained at overpotentials of 800 and 1,000 mV. As already mentioned, at these overpotentials hydrogen evolution is vigorous enough to strongly affects hydrodynamic conditions in the near-electrode layer. From Fig. 3.16, it can be seen that there is no maximum of the specific surface in a function of quantity of the passed electricity. Due to the overlap of the agglomerates of grains during prolonged electrodeposition, the decrease of the specific surface of the electrodeposited copper with the increasing quantity of the electricity was observed. With the short electrodeposition times, $S_{sp,m}$ was very large. Also, it can be observed from Fig. 3.16 that there is not any difference between the specific surfaces obtained at overpotentials of 800 and 1,000 mV. SEM analysis of copper deposits obtained at these overpotentials showed that the honeycomb-like structures are formed at these overpotentials [30, 38]. The typical agglomerates of copper grains formed around holes are shown in Fig. 3.19. The decrease of grain size obtained at an overpotential of 1,000 mV in relation to those obtained at an overpotential of 800 mV is due to the increase of nucleation rate with the increase of overpotential. Anyway, copper dendrites were not formed during electrodepositions at overpotentials of 800 and 1,000 mV.

Figure 3.20 shows the dependence of the specific surface of the electrodeposited copper on overpotential of the electrodeposition for

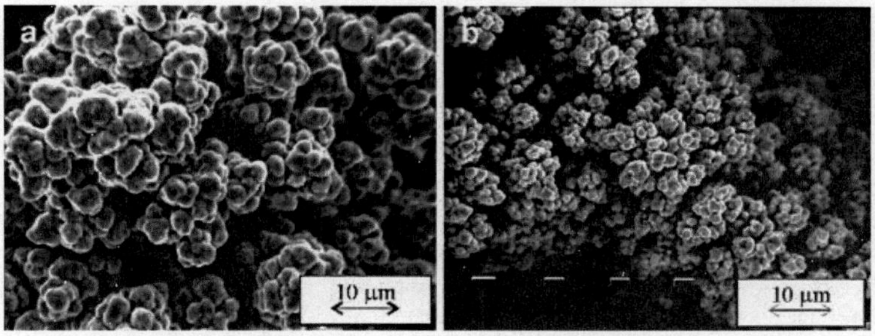

Fig. 3.19 Copper grain agglomerates obtained with the quantity of the electricity of 20 mAh cm^{-2} at an overpotential of (**a**) 800 mV, and (**b**) 1,000 mV

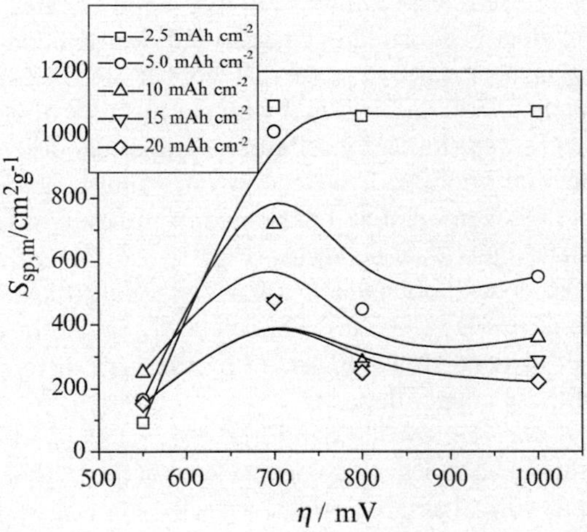

Fig. 3.20 The dependence of the specific surface of the electrodeposited copper on overpotential of the electrodeposition

different quantities of the electricity. Aside from for quantity of the electricity of 2.5 mAh cm^{-2}, the maximum of $S_{sp,m}$ is obtained at an overpotential of 700 mV with all the analyzed quantities of electricity. The formation of dendrites corresponds to these maximal values of $S_{sp,m}$.

The decrease of the specific surface with the increase of the overpotential of the electrodeposition from 700 to 800 mV can be explained by the change of the hydrodynamic conditions in the near-electrode layer caused by hydrogen evolution. Due to the change of hydrodynamic conditions in the near-electrode layer, copper electrodeposition occurs at an overpotential which is effectively lower than the specified one. For that reason, morphologies of electrodeposited copper obtained at overpotentials of 800 and 1,000 mV are similar to ones obtained at some lower overpotentials before the initiation of dendritic growth (the concept of "effective overpotential") [30, 38].

The absence of maximum of the $S_{sp,m}$ for powdered deposits obtained with the quantity of the electricity of 2.5 mAh cm^{-2} can be explained by the fact that morphologies of copper deposits obtained with this quantity of the electricity did not depend on the overpotential used. The cauliflower-like agglomerates of copper grains were electrodeposited at all analyzed overpotentials.

Also, it is necessary to note that copper dendrites formed at overpotentials of 550 and 700 mV with the quantity of electricity of 20 mAh cm^{-2} were mutually different. Copper dendrites formed at an overpotential of 550 mV were very rare, and the copper deposit obtained at this overpotential represented the mixture of cauliflower-like forms and individual dendrites (Fig. 3.18a). On the other hand, the powdered deposit obtained at 700 mV was constructed from very branchy dendrites (Fig. 3.18b). This clear difference in the surface morphology of deposits is confirmed by the very different specific surfaces of these copper deposits.

Hence, it was shown that dendritic deposits showed the largest electrode surface area, i.e., these deposits are of the largest $S_{sp,m}$. This is consistent with Chassaing et al. [80] who showed by impedance spectroscopy that the electrodeposition of ramified deposits was accompanied by the large increase of the deposit surface area.

3.4.2 *The Representative Powder Particle and the Particle Size Distribution Curves*

As already mentioned, a copper powder is not formed of particles of identical size and morphology; the individual particles may assume various forms and have very different surface areas for the same average size of granule [3]. For that reason, to relate the powder properties with the deposition process parameters and the deposition conditions, a representative particle of metal powder must be defined, and then, metal powder is defined as a group of identical (i.e. representative) powder particles.

The representative powder particle must have at least one common property with powder as a whole. The specific surface is one such common property characterizing both a metal powder and an individual (representative) powder particle. The specific surface of an individual powder particle can be determined only if its form is approximated by some regular geometric form, such as a cube [60]. On the other hand, numerous methods are available in the literature for the determination of the specific surface of powder, S_{sp} [81].

The representative powder particle is characterized by the same specific surface as a powder consisting of a mixture of different particles. The specific surface of a powder particle, $S_{sp,p}$, is given by

$$S_{sp,p} = \frac{S_{par}}{m_p}, \tag{3.33}$$

where S_{par} and m_p are the surface and mass of particle, respectively.

The shape of particle size distribution curve can be calculated assuming that the largest fraction of particles corresponds to the representative ones [60, 67], and one such calculated curve is shown in Fig. 3.21. The shape of this curve was in good agreement with those found in the literature [1, 82].

3.4.2.1 Real Systems

In real conditions, the size and shape of powder particles and hence the particle size distribution curve depend on a regime of electrolysis, a composition of solution, deposition time, cathodic material, temperature, hydrodynamic regime, *etc*.

During electrochemical deposition processes of copper at overpotentials belonging to the plateau of the limiting diffusion

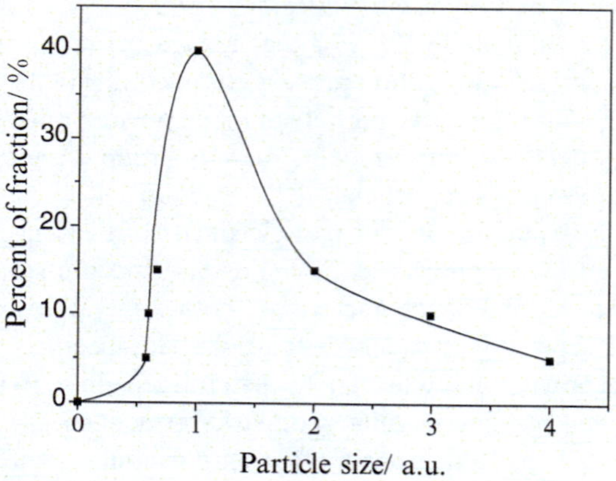

Fig. 3.21 The shape of the typical calculated particle size distribution curve (according to [60, 67])

current density, the shape of powder particles strongly depended on the applied overpotential of electrodeposition. The reason for it is relatively wide range of overpotentials belonging to the plateau of the limiting diffusion current density of about 500 mV. The characteristic of electrodeposition processes in this range of overpotentials is the absence of vigorous hydrogen evolution. The typical particle size distribution curves for copper powders obtained by electrodeposition at overpotentials of 600, 700, and 800 mV on platinum electrodes are shown in Fig. 3.22 [83]. In all results presented here, copper was electrodeposited from 0.10 M $CuSO_4$ in 0.50 M H_2SO_4 at the room temperature. From Fig. 3.22, it can be seen that the increase of overpotential leads to the formation of smaller particles and to narrower particles size distribution curves.

The type of cathodic materials used had strong effect on the shape of particle size distribution curves [6, 83]. For example, at overpotentials of 600 and 700 mV (i.e., at overpotentials belonging to the plateaus of the limiting diffusion current density), smaller particles and narrower distribution curves were obtained for the electrodeposition on a platinum electrode than on an aluminum electrode. This is due to fact that aluminum is covered with a relatively thick oxide film which causes an enlarged ohmic resistance

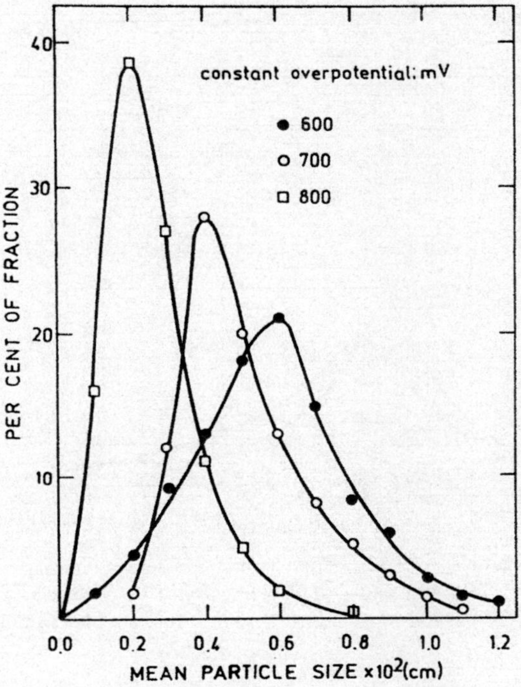

Fig. 3.22 Particle size distribution curves for copper powders obtained by potentiostatic electrodeposition on platinum electrodes (Reprinted from [6, 83] with permission from Springer.)

of the electrode–solution interface of aluminum than the one for platinum [83]. At an overpotential of 800 mV (this overpotential is outside the plateau of the limiting diffusion current density), there was not any difference in particle size distribution curves obtained on platinum and aluminum electrodes. The reason for it is the fact that an overpotential of 800 mV is situated in the hydrogen codeposition range where the process is dominantly controlled by hydrogen evolution reaction.

Anyway, increasing overpotential leads to the formation of a more disperse deposit characterized by the decreased particle size. This can be explained by the fact that increasing overpotential leads to the decrease of the height of protrusion, $h_{0,i}$, at which dendrites start to

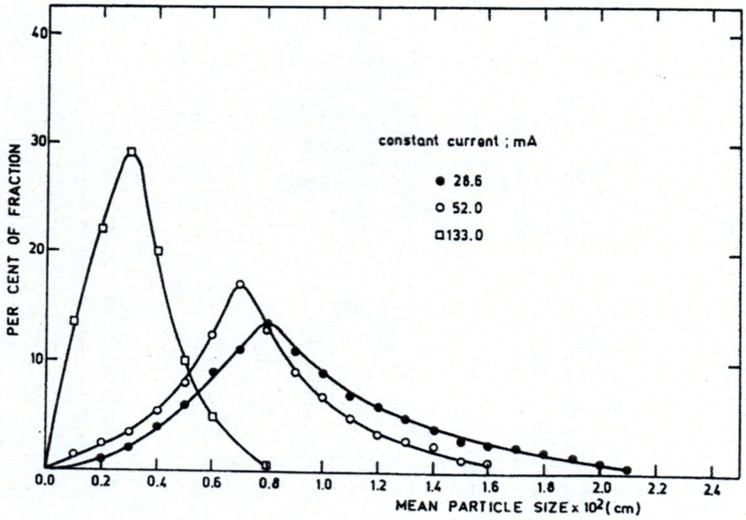

Fig. 3.23 Particle size distribution curves for copper powders obtained by galvanostatic electrodeposition on platinum electrodes. Surface area of the electrode: 0.63 cm^2 (Reprinted from [6, 83] with permission from Springer.)

grow instantaneously. Hence, increasing overpotential means a larger number of growth sites suitable for the growth of dendrites. On the other hand [19], the velocity of dendrite growth is maximum for some optimal value of the dendrite tip radius. The optimal tip radius decreases with increasing overpotential. With the dendrite tip radii larger than the optimal value, the difference between maximal and actual velocities of dendrite growth increases with the increasing overpotential. Hence, smaller particles and narrower particle size distribution curves are expected with the increasing overpotential of powder formation [6, 83].

The particle size distribution curves for copper powders obtained on the platinum electrodes in galvanostatic regime at currents of 28.6, 52.0, and 133 mA are shown in Fig. 3.23. The selected currents corresponded to average currents recorded in potentiostatic electrodepositions at overpotentials of 600, 700, and 800 mV, respectively [83]. The formation of larger particles and less narrow distribution curves in the galvanostatic regime than those formed in the potentiostatic

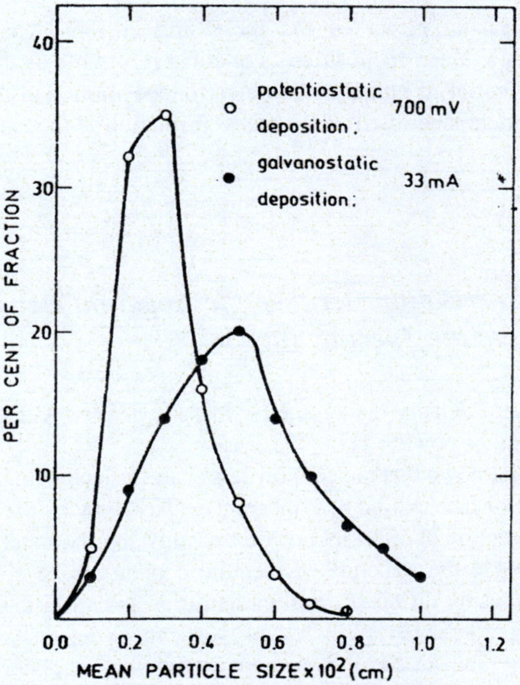

Fig. 3.24 Particle size distribution curves for copper powders obtained by the potentiostatic and galvanostatic (the average current in the potentiostatic regime) electrodepositions on copper electrodes. Surface area of the electrode: 0.63 cm^2 (Reprinted from [83] with permission from Springer.)

regime (Fig. 3.22) can be considered as follows: during electrodeposition in the galvanostatic regime in the hydrogen codeposition range, overpotential is determined by hydrogen reduction, and for the difference of overpotentials of the order 100 mV, ten times larger current of electrodeposition is required. Hence, in real conditions, smaller differences in size of the particles are expected during electrodeposition by different current densities, as well as less narrow particle size distribution curves relative to the ones obtained in potentiostatic electrodeposition. A similar situation was observed when copper was used as a cathode material (Fig. 3.24).

Anyway, the effect of the increasing current density in the galvanostatic electrodeposition is qualitatively same as the increase of overpotential in potentiostatic electrodeposition, and the essence of the particle size distribution curve formation is the same in both the cases.

3.4.3 Correlation Between the Apparent Density and the Specific Surface

The apparent density or volumetric mass is defined as the mass per unit volume of powder [1].

It is well known that copper powders characterized with high values of specific surface exhibit low apparent density. Powder particles from the same fraction of different powders occupy approximately the same volume, while the structure of metallic copper can be considerably different causing different apparent densities and specific surfaces of powder (Fig. 3.15) [61, 62]. Obviously, the more disperse the powder particles are, the smaller the apparent density of the copper powder is and the larger the specific surface is. Factors affecting the apparent density of copper powders are electrolyte composition (copper and acid content), electrolyte temperature, electrolyte circulation rate, current density, and brush-down interval [61].

Using the data of Calusaru [3], the dependence of the apparent density on the specific surface of a copper powder can be determined [60, 67]. The correlation between the apparent density ρ' and the specific surface of a copper powder can be given by (3.34):

$$\rho' = \frac{K}{S_{\mathrm{sp}}}, \tag{3.34}$$

where K is a constant. This constant is determined from the slope of the dependence of $\rho' - 1/S_{\mathrm{sp}}$ as $K \approx 1{,}000\ \mathrm{cm}^{-1}$ [67]. The maximal difference between values of the apparent density determined experimentally and the one calculated using Eq. (3.34) is 20% and less than 10% in other cases. Hence, the apparent density is well correlated with the specific surface of a copper powder by the use of Eq. (3.34). On the other hand, Eq. (3.34) can be rewritten in the form

$$\rho' S_{sp} = K, \tag{3.35}$$

which means that the value of K can be estimated using the values of specific surface and apparent density of each particular powder.

The relation between the apparent density and the specific surface of powder is very important from the practical point of view. The experimental determination of S_{sp} requires a relatively complicated equipment and experimental procedure, while ρ' can be easily determined. S_{sp} and ρ' depend on many variables: concentration of depositing ion, concentration of the supporting electrolyte, temperature, and stirring rate, and for example, using data from [84], S_{sp} can be satisfactory estimated by the use of Eq. (3.34).

3.4.3.1 The Effect of the Regime of Reversing Current on the Apparent Density of Copper Powder

The strong effect on the apparent density of copper powder can be achieved by the application of periodically changing regimes of electrolysis, such as the reversing current (RC) regime. Generally, the apparent density of powder increases by the selection of appropriate parameters of the RC regimes [62]. Figures 3.25–3.27 show comparative inspection of morphologies of powder particles obtained in the constant galvanostatic regime (DC regime; Fig. 3.25) and by the different RC regimes (Figs. 3.26 and 3.27). It is necessary to note that the current density amplitude in the RC regimes corresponded to the selected current density in the DC regime (j = 3,600 A/m^2). The other parameters of electrolysis were same (15 g dm^{-3} $CuSO_4{\cdot}5H_2O$ in 140 g dm^{-3} H_2SO_4; temperature: (50 ± 2.0)°C; electrolyte circulation rate: 0.11 dm^3 min^{-1}) [61]. The decrease of dendritic character and the increase of compactness of powder particles were observed by the application of RC regimes. The effect of this regime is less pronounced in the minute range (Fig. 3.26) than in the second one when agglomerates of monocrystal subparticles were formed (Fig. 3.27). The powder particles obtained by the RC regime in the

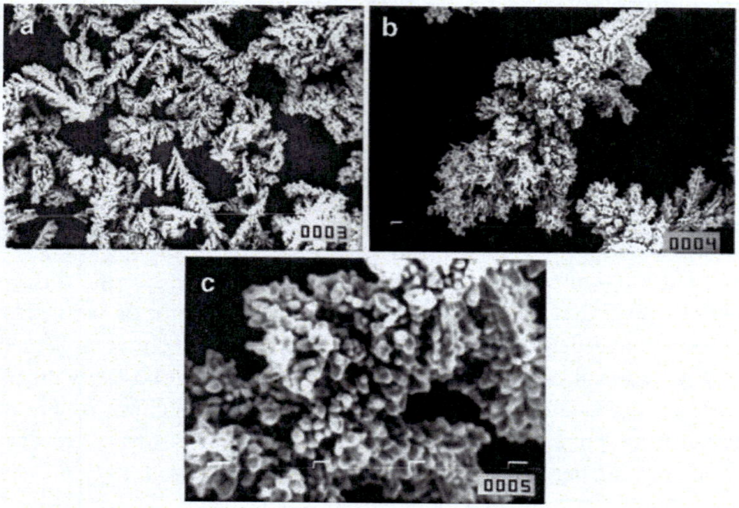

Fig. 3.25 The powder particles obtained by the galvanostatic electrodeposition. The current density: 3,600 A/m^2. Fraction (149–177) μm and apparent density: 0.524 g cm^{-3}. (**a**) ×200; (**b**) ×1000, and (**c**) ×3500. The powder was not sieved (Reprinted from [62] with permission from Serbian Chemical Society.)

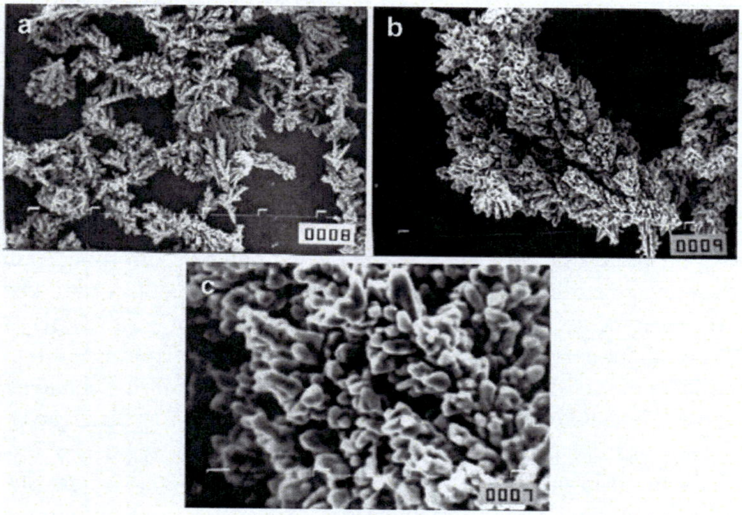

Fig. 3.26 The powder particles obtained by the RC regime. Amplitude current density: 3,600 A/m^2. Cathodic to anodic time ratio: 5. Cathodic pulse duration: 5 min. Apparent density: 0.644 g/cm^3. (**a**) ×200; (**b**) ×1000, and (**c**) ×3500. The powder was not sieved (Reprinted from [62] with permission from Serbian Chemical Society.)

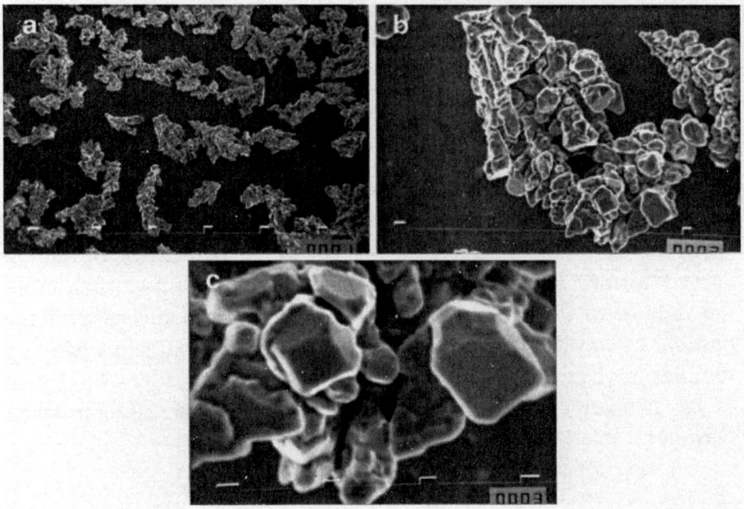

Fig. 3.27 The powder particles obtained by the RC regime. Amplitude current density: 3,600 A/m^2. Cathodic to anodic time ratio: 2.5. Cathodic pulse duration: 1 s. Apparent density: 1.624 g cm^{-3}. (**a**) ×200; (**b**) ×1000, and (**c**) ×3500. The powder was not sieved (Reprinted from [62] with permission from Serbian Chemical Society.)

second range were of considerably larger apparent density (1.624 g/cm^3) than those formed in the minute range (0.644 g/cm^3) and in the constant galvanostatic regime (0.524 g/cm^3).

The increase of the apparent density of powder particles by the application of RC regimes can be explained by the effect of the anodic time in square-waves RC on selective dissolution of the electrode surface [62]. The selective dissolution of the electrode surface during the anodic time only occurs at points with a very small radii of curvature, which dissolve faster than flat parts of the surface or of points with larger tip radii. In the minute range, the duration of selective dissolution must be shorter compared to the overall anodic dissolution time, because the tip radii of dendrites or dendrite branches very quickly become sufficiently large to make the effect of selective dissolution negligible and the particles dissolve uniformly. A decrease of the overall dissolution time leads to a decrease of the time in which the particles dissolve uniformly and the effect of selective dissolution is more pronounced from the point of view of the Kelvin effect, i.e., the selective dissolution on the particle "macrolevel," making the particles less branched.

On the other hand, the adatoms which are not included completely in the metal lattice will be dissolved faster than those which are included in it, which has the effect of selective dissolution on the "microlevel" of the particle which results in the formation of regular crystal forms. The effect of the decreased dissolution time from the minute to the second range is the same as in the case of the particle "macrolevel." More about the effect of periodically changing regimes of electrolysis on morphology of powder particles was given in the chapter considering morphology of lead and silver powder particles.

3.4.4 *The Effect of the Shape and Structure of the Particle on the Flowability of Electrolytic Copper Powder*

The flowability of a copper powder depends on the interparticle friction, which is determined by the surface area and surface roughness of the particles. As the surface area and surface roughness increase, the amount of friction in the powder mass increases and the powder exhibits less efficient flow. The same appears with the shape of particle.

The more irregular the shape of particles is, the less efficient the flow of powder is. Resistance to flow is the main feature of friction and it decreases as the particles approach to a smooth spherical shape. The effect of particle size distribution on the powder flowability is also important. If the powder consists of monosized particles which are more or less in mutual point contact, making the contact surface as low as possible, even dendritic deposits can flow. If the powder consists of different particles, the interstitial voids of the larger particles can be filled by the smaller ones, the contact surface area increases, and the flow of the powder is less efficient [1]. As a result of this, a nonsieved powder often does not flow, while the fractions of the same powder flow [61, 62]. Hence, the best conditions for the free flow of the powder are fulfilled if the powder consists of mono-sized particles of spherical shape with a surface structure approaching to the structure of a smooth metal surface.

For the analysis of the flowability of the powder, due to the existence of nonsieved powders which can flow, the shape and the structure of the powder particles are more important than particle size distribution. Flowability of nonsieved powders occurs when the surface parts of the particles corresponding to the metal segments are larger than or equal to the pores between them [65].

Typical particles of the fraction 149–177 μm of *powders A* and *B* are shown in Fig. 3.15. (*Powder A* was obtained at 1,800 A m^{-2}, while *Powder B* was obtained at 3,600 A/m^2.) The fraction of *powder A* exhibits excellent flowability, while the fraction of *powder B* does not flow. Although the surface structure is very different, in the first approximation, the shape of both particles can be taken as spherical.

Hence, the powder particles can be approximated to be mono-sized and spherical ones. The flow time of fraction 149–177 μm of *powder A* was about 20 s, which corresponds to excellent flowability [61] but the nonsieved powder does not flow due to the fact that the surface structure can allow the jamming of the particles of different fractions. On the other hand, the structure of the particle of fraction 149–177 μm of *powder B* from Fig. 3.15 is very porous and such particles can interweave. Obviously, such behavior leads to the nonflowing powder.

As already shown, the flowability of copper powder is mainly determined by the structure of the surface of the powder particles. The effect of the particle shape is also important, but probably it is not the decisive factor. If the surface structure of powders approaches the structure of the surface of bulk copper and if the shape of the particles approaches a sphere, the friction in the powder mass is low and the flow of the powder is efficient. Besides, in these cases, the particle size distribution will not have an effect on the flowability of the copper powder, so nonsieved powders exhibit free flow.

3.5 Conclusions

There is an important difference between disperse deposits formed in galvanostatic and potentiostatic conditions. In potentiostatic electrodeposition, the properties of disperse deposits mainly depend on overpotential of electrodeposition. The electrodeposition overpotential remains constant during deposition time, as well as the real current density, and the structure of disperse deposits does not change with the electrodeposition time. Hence, it can be expected that both the structure of powder particles and the properties of powder (as association of different powder particles) on the macrolevel do not depend on the electrodeposition time.

In the formation of disperse metals by galvanostatic electrodeposition [75], the apparent current density is constant, while the surface area of a deposit increases during electrodeposition leading to the decrease of both the real current density and the overpotential of electrodeposition. Then, the change of the structure of disperse deposits is observed. In this way, in galvanostatic conditions, the structure of particles can be different in different parts of the one and the same powder particle [85]. Naturally, due to the change of the structure of particles, the properties of powder obtained in galvanostatic electrodeposition will depend on the electrodeposition time. The dependence of the properties of metal powders on the structure of powder particles, and the dependence of metal coatings on the structure of the surface [77, 78, 86–88] are examples of the effects of micro- and submicrostructures on the behavior of the macrosystems.

Electrodeposition at a periodically changing rate offers a number of possibilities for changing the deposition conditions at one and the same deposition rate [43]. In this way, powder particles with different grain size and morphology can be obtained by varying the wave of periodically changing current or overpotential [6]. Electrodeposition at a periodically changing rate is based on the periodic repetition of current or overpotential waves [27, 43]. The most important regime from the theoretical point of view is the regime of pulsating overpotential (PO). On the other hand, the most important regime from a practical point of view is the regime of reversing current (RC).

Copper powders with different apparent densities and related properties were obtained by the change of conditions of electrolysis such as: electrolyte composition (acid and copper content), electrolyte temperature, electrolyte circulation rate, current density, and brush-down interval. Similar effects are expected by changing the parameters which determine the shape of the deposition reversing current wave [6]. Besides, it seems that the surface structure of powder particles obtained in reversing current (RC) electrodeposition is more compact than in the constant galvanostatic regime permitting the free flow of powders with considerably lower apparent densities [62]. This will be considered in more detail in the future.

The relations between the decisive characteristics of metal powder and the conditions of electrodeposition have not been established so far in a semiquantitative way. Hence, electrodeposition of metal powders can be regarded as largely empirical, an activity in which there is much art and little science. We hope that this will change with the publication of this chapter.

The specific surface of copper powder is related to the overpotential of electrodeposition. Also, it is shown that the apparent density is a function of the specific surface of powder, while the flowability of powder is a function of the apparent density. In this way, the most important properties of the powder are related to both the specific surface and the conditions of electrodeposition. Additionally the shape of particle size distribution curve is also discussed qualitatively.

The importance of hydrogen codeposition in copper powder formation is also analyzed in this chapter. Regardless of the fact that all above relations are done, more or less, in a semiquantitative way, this chapter can be considered as the initiation of the qualitative analysis of decisive characteristics of metal powders and their relation with conditions of electrodeposition.

Acknowledgments The authors are grateful to Dr. Goran Branković and Dr. Ljubica Pavlović for SEM analysis of investigated systems, as well as to Dr. Snežana Krstić and Dr. Miomir Pavlović for helpful discussions.

The work was supported by the Ministry of Education and Science of the Republic of Serbia under the research project "Electrochemical synthesis and characterization of nanostructured functional materials for application in new technologies" (No. 172046).

References

1. German RM (1994) Powder metallurgy science, 2nd edn. Metal Powder Industries Federation, Princeton, NJ
2. Pavlović MG, Popov KI (2005) Electrochemistry encyclopedia. http://electrochem.cwru.edu/ed/encycl/
3. Calusaru A (1979) Electrodeposition of metal powders, Materials science monographs. Elsevier, New York
4. Orhan G, Hapci G (2010) Powder Technol 201:57
5. Popov KI, Djokić SS, Grgur BN (2002) Fundamental aspects of electrometallurgy. Kluwer Academic/Plenum, New York
6. Popov KI, Pavlović MG (1993) Electrodeposition of metal powders with controlled grain size and morphology. In: White RE, Bockris JO'M, Conway BE (eds) Modern aspects of electrochemistry, vol 24. Plenum, New York, pp 299–391
7. Hirakoso K (1935) Denkikogaku Kyokoishi 3:7
8. Hirakoso K (1935) Chem Abst 29:5749u
9. Ibl N (1962) Advances in electrochemistry and electrochemical engineering, vol 2. Interscience, New York
10. Ibl N (1954) Helv Chim Acta 37:1149
11. Kudra O, Lerner ME (1951) Ukrain Khim Zh 17:890
12. Kudra O, Gitman E (1952) Elektroliticheskoe Poluchenie Metallicheskiekh Poroshkov, Izd. Akad. Nauk Ukr. SSR, Kiev
13. Ibl N, Schadegg K (1967) J Electrochem Soc 114:54
14. Calusaru A (1957) Revista de Chemie Bucuresti 8:369
15. Atanasiu I, Calusaru A (1957) Studii Cercet Met Bucuresti 2:237
16. Russev D (1981) J Appl Electrochem 11:177
17. Theis G, Fassler C, Robertson PM, Dossenbach O, Ibl N (1981) 32nd ISEMeeting, Dubrovnik/Cavtat, vol 1, p 383
18. Popov KI, Maksimović MD, Trnjančev JD, Pavlović MG (1981) J Appl Electrochem 11:239
19. Barton JL, Bockris JO'M (1962) Proc Roy Soc A268:485
20. Maksimović MD, Popov KI, Pavlović MG (1979) Bull Soc Chim 44:687
21. Maksimović MD, Popov KI, Jović LjJ, Pavlović MG (1979) Bull Soc Chim 44:47
22. Krichmar SI (1965) Electrokhim 1:609
23. Diggle JW, Despić AR, Bockris JO'M (1969) J Electrochem Soc 116:1503
24. Despić AR, Diggle JW, Bockris JO'M (1968) J Electrochem Soc 115:507
25. Popov KI, Despić AR (1971) Bull Soc Chim 36:173
26. Despić AR (1970) Croat Chim Acta 42:265
27. Despić AR, Popov KI (1972) Transport controlled deposition and dissolution of metals. In: Conway BE, Bockris JO'M (eds) Modern aspects of electrochemistry, vol 7. Plenum, New York, pp 199–313
28. Popov KI, Pavlović MG, Maksimović MD (1982) J Appl Electrochem 12:525

29. Popov KI, Krstajić NV, Čekerevac MI (1996) The mechanism of formation of coarse and disperse electrodeposits. In: White RE, Conway BE, Bockris JO'M (eds) Modern aspects of electrochemistry, vol 30. Plenum, New York, pp 261–312
30. Nikolić ND, Popov KI, Pavlović LjJ, Pavlović MG (2006) J Electroanal Chem 588:88
31. Nikolić ND, Popov KI, Pavlović LjJ, Pavlović MG (2007) Sensors 7:1
32. Nikolić ND, Pavlović LjJ, Pavlović MG, Popov KI (2008) Powder Technol 185:195
33. Lowenheim FA (1978) Electroplating. McGraw-Hill Book, New York, St. Louis
34. Wolery TJ (1992) EQ3NR – a computer program for geochemical aqueous speciation-solubility calculations: theoretical manual and user's guide, version 7.0. Lawrence Livermore National Laboratory, Livermore, CA
35. Roine A (1999) HSC chemistry: chemical reaction and equilibrium software with extensive thermochemical database. 4.0. Outokumpu Research Oy, Finland
36. Casas JM, Alvarez F, Cifuentes L (2000) Chem Eng Sci 55:6223
37. Pitzer KS (1991) Activity coefficients in electrolyte solutions, 2nd edn. CRC, Boca Raton, FL
38. Nikolić ND, Popov KI (2010) Hydrogen co-deposition effects on the structure of electrodeposited copper. In: Djokić SS (ed) Electrodeposition: theory and practice. Modern aspects of electrochemistry, vol 48. Springer, Berlin, pp 1–70
39. Nikolić ND, Pavlović LjJ, Krstić SB, Pavlović MG, Popov KI (2008) Chem Eng Sci 63:2824
40. Nikolić ND, Pavlović LjJ, Branković G, Pavlović MG, Popov KI (2008) J Serb Chem Soc 73:753
41. Nikolić ND, Branković G, Pavlović MG, Popov KI (2008) J Electroanal Chem 621:13
42. Nikolić ND, Pavlović LjJ, Pavlović MG, Popov KI (2007) Electrochim Acta 52:8096
43. Popov KI, Maksimović MD (1989) Theory of the effect of electrodeposition at periodically changing rate on the morphology of metal deposition. In: Conway BE, Bockris JO'M, White RE (eds) Modern aspects of electrochemistry, vol 19. Plenum, New York, pp 193–250
44. Li Y, Jia W-Z, Song Y-Y, Xia XH (2007) Chem Mater 19:5758
45. Shin H-C, Liu M (2004) Chem Mater 16:5460
46. Kim J-H, Kim R-H, Kwon H-S (2008) Electrochem Commun 10:1148
47. Nikolić ND, Branković G (2010) Electrochem Commun 12:740
48. Shin H-C, Dong J, Liu M (2003) Adv Mater 15:1610
49. Everhart JL (n.d) Copper and copper alloy powder metallurgy properties and applications. http://www.copper.org/resources/properties/129_6/homepage.html
50. Walker R, Duncan SJ (1984) Surf Technol 23:301

51. Maksimović VM, Pavlović LjJ, Pavlović MG, Tomić MV (2009) J Appl Electrochem 39:2545
52. Owais A (2009) J Appl Electrochem 39:1587
53. Pavlović MG, Pavlović LjJ, Maksimović VM, Nikolić ND, Popov KI (2010) Int J Electrochem Sci 5:1862
54. Djokić SS, Nikolić ND, Živković PM, Popov KI, Djokić NS (2011) ECS Trans 33:7
55. Nikolić ND, Branković G, Pavlović MG (2012) Powder Technol. http://dx.doi.org/10.1016/j.powtec.2012.01.014
56. Nikolić ND, Branković G, Maksimović VM, Pavlović MG, Popov KI (2009) J Electroanal Chem 635:111
57. Nikolić ND, Popov KI, Pavlović LjJ, Pavlović MG (2007) J Solid State Electrochem 11:667
58. Popov KI, Nikolić ND, Rakočević Z (2002) J Serb Chem Soc 67:635
59. Popov KI, Nikolić ND, Rakočević Z (2002) J Serb Chem Soc 67:769
60. Nikolić ND, Krstić SB, Pavlović LjJ, Pavlović MG, Popov KI (2008) The mutual relation of decisive characteristics of electrolytic copper powder and effect of deposition conditions on them. In: Hayashi K (ed) Electroanalytical chemistry research trends. Nova, New York, pp 185–209
61. Pavlović MG, Pavlović LjJ, Ivanović ER, Radmilović V, Popov KI (2001) J Serb Chem Soc 66:923
62. Popov KI, Pavlović LjJ, Ivanović ER, Radmilović V, Pavlović MG (2002) J Serb Chem Soc 67:61
63. Popov KI, Nikolić ND, Rakočević Z (2002) J Serb Chem Soc 67:861
64. Popov KI, Krstić SB, Pavlović MG (2003) J Serb Chem Soc 68:511
65. Popov KI, Krstić SB, Obradović MČ, Pavlović MG, Pavlović LjJ, Ivanović ER (2003) J Serb Chem Soc 68:771
66. Popov KI, Pavlović MG, Pavlović LjJ, Ivanović ER, Krstić SB, Obradović MČ (2003) J Serb Chem Soc 68:779
67. Popov KI, Živković PM, Krstić SB (2003) J Serb Chem Soc 68:903
68. Popov KI, Krstić SB, Obradović MČ, Pavlović MG, Pavlović LjJ, Ivanović ER (2004) J Serb Chem Soc 69:43
69. Popov KI, Krstić SB, Pavlović MG, Pavlović LjJ, Maksimović VM (2004) J Serb Chem Soc 69:817
70. Popov KI, Nikolić ND, Krstić SB, Pavlović MG (2006) J Serb Chem Soc 71:397
71. Nikolić ND, Popov KI, Pavlović LjJ, Pavlović MG (2007) Mater Prot 48:3 (in Serbian)
72. Pavlović MG, Nikolić ND, Popov KI (2003) J Serb Chem Soc 68:649
73. Pavlović MG, Pavlović LjJ, Doroslovački ID, Nikolić ND (2004) Hydrometallurgy 73:155
74. Popov KI, Pavlović LjJ, Pavlović MG, Čekerevac MI (1988) Surf Coat Technol 35:39
75. Popov KI, Pavlović MG, Pavlović LjJ, Čekerevac MI, Remović GŽ (1988) Surf Coat Technol 34:355

76. Popov KI, Maksimović MD, Pavlović MG, Lukić DT (1980) J Appl Electrochem 10:299
77. Nikolić ND, Rakočević Z, Popov KI (2005) Nanostructural analysis of bright metal surfaces in relation to their reflectivities. In: Conway BE, Vayenas CG, White RE, Gamboa-Adelco ME (eds) Modern aspects of electrochemistry, vol 38. Kluwer Academic/Plenum, New York, pp 425–474
78. Nikolić ND, Rakočević Z, Popov KI (2001) J Electroanal Chem 514:56
79. Nikolić ND, Popov KI, Pavlović LjJ, Pavlović MG (2006) Surf Coat Technol 201:560
80. Chassaing E, Rosso M, Sapoval B, Chazalviel J-N (1993) Electrochim Acta 38:1941
81. Schatt W, Wierters KP (1997) Powder metallurgy – processing and materials. European Powder Metallurgy Association, Technical University Dresden, Dresden, p 8
82. Peissker E (1984) Int J Powder Metallurgy Powder Technol 20:87
83. Popov KI, Pavlović MG, Maksimović MD, Krstajić SS (1978) J Appl Electrochem 8:503
84. Pavlović LjJ, Nikolić ND, Popov KI (2000) Mater Sci Forum 352:65
85. Murashova I, Pomosov B (1989) In: Polukarov YuM (ed) Itogi nauki i tehniki, Seria Elektrokhimiya, vol 30. Acad. Sci., Moscow, p 90
86. Nikolić ND, Novaković G, Rakočević Z, Djurović DR, Popov KI (2002) Surf Coat Technol 161:188
87. Nikolić ND, Rakočević Z, Popov KI (2004) J Solid State Electrochem 8:526
88. Nikolić ND, Rakočević Z, Djurović DR, Popov KI (2006) Russ J Electrochem 42:112

Chapter 4
Porous Copper Electrodes Formed by the Constant and the Periodically Changing Regimes of Electrolysis

Nebojša D. Nikolić

4.1 Introduction

The formation of open and porous structures with extremely large surface area is of high technological significance, because this structure type is very suitable for electrodes in many electrochemical devices, such as fuel cells, batteries and sensors [1, 2], and in catalysis applications [3]. The template-directed synthesis method is most commonly used for the preparation of such electrodes. This method is based on a deposition of desired materials in interstitial spaces of disposable hard template. When interstitial spaces of template are filled by deposited material, the template is removed by combustion or etching, and then the deposited material with the replica structure of the template is obtained [4, 5]. The most often used hard templates are porous polycarbonate membranes [6, 7], anodic alumina membrane [8–10], colloidal crystals [11, 12], echinoid skeletal structures [13], and polystyrene spheres [14, 15].

N.D. Nikolić (✉)
ICTM-Institute of Electrochemistry, University of Belgrade,
Njegoseva 12, P.O.B. 473,11001 Belgrade, Serbia
e-mail: nnikolic@tmf.bg.ac.rs

Stojan S. Djokić (ed.), *Electrochemical Production of Metal Powders*, Modern Aspects of Electrochemistry 54, DOI 10.1007/978-1-4614-2380-5_4,

Fig. 4.1 The honeycomb-like structure formed by electrodeposition from 0.15 M $CuSO_4$ in 1.0 M H_2SO_4 at an overpotential of 1,000 mV with a quantity of the electricity 10 mAh/cm^2 (Reprinted from [22] with permission from Elsevier and [23] with permission from Springer.)

Electrodeposition technique showed as very favorable way for the production of porous electrodes [1, 2, 4]. The open porous copper and copper–tin alloys electrodes, denoted as both 3-D foam [1, 2, 4] and honeycomb-like ones [16–23], are formed by electrodeposition at high current densities and overpotentials, where parallel to electrodeposition process, the hydrogen evolution reaction occurs. The main characteristics of these electrodes are holes or pores formed by attached hydrogen bubbles surrounded by metals grain agglomerates or dendritic particles (Fig. 4.1). This way of preparing porous electrodes is denoted as gas bubble dynamic template method, where the hydrogen bubbles function as a dynamic template for the formation of this type of deposits. The advantage of producing porous materials by this hydrogen bubble dynamic template method when compared with hard template methods is its low cost, ease of preparation, facile control of structure, and facile one-step synthesis process, including preparation of the template, metal deposition, and elimination of the template [5].

4.2 Potentiostatic Regime of Electrolysis

In potentiostatic regimes of electrolysis, honeycomb-like copper electrodes are formed by electrochemical deposition at overpotentials outside the plateau of the limiting diffusion current density where the parallelism between copper electrodeposition rate and hydrogen evolution reaction is evident [16, 23]. Hydrogen evolution responsible for the formation of the honeycomb-like electrodes was vigorous enough to cause such stirring of the copper solution which leads to the decrease of the cathode diffusion layer thickness and to the increase of the limiting diffusion current density and hence to the change of the hydrodynamic conditions in the near-electrode layer [16]. For copper solutions containing 0.15 M $CuSO_4$ and less (in 0.50 M H_2SO_4), the critical quantity of evolved hydrogen leading to a change of hydrodynamic conditions in the near-electrode layer was estimated to correspond to the average current efficiency of hydrogen evolution of 10.0% [20].

4.2.1 Phenomenology of Formation of the Honeycomb-Like Structures

The formation of porous deposits by dynamic template method can be briefly presented as follows: in the initial stage of the electrodeposition process, both the nuclei of deposited metal and the "nuclei" of hydrogen bubbles are formed at the active sites of the electrode surface (Fig. 4.2a) [18]. The hydrogen bubbles isolate the substrate and then the current lines are concentrated around them making rings consisted of agglomerates of grains of deposited metal. The current lines are also concentrated at the metal nuclei formed in the initial stage between the hydrogen bubbles forming copper grains agglomerates of them. In the growth process, due to current density distribution effect, both hydrogen evolution and copper nucleation primarily occur at top of these agglomerates. Some of the new, freshly formed hydrogen bubbles will coalesce with hydrogen bubbles formed in the initial stage of electrodeposition process, leading to their growth with

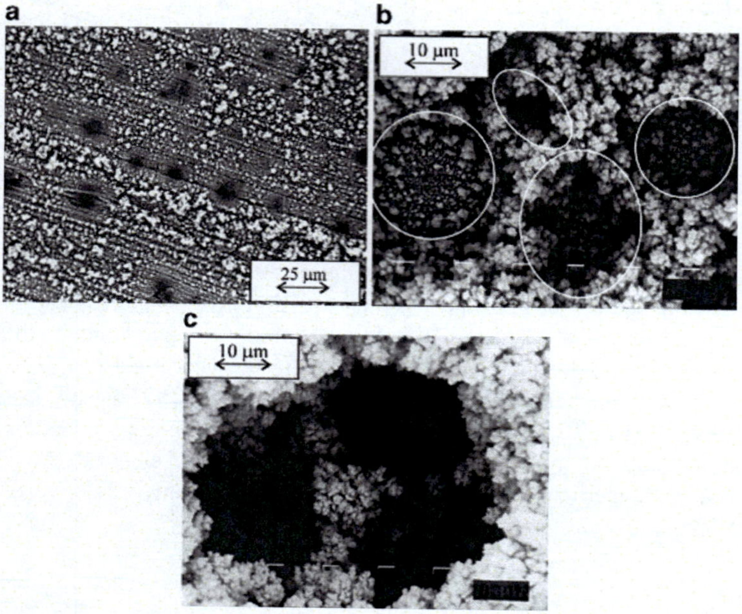

Fig. 4.2 Copper deposits electrodeposited from 0.15 M $CuSO_4$ in 0.50 M H_2SO_4 at an overpotential of 1,000 mV. Time of electrolysis: (**a**) 10 s; (**b**) 30 s, and (**c**) 120 s (Reprinted from [18, 23] with permission from Springer.)

electrolysis time. When the critical size of these hydrogen bubbles to detach from electrode surface is reached, they will detach from electrode surface forming holes of regular shapes at electrode surface. This "regular" type of holes is shown in Fig. 4.2b in circle. Simultaneously, holes of irregular shape are formed at electrode surface of agglomerate copper grains formed between hydrogen bubbles [18]. These "irregular" holes are situated between regular holes, and they are shown in Fig. 4.2b in ellipse. For longer electrodeposition time, coalescence of closely formed hydrogen bubbles occurs, leading to the formation of large so-called coalesced holes (Fig. 4.2c) [18].

Meanwhile, some of the new, freshly formed hydrogen bubbles will not coalesce with the previously formed hydrogen bubbles because

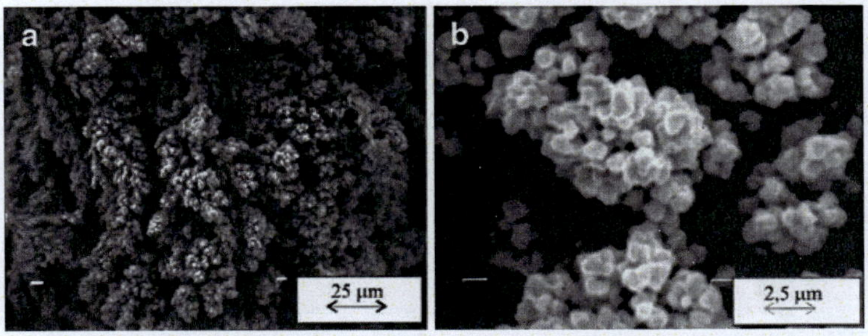

Fig. 4.3 Copper deposits electrodeposited from 0.075 M $CuSO_4$ in 0.50 M H_2SO_4 at an overpotential of 1,000 mV. Quantity of electricity: 20 mAh/cm^2 (Reprinted from [19] with permission from Elsevier and [23] with permission from Springer.)

they are situated between the freshly formed copper nucleus and these hydrogen bubbles have not enough place to develop in large hydrogen bubbles. These hydrogen bubbles will detach very fast from the electrode surface forming a channel structure through the interior of the deposit [19]. The typical channel structure formed by simultaneous hydrogen evolution and copper nucleation is shown in Fig. 4.3a. The "top view" of the part of the honeycomb-like structure shown in Fig. 4.3a is given in Fig. 4.3b; from it can be seen cauliflower-like agglomerates of copper grains surrounded by irregular channels for which the origin is of evolved hydrogen.

All elements constructing the honeycomb-like structure can also be seen from Fig. 4.4 showing a cross section of this type of deposit [24].

The "regular holes" formed by both the attached hydrogen bubbles (part in circle denoted with *A* in Fig. 4.4a) and the coalescence of neighboring hydrogen bubbles (part in circle denoted with *B* in Fig. 4.4a) and "irregular holes" formed due to the effect of current distribution at the growing surface (parts denoted by arrow labeled *C* in Fig. 4.4a) are shown in Fig. 4.4a [24]. The presence of channel structures formed through the interior of the deposit can be easily observed by cross section analysis of this deposit at a higher magnification (Fig. 4.4b).

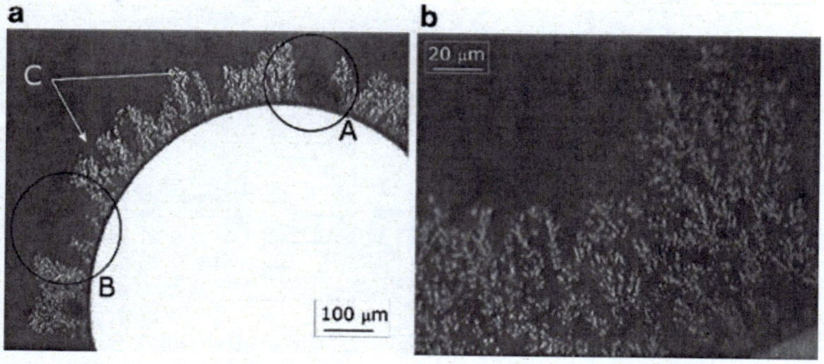

Fig. 4.4 Cross section of copper deposit electrodeposited from 0.15 M $CuSO_4$ in 0.50 M H_2SO_4, at an overpotential of 1,000 mV with a quantity of the electricity of 10 mAh/cm^2 (Reprinted from [24] with permission from Serbian Chemical Society.)

4.2.2 Parameters Affecting Number, Size and Distribution of Holes in the Honeycomb-Like Structures

Electrodeposition technique is a suitable way to get open and porous structure because it is very easy to control number, size and distribution of holes by the choice of appropriate electrodeposition conditions [19].

Factors affecting number, size, and distribution of holes are:

(a) Overpotential of electrodeposition
(b) Preparation of working electrode
(c) Concentration of Cu(II) ions
(d) Concentration of sulfuric acid
(e) Temperature of electrolysis
(f) Time of electrolysis

4.2.2.1 Overpotential of Electrodeposition

Increasing the overpotential intensifies hydrogen evolution reaction [16, 23]. For copper solution containing 0.15 M $CuSO_4$ in 0.50 M H_2SO_4, the average current efficiency of hydrogen evolution, $\eta_{I,av}(H_2)$, was about three times larger at an overpotential of 1,000 mV ($\eta_{I,av}(H_2) = 30.0\%$) than at 800 mV ($\eta_{I,av}(H_2) = 10.8\%$) [16]. It is manifested by the formation of the honeycomb-like structures with the considerably larger number of holes formed of detached hydrogen bubbles at 1,000 mV than at 800 mV.

4.2.2.2 Preparation of Working Electrode

The number, size, and distribution of holes in the honeycomb-like electrodes strongly depended on the type of working electrode used for copper electrodeposition [17, 18]. The strong difference in the initial stage of their formation, as well as in the formed honeycomb-like structures, was observed in the dependence of the type of used working electrode. The number of hydrogen bubbles formed at the electrode with large number of active centers, where irregularities at electrode surface represent active centers for the formation of both the hydrogen bubbles and agglomerates of copper grains, was several times higher than the number of holes formed at the electrode with "killed" active centers, where active centers were removed by the formation of uniform thin copper film by electrodeposition at some lower overpotential [18].

4.2.2.3 Concentration of Cu(II) Ions

The increase of concentration of Cu(II) ions causes a sharp decrease of the quantity of evolved hydrogen and hence the decrease of the average current efficiencies for hydrogen evolution reaction [19, 20], what is in a good agreement with the prediction of the ionic equilibrium of the species in the $CuSO_4$–H_2SO_4–H_2O system [21, 25, 26]. Electrodeposition processes from copper solutions containing the concentration of Cu(II) ions above 0.15 M $CuSO_4$ (in 0.50 M H_2SO_4) lead to the formation of new type of holes, denoted as dish-like hole [19]. The typical dish-like holes obtained by electrodeposition from 0.60 M $CuSO_4$ in 0.50 M H_2SO_4 at an overpotential of 1,000 mV are shown in Fig. 4.5a. The appearance of very branchy dendrites between dish-like holes (Fig. 4.5b), and at shoulders of holes with longer electrodeposition times (Fig. 4.5c) clearly points out that the diffusion layer of the macroelectrode is not disturbed during copper electrodeposition from this solution, and that the quantity of evolved hydrogen

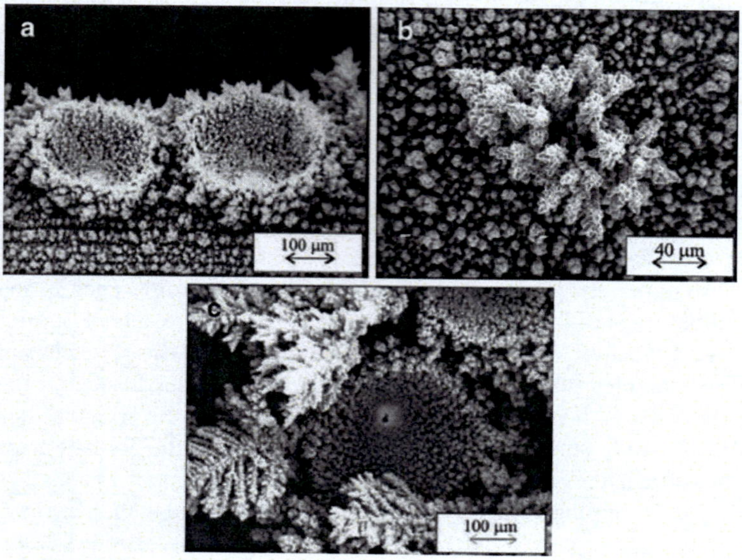

Fig. 4.5 Copper deposits electrodeposited from 0.60 M $CuSO_4$ in 0.50 M H_2SO_4 at an overpotential of 1,000 mV. Quantity of electricity: (**a**) and (**b**) 2.5 mAh/cm^2 and (**c**) 20 mAh/cm^2 (Reprinted from [19] with permission from Elsevier and [23] with permission from Springer.)

was not enough to cause stirring of solution in the near-electrode layer. Honeycomb-like structures were formed during copper electrodeposition from solutions with concentrations of Cu(II) ions less than 0.15 M $CuSO_4$ (for example, from 0.075 M $CuSO_4$ in 0.50 M H_2SO_4) [19].

The concentration of 0.30 M $CuSO_4$ (in 0.50 M H_2SO_4) is the transitional concentration between lower and higher concentrations of Cu(II) ions. The mixture of holes forming the honeycomb-like structure and dish-like holes was obtained by electrodeposition from this solution at an overpotential of 1,000 mV (Fig. 4.6) [19].

4.2.2.4 Concentration of Sulfuric Acid

The effect of H_2SO_4 on copper electrodeposition processes was examined keeping the concentration of Cu(II) ions constant

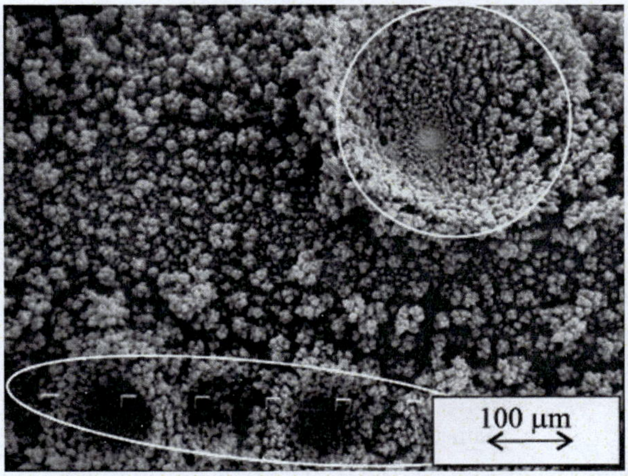

Fig. 4.6 Copper deposit obtained at an overpotential of 1,000 mV from 0.30 M $CuSO_4$ in 0.50 M H_2SO_4 with a quantity of the electricity of 2.5 mAh/cm^2 (Reprinted from [19] with permission from Elsevier and [23] with permission from Springer.)

(0.15 M $CuSO_4$), while the concentration of H_2SO_4 was varied, and they were 0.125, 0.25, and 1.0 M H_2SO_4 [22]. As expected, the increasing H_2SO_4 concentration led to the increase of the average current efficiencies of hydrogen evolution. The main characteristics of electrodeposition processes at an overpotential of 1,000 mV from the solutions containing 0.15 M $CuSO_4$ in both 0.25 and 1.0 M H_2SO_4 were holes or pores surrounded by agglomerates of copper grains. Aside from holes and cauliflower-like agglomerates of copper grains between them, degenerate dendrites, a channel structure around dendritic and cauliflower-like particles and holes with the shoulders of degenerate dendrites were electrodeposited at 1,000 mV from 0.15 M $CuSO_4$ in 0.125 M H_2SO_4 [22, 23]. These morphological forms were obtained in spite of relatively high average current efficiency of hydrogen evolution of 20.3% by which this deposit was formed, and their formation can be explained by the effect of a density and a surface tension of the electroplating solution on the bubble break-off diameter [22, 23]. The number of holes increased with the increasing H_2SO_4 concentration, while the hole size decreased with the increasing H_2SO_4 concentration. Also, the orientation of holes was changed from random oriented holes to holes which were lined up in parallel rows [22].

4.2.2.5 Temperature of Electrolysis

The increase of temperature of electrolysis leads to an intensification of hydrogen evolution reaction [27]. Meanwhile, despite intensification of hydrogen evolution with the increasing temperature, the decrease of the number of holes formed per mm^2 surface area of electrodes and the increase of their average diameter were observed during copper electrodeposition at an overpotential of 800 mV (Fig. 4.7). To explain this unexpected trend in the development of morphology of electrodeposited copper, the effect of temperature on some properties of electroplating solution, such as viscosity and surface tension, is considered [27]. The increase of temperature causes the decrease of both the viscosity [28] and the surface tension of this solution [29]. The decrease of the surface tension of the solution lowers the break-off diameter of the hydrogen bubble from the electrode surface [29], while the decreased viscosity of the solution probably facilitates a transport of the detached hydrogen bubbles through the interior of the deposit forming the channel structure through it. Anyway, increasing temperature leads to redistribution of evolved hydrogen from those creating honeycomb-like structure (Fig. 4.7a, b) to structure with dish-like holes (Fig. 4.7c) and by the dominant presence of agglomerates of copper grains surrounded by irregular channels of evolved hydrogen (that is a channel structure) (Fig. 4.7d).

4.2.2.6 Time of Electrolysis

The size of holes increases with electrolysis time due to the growth of the hydrogen bubbles with time, as well as due to a coalescence of neighboring hydrogen bubbles. In the growth process, due to the current density distribution effect, some hydrogen bubbles can remain captive in the interior of deposit making the honeycomb-like structure very porous (Fig. 4.8; part in circle denoted with *D*) [24].

Fig. 4.7 Copper deposits electrodeposited at an overpotential of 800 mV from 0.15 M $CuSO_4$ in 0.50 M H_2SO_4 at temperatures of: (**a**) 14.0; (**b**) 35.0; (**c**) and (**d**) 50.0 ± 0.5°C. Quantity of electricity: 10 mAh/cm^2 (Reprinted from [27] with permission from Serbian Chemical Society and [23] with permission from Springer.)

4.3 Galvanostatic Regime of Electrolysis

In galvanostatic regimes of electrolysis, the honeycomb-like structures are formed at current densities larger than the limiting diffusion current density [30]. The typical honeycomb-like structure electrodeposited at a current density of 0.44 A/cm^2, which is about 27.5 larger than the limiting diffusion current density, is shown in Fig. 4.9a. From Fig. 4.9b–d, all elements of which this structure type is constructed can be seen: individual hole formed by attached hydrogen bubble (this type of hole is denoted as noncoalesced hole in the future text; Fig. 4.9b), hole formed by coalescence of closely formed hydrogen bubbles (this type of hole is denoted as coalesced

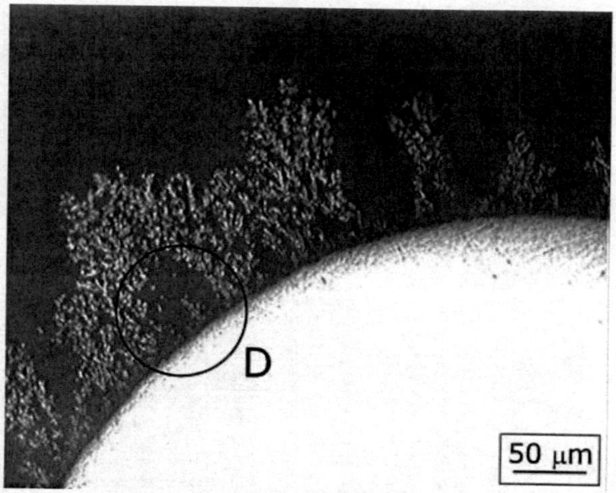

Fig. 4.8 Cross section of copper deposit electrodeposited from 0.15 M $CuSO_4$ in 0.50 M H_2SO_4 at an overpotential of 1,000 mV with a quantity of the electricity of 10 mAh/cm^2 (Reprinted from [24] with permission from Serbian Chemical Society.)

hole in the future text; Fig. 4.9c), and cauliflower-like agglomerates of copper grains formed around holes (Fig. 4.9d).

The hole size decreases while the number of holes increases with the increasing current density of electrodeposition [5].

4.4 A Brief Review of Previous Results

The increasing overpotential, the decreasing concentration of Cu(II) ions, and the increasing H_2SO_4 concentration intensify the hydrogen evolution reaction. The intensification of hydrogen evolution leads to an increase of the number of formed holes, as well as to a decrease of hole size. Meanwhile, the ratio of the coalesced holes to the overall number of formed holes increased with the intensification of hydrogen evolution.

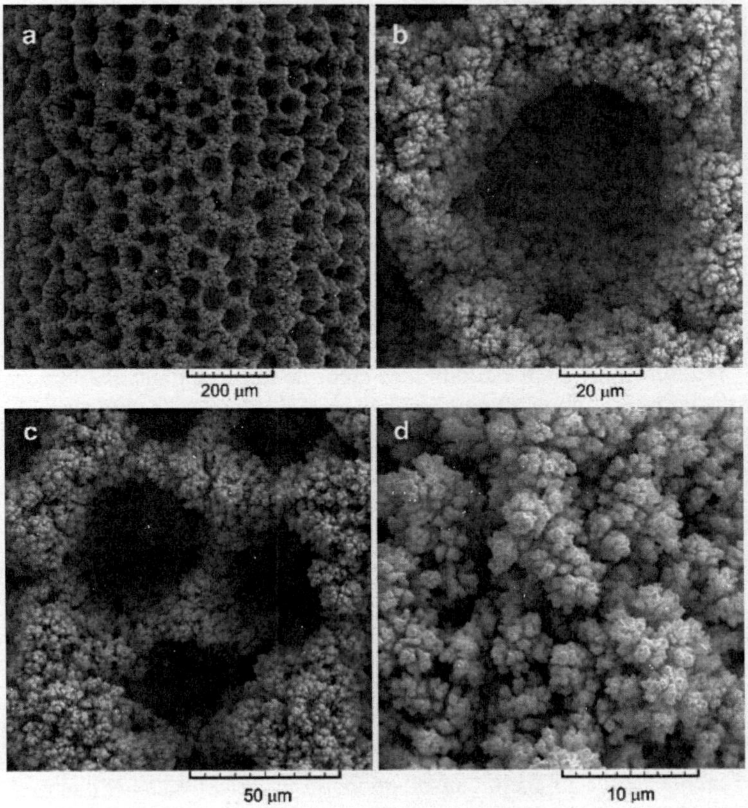

Fig. 4.9 (**a**) The honeycomb-like structure formed at a current density of 0.44 A/cm^2 and the typical elements constructing this type of structure; (**b**) noncoalesced hole; (**c**) coalesced hole; and (**d**) cauliflower-like agglomerates of copper grains formed around holes (Reprinted from [30] with permission from Elsevier.)

The process of a coalescence of closely formed hydrogen bubbles should be avoided during the formation of the honeycomb-like deposits because this process causes both the decrease of the overall number of the formed holes and the increase of the hole size and hence leads to the decrease of the specific surface area of these electrodes.

To increase the specific surface area and enhance the effectiveness/activity of the porous electrodes, it is necessary to reduce the size of the pores, as well as the branches in the foam or agglomerates of copper grains in the honeycomb-like structures [4].

The two ways are proposed to increase the specific surface area of open porous copper electrodes and to improve micro- and nanostructural characteristics of these electrodes. The first way is the addition of specific substances, known as additives, to the electroplating solution [4]. So, the decrease of the diameter of holes, as well as the increase of their number in 3D foam copper structures, can be realized by the addition of acetic acid to the copper sulfate solution [4]. Also, the addition of chloride ions dramatically reduces the size of the copper branches in the walls of holes. The reduction in pore size is a result of lowering hydrophobic force of the generated hydrogen gas by adding bubble stabilizer (e.g., acetic acid) that suppresses the coalescence of bubbles, while the decrease in branch size in the foam wall is a consequence of the catalytic effect of chloride ions on the copper deposition reaction. Mechanical strength of the foam structure can be improved by the addition of $(NH_4)^+$, Cl^-, polyethylene glycol, and 3-mercapto-1-propane sulfonic acid to the deposition bath [31]. The foam structure obtained by a combination of these additives was a highly porous with better mechanical strength than the one obtained without additives, owing to higher compactness of crystallites. Meanwhile, the use of additives in electroplating practice leads to their consumption during electrodeposition processes and the requirement for their permanent control is necessary. The consumption of additives occurs due to removal with the plated objects, by their incorporation in the deposit (codeposition) and by reaction on the plated object [32, 33].

The second way for the increase of the specific surface area of copper electrodes is the application of periodically changing regimes of electrolysis, such as pulsating overpotential (PO), pulsating current (PC), and reversing current (RC). The application of PO regime is primarily important from academic point of view for understanding mechanism of electrodeposition processes at periodically changing rate. For technological purposes, pulse and reverse plating techniques, such as pulsating current (PC) and reversing current (RC), are more important [34, 35].

4.5 The Regime of Pulsating Overpotential

Pulsating overpotential (PO) consists of a periodic repetition of overpotential pulses of different shapes [34, 35]. Square-wave PO is defined by the overpotential amplitude, η_A, deposition pulse, t_c, and pause, t_p. The pause to pulse ratio is defined as $p = t_p/t_c$.

4.5.1 Characteristics of the Honeycomb-Like Structures Obtained by the PO Regime and Their Comparison with Those Obtained by the Constant Regimes of Electrolysis

Figure 4.10 shows the honeycomb-like electrodes obtained at a constant overpotential of 1,000 mV (Fig. 4.10a) and by the PO regime with the overpotential amplitude of 1,000 mV, deposition pulse, t_c, of 10 ms, and pause duration, t_p, of 50 ms (Fig. 4.10b). The difference in the number of holes, as well as in their size, can be clearly seen from this figure. In all experiments for which results are presented in this section,

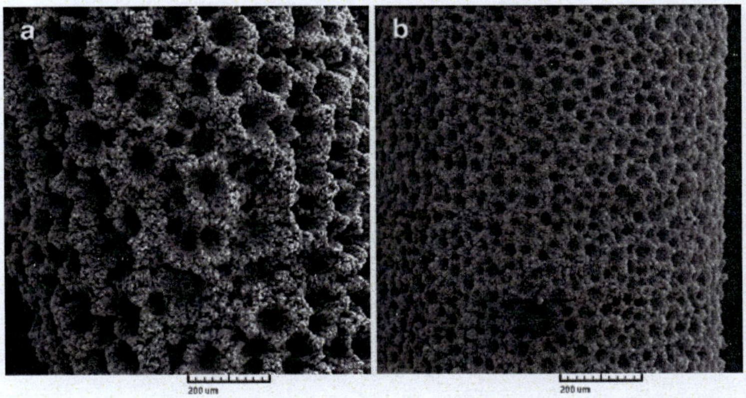

Fig. 4.10 Honeycomb-like structures obtained: (**a**) at a constant overpotential of 1,000 mV and (**b**) by PO regime with deposition pulse of 10 ms and a pause of 50 ms (Reprinted from [36] with permission from Elsevier.)

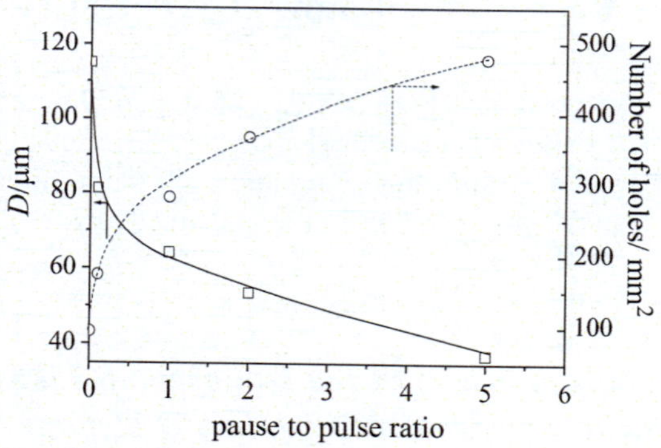

Fig. 4.11 The dependences of the average diameter of the surface holes, *D*, (*square*) and the number of holes per mm^2 surface area of the copper electrode (*circle*) on the pause to pulse ratio (Reprinted from [36] with permission from Elsevier.)

electrodeposition of copper was performed from 0.15 M $CuSO_4$ in 0.50 M H_2SO_4 at room temperature using cylindrical copper electrodes [24, 36, 37]. In pulsating overpotential (PO) deposition the overpotential amplitude of 1,000 mV and pulse duration of 10 ms were applied. A pause duration was selected to be 5, 10, 20, 50, and 100 ms (the pause to pulse ratios were 0.5, 1, 2, 5, and 10, respectively).

The dependences of the average diameter of the holes and of the number of holes formed per mm^2 surface area of the honeycomb-like copper electrodes on the pause to pulse ratio are shown in Fig. 4.11.

The decrease of the hole size and the increase of the number of formed holes with the increasing pause to pulse ratio can be primarily ascribed to the suppression of a coalescence of closely formed hydrogen bubbles [36]. The coalesced holes are observed in the honeycomb-like structures formed with pause to pulse ratios up to 2. Holes formed with the pause durations shorter than the deposition pulse were similar to those obtained at constant overpotential [36]. The bottom of these holes was very compact (Fig. 4.12a, b). The prolonging pause duration led to the change of the bottom of holes from compact to the one constructed of very disperse agglomerates of copper grains (Fig. 4.12c). The significantly smaller number of holes was formed with a pause duration of 100 ms and

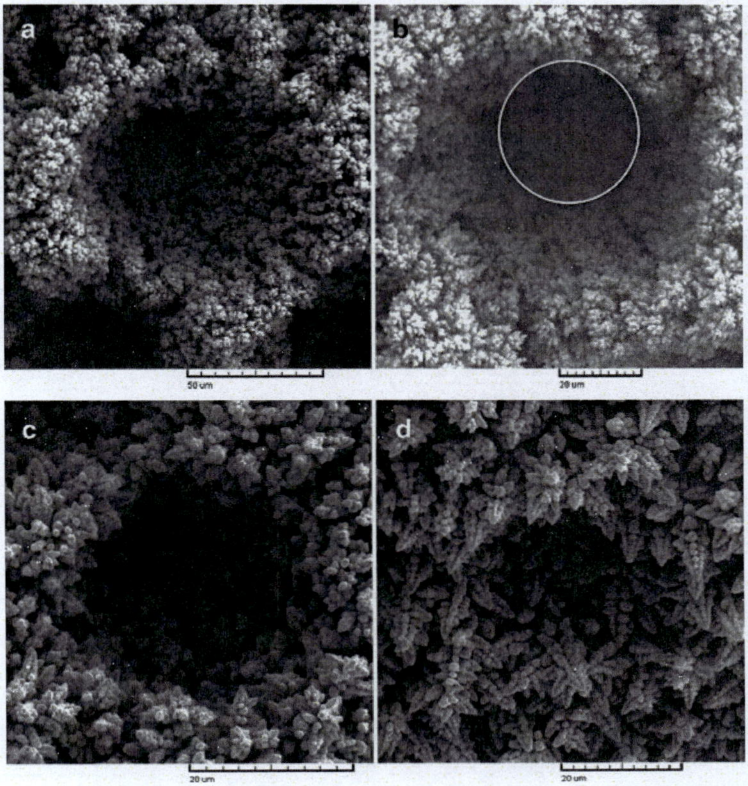

Fig. 4.12 Holes formed by attached hydrogen bubbles obtained at: (**a**) a constant overpotential of 1,000 mV and by PO regimes with a pause duration of: (**b**) 5 ms; (**c**) 50 ms; and (**d**) 100 ms. Deposition pulse: 10 ms (Reprinted from [36] with permission from Elsevier and [37] with permission from Springer.)

these holes were completely different from those obtained with the smaller pause to pulse ratios [37]. This type of holes was constructed from dendrites (Fig. 4.12d).

Simultaneously, the morphology of electrodeposited copper formed around holes, as well as inside holes, changed with the increasing pause to pulse ratio from cauliflower-like agglomerates of copper grains to dendrites. Very disperse agglomerates of copper grains were formed at the constant overpotential and by the PO

Fig. 4.13 Morphologies of electrodeposited copper formed around holes: (**a**) at a constant overpotential of 1,000 mV and by the PO regimes with pause duration of: (**b**) 50 ms and (**c**) 100 ms. Deposition pulse: 10 ms (Reprinted from [36, 38] with permission from Elsevier.)

regimes with pause durations shorter than the deposition pulse (Fig. 4.13a). Copper dendrites are noticed at shoulders of holes electrodeposited with pause duration of 20 ms and their number and size increased with the prolonging pause durations, as shown in Fig. 4.13b for the morphology of electrodeposited copper obtained with a pause duration of 50 ms. Finally, the walls of holes obtained with pause duration of 100 ms were only composed of dendrites (Fig. 4.13c). The appearing of dendritic forms clearly indicates a decrease of effectiveness of stirring of copper solution by evolved hydrogen with the increase of pause duration.

The analysis of the interior of holes obtained with pause durations up to 50 ms showed that the prolongation of pause duration leads to a reduction of the size of agglomerates of copper grains of which the walls of holes are constructed. With the prolonging pause duration up to 50 ms, holes became closer to each other, while compactness of the formed agglomerates between them was also increased [36]. The size of grains of which both cauliflower-like agglomerates and dendrites are composed increased with increasing pause to pulse ratio due to the selective dissolution of grains during the pauses. It was shown [39] that the smaller grains would dissolve faster than the larger ones due to the Kelvin effect [40]. In addition, the structure of the grains becomes more regular with increasing pause duration due to the fact that the adatoms in nonstable positions dissolve faster than the atoms in a stable position in lattice [35]. Finally, the deposit at the shoulders of the holes dissolve faster due to the edge effect, which also leads to the formation of a more homogenous distributed deposit with increasing pause duration and to an increased number of less deep holes.

The increased compactness of the copper deposits, the suppression of coalescence of neighboring hydrogen bubbles, and a decrease of the depth of the holes can be clearly seen from Fig. 4.14 showing cross section of copper deposits obtained with different pause to pulse ratios [24]. The compactness of the formed deposits increased with the increasing pause duration [24], and it was larger than the compactness of the deposit obtained by the constant regime of electrolysis (Fig. 4.4a). From Figs. 4.8 and 4.14, it can be seen that the increase of the number of holes by the application of PO regime can be ascribed not only to supressed coalescence of neighboring hydrogen bubbles but also to the improved current distribution at growing copper surface by which an inclusion of hydrogen bubbles in deposit was prevented. Due to the current density distribution effect, the loosing of "irregular holes" was also observed by the application of PO regimes. With the prolonging pause duration, pores or channels formed through the interior of deposits were mutually coalesced forming larger pores. In this way, a transport of electroactive species through the interior of structures was facilitated, what is very desirable for evaluation of electrochemical reactions [1].

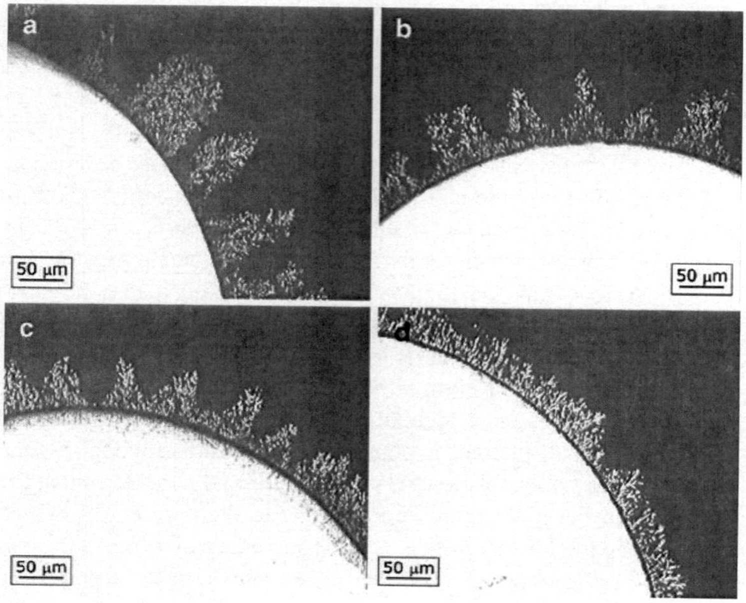

Fig. 4.14 Cross section of copper deposits electrodeposited by pulsating overpotential (PO) regime with pause duration of: (**a**) 5 ms; (**b**) 20 ms; (**c**) 50 ms; and (**d**) 100 ms. Deposition pulse: 10 ms (Reprinted from [24] with permission from Serbian Chemical Society.)

Anyway, the effects observed by the application of PO regime are ascribed to a current density during "off" periods (i.e., during duration of pause). Although this current density can be neglected in comparison with the current density during "on" periods (i.e., during the duration of deposition pulse), it is clear that its effect on the formation of these deposits is very important [36, 37].

4.5.2 Formation of the Honeycomb-Like Structures by the PO Regime

The values of the average current efficiencies of hydrogen evolution, $\eta_{I,av}(H_2)$, obtained for different pause to pulse ratios are given in Table 4.1 [41]. The following parameters of square-waves PO were

Table 4.1 The values of the average current efficiencies of hydrogen evolution, $\eta_{I,av}(H_2)$, in %, obtained for different pause to pulse ratios (t_c deposition pulse; t_p pause duration)

$t_c{:}t_p$	1:10	3:10	5:10	10:10	20:10
$\eta_{I,av}(H_2)$ (%)	0	16.4	22.4	27.2	28.1

Reprinted from [41] with permission from Elsevier

analyzed: the overpotential amplitude of 1,000 mV, pause duration of 10 ms, and deposition pulses of 1, 3, 5, 10, and 20 ms.

The pyramid-like forms were electrodeposited by square-wave PO with a deposition pulse of 1 ms (Fig. 4.15a, b). These pyramid-like forms represent precursors of dendrites, what is concluded by the comparison with the top of copper dendrites obtained by copper electrodeposition at a constant overpotential of 650 mV (Fig. 4.15c). Copper pyramids of different shapes are also obtained by electrodeposition in a constant regime of electrolysis, where the morphology of pyramidal nanoparticles depended on the ratio of the concentration of surfactant/precursor and deposition time [42]. Despite the fact that the overpotential amplitude of 1,000 mV was used, holes for which the origin was of attached hydrogen bubbles were not formed (Fig. 4.15a). The absence of holes clearly indicates that a deposition pulse of 1 ms was insufficient for the formation of hydrogen bubbles.

From Fig. 4.16, it can be seen that honeycomb-like structures were formed with deposition pulses of 3, 5, 10, and 20 ms. The analysis of the honeycomb-like structures showed that the number of holes formed by the attached hydrogen bubbles did not change considerably with the length of deposition pulses of 3, 5, and 10 ms (i.e., with pause to pulse ratios up to 1). The mild decrease of the number of the formed holes was observed with a deposition pulse of 20 ms, what can be ascribed to the enhanced coalescence of neighboring hydrogen bubbles as well as to effects related to the current density distribution at the growing electrode by which some of the hydrogen bubbles remained captive in the interior of deposits [41].

Meanwhile, the strong effect of the length of deposition pulse on the morphology of electrodeposited copper formed around holes was achieved (Fig. 4.17). Copper dendrites were formed with a deposition

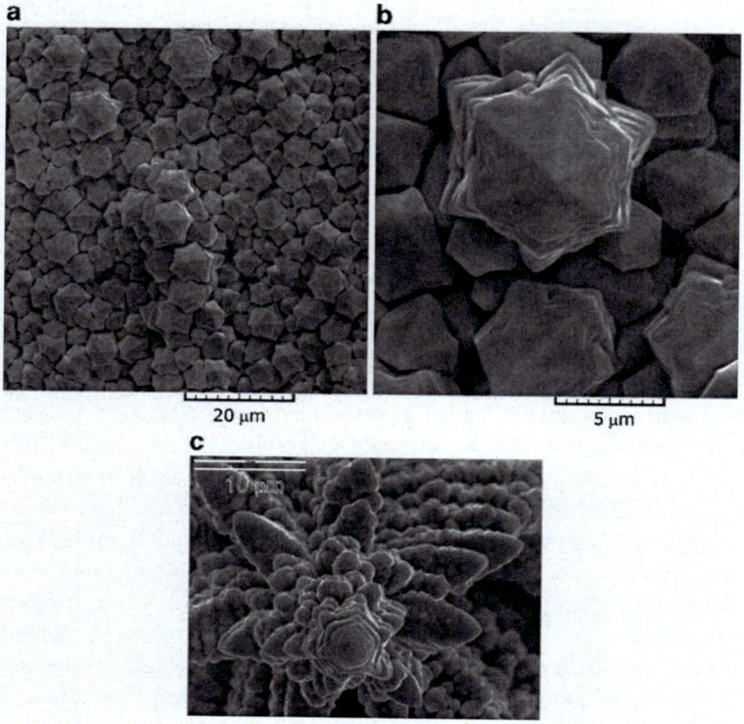

Fig. 4.15 (**a**) and (**b**) Pyramid-like precursors of dendrites obtained by square-wave PO with a deposition pulse of 1 ms and a pause of 10 ms; deposition time: 1,800 s, (**c**) top view of copper dendrite obtained by electrodeposition at an overpotential of 650 mV (Reprinted from [41] with permission from Elsevier.)

pulse of 3 ms (Fig. 4.17a). Although dendrites were also formed with a deposition pulse of 5 ms (Fig. 4.17b), there were more branchy structures than those obtained with a deposition pulse of 3 ms (Fig. 4.17a). Agglomerates of copper grains were mainly formed with a deposition pulse of 10 ms, while the presence of developed dendrites was very rare (Fig. 4.17c). Finally, agglomerates of copper grains were formed between holes when deposition pulse of 20 ms was applied (Fig. 4.17d).

Figure 4.18 shows the cross section of the copper deposits obtained with deposition pulses of 3, 5, 10, and 20 ms. Dendrites formed around

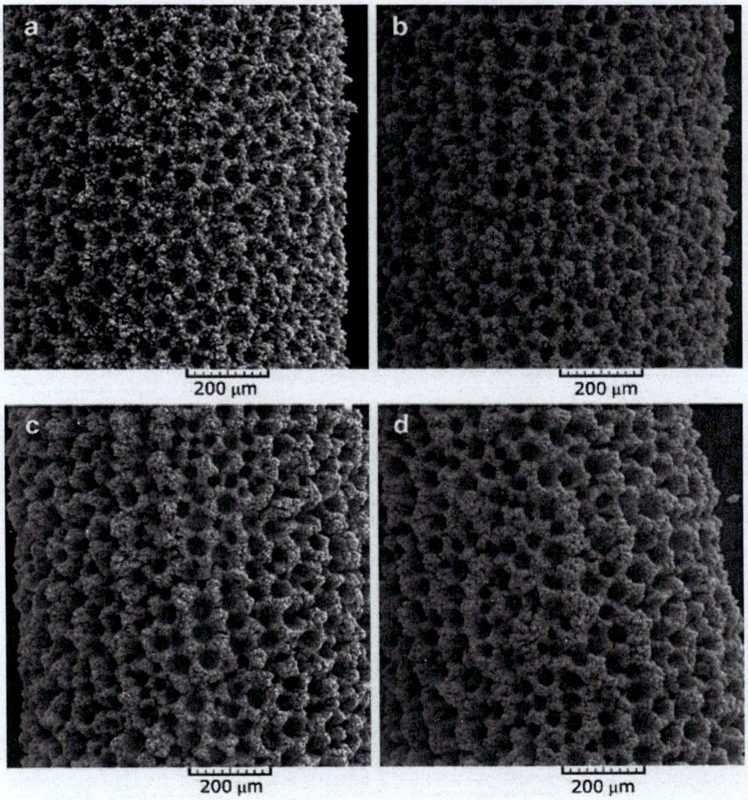

Fig. 4.16 Honeycomb-like copper structures obtained by the PO regimes with deposition pulses of: (**a**) 3 ms; (**b**) 5 ms; (**c**) 10 ms; and (**d**) 20 ms. Pause duration: 10 ms (Reprinted from [41] with permission from Elsevier.)

the hydrogen bubbles in the copper deposit obtained with a deposition pulse of 3 ms are clearly seen from Fig. 4.18a. The analysis of Fig. 4.18b confirmed that more branchy dendrites are formed with a deposition pulse of 5 ms than with 3 ms. Further increase of the deposition pulse led to the increase of a dispersity of the internal structures of the copper deposits, which is observed from Fig. 4.18c, d showing the cross sections of the copper deposits obtained with deposition pulses of 10 and 20 ms. The numbered channels formed

Fig. 4.17 Morphology of electrodeposited copper formed around holes by the PO regimes with deposition pulses of: (**a**) 3 ms; (**b**) 5 ms; (**c**) 10 ms; and (**d**) 20 ms. Pause duration: 10 ms (Reprinted from [41] with permission from Elsevier.)

around relatively small copper particles and degenerate dendrites are easily observed from Fig. 4.18c, d. From Fig. 4.18, it can also be seen that the depth of holes did not change noticeably when deposition pulses of 3, 5, and 10 ms were applied. The depth of the hole obtained with a deposition pulse of 20 ms was larger than those obtained with 3, 5, and 10 ms, what can be ascribed to the enhanced dispersity of this deposit. The analysis of holes shown in Fig. 4.18 showed that there

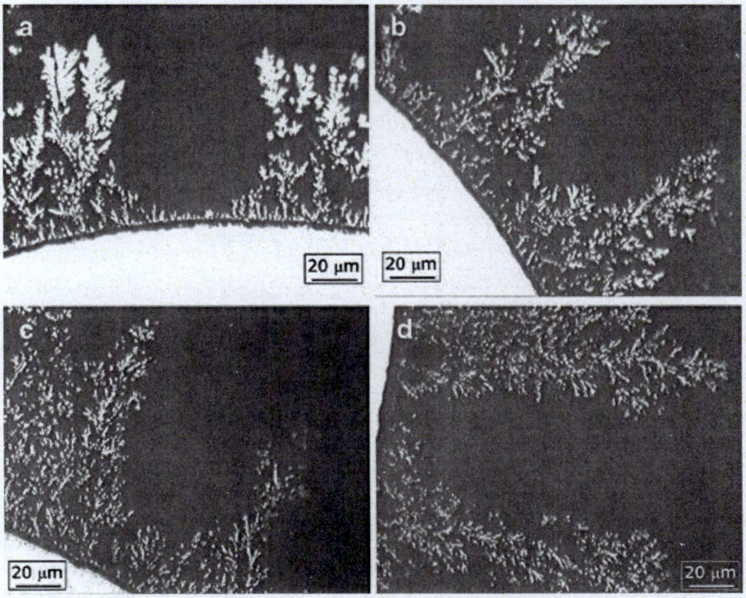

Fig. 4.18 Cross section of the copper deposits obtained with deposition pulses of: (**a**) 3 ms; (**b**) 5 ms; (**c**) 10 ms; and (**d**) 20 ms. Pause duration: 10 ms (Reprinted from [41] with permission from Elsevier.)

was not any effect of deposition pulse duration on the size of holes. The average diameter of holes obtained with deposition pulses of 3, 5, 10, and 20 ms (for pause duration of 10 ms) was estimated to be 65 ± 2.0 μm [41].

Then, the formation of the honeycomb-like structures by the PO regimes, as well as the change of morphology of the electrodeposited copper from dendrites to degenerate dendrites and cauliflower-like agglomerates of copper grains, can be explained as follows: similar to the constant regimes of electrolysis, in the initial stage of the electrodeposition process, both the nuclei of copper and the "nuclei" of the hydrogen bubbles are formed at the active sites of the electrode surface [18]. The number of the formed hydrogen bubbles is determined by the overpotential amplitude. Since the formed bubbles isolate electrode surface, electrodeposition of copper occurs around them forming rings consisted of agglomerates of copper grains. In the growth process, during deposition pulse, due to a concentration of current lines, both copper nucleation and hydrogen evolution primarily occur at these agglomerates of copper grains. Some of the new, freshly formed hydrogen bubbles coalesce with the hydrogen bubbles formed in the initial stage, leading to their growth with electrolysis time.

Meanwhile, some of the freshly formed hydrogen bubbles cannot find a path to coalesce with them, and these hydrogen bubbles start to grow independently making an interior of deposit to be porous. Due to a constant value of pause duration of 10 ms, the effect of pause duration was the same for all deposition pulses. During the pause, the dissolution of both copper particles and the formed hydrogen bubbles occur.

The prolonging duration of deposition pulse increases both copper deposition and hydrogen evolution rates. This means a greater number of both new hydrogen bubbles and copper "nuclei" is formed with the prolonging deposition pulses. The greater part of these freshly formed bubbles will not find a path to coalesce with the previously formed hydrogen bubbles, because they are surrounded by freshly formed copper grains. Also, they will not be able to develop into large hydrogen bubbles the same reason. Due to impossibility for their further growth, these hydrogen bubbles will detach very fast from a growing electrode surface forming a channel structure through the interior of deposit and causing stirring of solution in the near-electrode layer.

The analysis of Figs. 4.15–4.18 confirms that the effectiveness of a stirring of solution increased with the increasing duration of deposition pulse. Because of the increased effectiveness of a stirring of solution by evolved hydrogen, the change of morphology of the electrodeposited copper from dendrites to degenerate dendrites and cauliflower-like agglomerates was observed.

On the basis of the same hole sizes and their unchanged number, as well as of the same depth of holes (Figs. 4.16–4.18), it can be concluded that the critical size of the hydrogen bubbles formed in the initial stage of electrodeposition process to be detached from electrode surface does not depend on the length of deposition pulse. It can be assumed that approximately a same quantity of evolved hydrogen is used for the formation of these holes. Then, the difference between this quantity and the overall quantity of evolved hydrogen is responsible for a stirring of solution and a change of hydrodynamic conditions in the near-electrode layer. Also, it is clear that the quality of deposits formed between holes is determined by this difference in the quantity of evolved hydrogen.

4.5.3 Energy Aspects of the Formation of the Honeycomb-Like Structures by the PO Regime

Energy aspects of the formation of these electrodes can be obtained by the analysis of the specific energy consumption, w, which can be an important energy parameter in the development of these electrodes for commercial purposes. For electrodeposition process in PO regime, the specific energy consumption can be presented by Eq. (4.1) [41]:

$$w = \frac{nF\eta_{\mathrm{A}}}{M(1 - \eta_{\mathrm{I,av}}(\mathrm{H_2}))}, \tag{4.1}$$

where η_{A} is the amplitude of overpotential, M is the molar mass of copper (63,55 g/mol), and the number of Faraday per mole of consumed ions of copper is 2×26.8 Ah/mol $= 53.6$ Ah/mol. Using the overpotential amplitude of 1,000 mV, and the values of $\eta_{\mathrm{I,av}}(\mathrm{H_2})$ obtained with deposition pulses of 3 and 10 ms (Table 4.1), it can be shown that the shortening of deposition pulse from 10 to 3 ms leads to the decrease of the specific energy consumption for about 15%. Simultaneously, the number, diameter, and depth of holes remained unchanged. Considering the unchanged number of holes with the approximate same diameter and depth, it is clear that the use of shorter deposition pulses showed as a valuable way for energy saving in a production of this structure type.

Aside from energy savings attained by the application of PO regime, the improvement of mechanical strength of the honeycomb-like structures was noticed in the process of development of these deposits as possible electrodes. It can be assumed that the increase of mechanical strength is related to the decrease of the quantity of evolved hydrogen needed for their formation.

4.6 The Regime of Pulsating Current

The regime of pulsating current (PC) consists of a periodic repetition of square pulses [34, 35], and it is characterized by the amplitude of the cathodic current density, j_A, the deposition pulse, t_c (on period), and the pause duration, t_p, in which the system relaxes (off period). The average current density, j_{av}, is given by Eq. (4.2):

$$j_{av} = \frac{j_A t_c}{t_c + t_p} \tag{4.2}$$

or

$$j_{av} = \frac{j_A}{1 + p}, \tag{4.3}$$

if

$$p = \frac{t_p}{t_c}, \tag{4.4}$$

where p is the pause to pulse ratio.

In the PC regimes, the surface concentration of depositing ions is only determined by the average current density [34]. Then, the overpotential amplitude, η_A, can be presented by Eq. (4.5) [43]:

$$\eta_A = \frac{b_c}{2.3} \ln \frac{j_{av}(p+1)}{j_0} + \frac{b_c}{2.3} \ln \frac{1}{1 - \frac{j_{av}}{j_L}}, \tag{4.5}$$

where b_c is the cathodic Tafel slope, j_L is the limiting diffusion current density, and j_0 is the exchange current density. The activation part of overpotential, η_{act}, is

$$\eta_{\mathrm{act}} = \frac{b_{\mathrm{c}}}{2.3} \ln \frac{j_{\mathrm{av}}}{j_0}(p+1) \tag{4.6}$$

while the diffusion part of overpotential, η_{diff}, is

$$\eta_{\mathrm{diff}} = \frac{b_{\mathrm{c}}}{2.3} \ln \frac{1}{1 - \frac{j_{\mathrm{av}}}{j_{\mathrm{L}}}}. \tag{4.7}$$

Equation (4.5) can be rewritten in the form

$$\eta_{\mathrm{A}} = \eta_{\mathrm{const}} + \frac{b_{\mathrm{c}}}{2.3} \ln(p+1), \tag{4.8}$$

where η_{const} is the overpotential in the constant regime of electrolysis defined by Eq. (4.9):

$$\eta_{\mathrm{const}} = \frac{b_{\mathrm{c}}}{2.3} \ln \frac{j_{\mathrm{av}}}{j_0} + \frac{b_{\mathrm{c}}}{2.3} \ln \frac{1}{1 - \frac{j_{\mathrm{av}}}{j_{\mathrm{L}}}}, \tag{4.9}$$

if $j_{\mathrm{av}} = j$, where j is the current density in the constant regime of electrolysis.

Equation (4.5) is valid in the frequency range 10–100 Hz, where the frequency is sufficiently high to produce constant concentration on the surface and sufficiently low that the effect of DC capacity can be neglected. Hence, in the analyzed frequency range from 10 to 100 Hz, the surface concentration of depositing ions is constant and equal to the one in the constant regime at the current density corresponding to the average current density in the PC regime [34, 35]. From the point of view of the average current density, it means that there is not any difference between electrochemical deposition processes in the constant regimes and PC conditions. On the other hand, it is very clear from Eq. (4.5) that at the fixed value of the average current density the amplitude of overpotential depends on pause to pulse ratio, and it increases with the increasing pause to pulse ratio.

According to Popov and Pavlović [43], the degree of diffusion control of electrodeposition process, ω, is defined by Eq. (4.10):

$$\omega = \frac{\ln \frac{1}{1 - \frac{j_{av}}{j_L}}}{\ln \frac{j_{av}}{j_0} + \ln(p+1) + \ln \frac{1}{1 - \frac{j_{av}}{j_L}}}, \tag{4.10}$$

and it represents a contribution of diffusion overpotential to total cathode overpotential. Hence, due to the increase of activation part of overpotential with the increasing pause to pulse ratio, the degree of diffusion control will decrease with the increasing pause to pulse ratio, resulting in the possible change of texture of deposit. It is noteworthy that it is valid if $j_{av} < j_L$ in the mixed controlled deposition.

In the hydrogen codeposition range, the current efficiency for metal electrodeposition is less than 1, and then the effective average current density, $j_{av}*$, can be presented by Eq. (4.11):

$$j_{av}* = \frac{\eta_{I,av}(M) j_A}{1+p}, \tag{4.11}$$

where $\eta_{I,av}(M)$ is the average current efficiency for metal electrodeposition.

Since

$$\eta_{I,av}(M) + \eta_{I,av}(H_2) = 1, \tag{4.12}$$

Eq. (4.11) can be rewritten as

$$j_{av}* = \frac{[1 - \eta_{I,av}(H_2)] j_A}{1+p}, \tag{4.13}$$

and, according to Eqs. (4.3) and (4.11),

$$\eta_{I,av}(H_2) = 1 - \frac{j_{av}*}{j_{av}}. \tag{4.14}$$

In the hydrogen codeposition range, if $j_{av}* > j_L$, the amplitude of overpotential is related to the hydrogen reduction, increasing with the current density of hydrogen evolution [43]. Equation (4.5) is still valid [43], but it must be modified and adapted to the effect of hydrogen evolution on metal electrochemical deposition process. Then, the modified Eq. (4.5) can be presented by Eq. (4.15):

$$\eta_{\mathrm{A,eff}} = \frac{b_{\mathrm{c}}}{2.3}\ln\frac{j_{\mathrm{av}}*(p+1)}{j_0} + \frac{b_{\mathrm{c}}}{2.3}\ln\frac{1}{1-\frac{j_{\mathrm{av}}*}{j_{\mathrm{L}}*}}, \tag{4.15}$$

where $\eta_{\mathrm{A,eff}}$ represents the effective overpotential amplitude and $j_{\mathrm{L}}*$ is the effective limiting diffusion current density which depends on the hydrodynamic conditions in the near-electrode layer caused by hydrogen evolution during electrodeposition process. Equation (4.15) is valid if the condition $j_{\mathrm{av}}*/j_{\mathrm{L}}* < 1$ is fulfilled. The effective average current density, $j_{\mathrm{av}}*$, can be calculated by the use of Eq. (4.13) if the value of the average current efficiency of hydrogen evolution is known. Meanwhile, the determination of $j_{\mathrm{L}}*$ when the change of hydrodynamic conditions in the near-electrode layer is caused by hydrogen evolution is not possible. For that reason, the analysis of morphologies of electrodeposited copper obtained in the hydrogen codeposition range showed an excellent tool for the estimation of the change of effectiveness of solution stirring by evolved hydrogen.

It is very clear from the above consideration that the quantities of evolved hydrogen and hence the morphologies of electrodeposited metal will depend strongly on the applied parameters of PC regimes, such as the amplitude of the current density, deposition pulse, and pause duration.

4.6.1 Formation of the Honeycomb-Like Structures by the PC Regime

In PC regimes, the intensification of hydrogen evolution reaction can be achieved by:

1. The increase of the current density amplitude and keeping durations of both the deposition pulse and pause constant

2. The prolonging of a deposition pulse duration and keeping both the current density amplitude and pause duration constant
3. The shortening of a pause duration and keeping both the current density amplitude and deposition pulse constant

Since the second and third ways are closely related, these two ways will be analyzed simultaneously.

4.6.1.1 The Effect of the Current Density Amplitude

Figure 4.19 shows SEM micrographs of copper deposits obtained with the current density amplitudes of 0.20 A/cm^2 (Fig. 4.19a) and of 0.44 A/cm^2 (Fig. 4.19b). In both cases, a deposition pulse of 1 ms and a pause duration of 10 ms were applied. In all experiments related to the analysis of the PC regimes, copper electrodeposition was performed from 0.15 M $CuSO_4$ in 0.50 M H_2SO_4 at room temperature using the vertical stationary cylindrical copper electrodes [30, 44, 45].

Very branchy dendrites, small cauliflower-like forms, and shallow holes formed from detached hydrogen bubbles are formed when the amplitude of current density of 0.20 A/cm^2 was applied (Fig. 4.19a). On the other hand, dish-like holes and small cauliflower-like agglomerates of copper grains were formed with a current density amplitude of 0.44 A/cm^2 (Fig. 4.19b). The formation of these morphological forms was accompanied by the quantity of evolved hydrogen which corresponded to the average current efficiency of hydrogen evolution, $\eta_{I,av}(H_2)$, of 5.5% with the applied current density amplitude, j_A, of 0.20 A/cm^2 [45], and 13.7% with j_A of 0.44 A/cm^2 [30].

4.6.1.2 The Effect of the Length of Deposition Pulse

The prolongation of a duration of deposition pulse leads to the formation of honeycomb-like structures with both the current density amplitudes applied [30, 45]. The formation of this structure type is analyzed applying the current density amplitude of 0.44 A/cm^2, deposition pulses of 1, 4, 7, 10, and 20 ms, and a pause duration of 10 ms [30]. The values of the average current efficiency of hydrogen evolution, $\eta_{I,av}(H_2)$, obtained for these parameters of PC regimes are

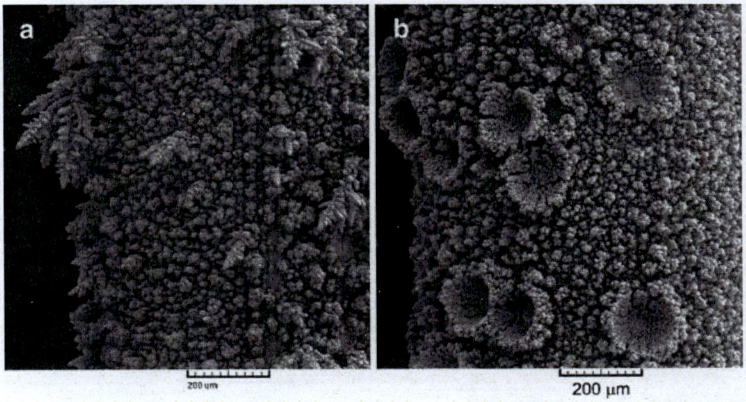

Fig. 4.19 Copper deposits obtained by the regime of pulsating current. The current density amplitude: (**a**) 0.20 A/cm^2 and (**b**) 0.44 A/cm^2. Deposition pulse: 1 ms. Pause duration: 10 ms (Reprinted from [30, 45] with permission from Elsevier.)

Table 4.2 The values of the average current efficiencies of hydrogen evolution, $\eta_{I,av}(H_2)$, in %, the average current efficiencies for copper electrodeposition, $\eta_{I,av}(Cu)$, in %, and the specific energy consumptions obtained for electrodeposition of copper with different pause to pulse ratios

t_c (ms)	1	4	7	10	20	Const. regime
$\eta_{I,av}(H_2)$ (%)	13.7	20.7	24.4	30.0	32.5	36.0
$\eta_{I,av}(Cu)$ (%)	86.3	79.3	75.6	70.0	67.5	64.0
w (kWh/kg)	1.075	1.170	1.228	1.326	1.375	1.450

Reprinted from [30] with permission from Elsevier
t_c deposition pulse in ms; t_p pause duration in ms ($t_p = 10$ ms); $j_A = 0.44$ A/cm^2

given in Table 4.2. Also, the value of $\eta_{I,av}(H_2)$ obtained for copper electrodeposition at the constant current density of 0.44 A/cm^2 is included in this table.

Using j_A of 0.44 A/cm^2, the honeycomb-like structures are formed with deposition pulses of 7 ms and longer and the typical honeycomb-like structure obtained with a deposition pulse of 10 ms is shown in Fig. 4.20a. The mixture of dish-like holes and holes constructing the honeycomb-like structure was formed with a deposition pulse of 4 ms (Fig. 4.20b). Finally, as already mentioned, dish-like holes and independently formed agglomerates of copper grains were formed with a deposition pulse of 1 ms (Fig. 4.19b).

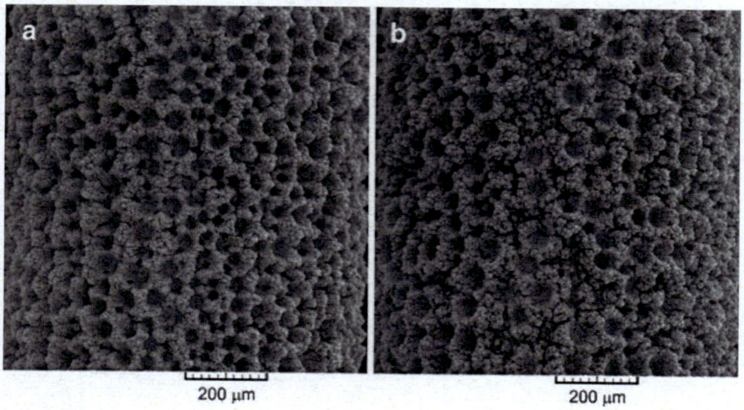

Fig. 4.20 Copper deposits obtained by the PC regimes with a deposition pulse of: (**a**) 10 ms and (**b**) 4 ms. Pause duration: 10 ms. The amplitude of current density: 0.44 A/cm^2 (Reprinted from [30] with permission from Elsevier.)

The shape, size, and number of holes strongly depended on the length of deposition pulse. The honeycomb-like structure electrodeposited with a deposition pulse of 20 ms was constructed of noncoalesced and coalesced holes (Fig. 4.21a, b, respectively). The decreasing length of deposition pulse led to the supression of the coalescence process. Noncoalesced holes formed with a deposition pulse of 10 ms were very similar to those obtained with a deposition pulse of 20 ms. Aside from the appearing of slightly larger holes than those formed with the deposition pulse of 10 ms, the decrease of deposition pulse to 7 ms did not have any effect on the shape and the size of holes. The mixture of dish-like holes and the noncoalesced holes constructing the honeycomb-like structure was formed with a deposition pulse of 4 ms [30]. The typical dish-like hole formed with a deposition pulse of 1 ms is shown in Fig. 4.21c.

Figure 4.22 shows the dependences of the average diameter of holes and the number of holes per mm^2 surface area of the copper electrode on pause to pulse ratio. In these dependences, the values obtained at the constant current density ($j = 0.44$ A/cm^2; $p = 0$) are also included. From Fig. 4.22 can be seen that the size of noncoalesced holes in the honeycomb-like structures did not depend on the length of deposition pulse. As expected, the size of dish-like

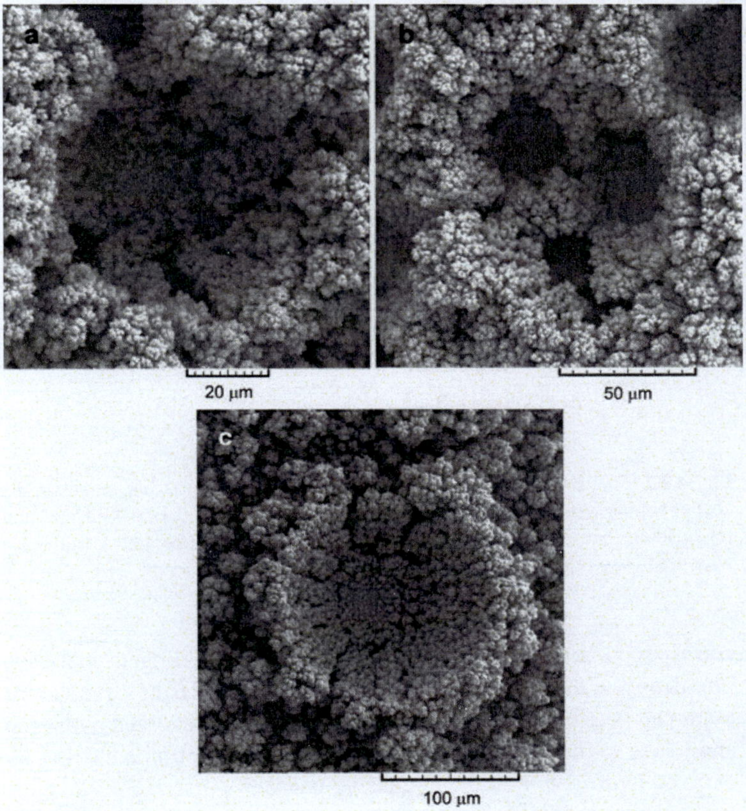

Fig. 4.21 The typical holes formed by the PC regimes with a deposition pulse of: (**a**) and (**b**) 20 ms; and (**c**) 1 ms. Pause duration: 10 ms. The amplitude of current density: 0.44 A/cm^2 (Reprinted from [30] with permission from Elsevier.)

holes was larger than those forming the honeycomb-like structures [19]. On the other hand, the dependence of the number of holes on the pause to pulse ratio shows maximal value for a pause to pulse ratio of 1 ($t_c = 10$ ms). The increase of number of holes obtained with a pause to pulse ratio of 0.5 ($t_c = 20$ ms) in relation to the one formed at the constant current density ($p = 0$) is due to the decreased coalescence of closely formed hydrogen bubbles. The maximum of the dependence obtained for a pause to pulse ratio of 1 ($t_c = 10$ ms)

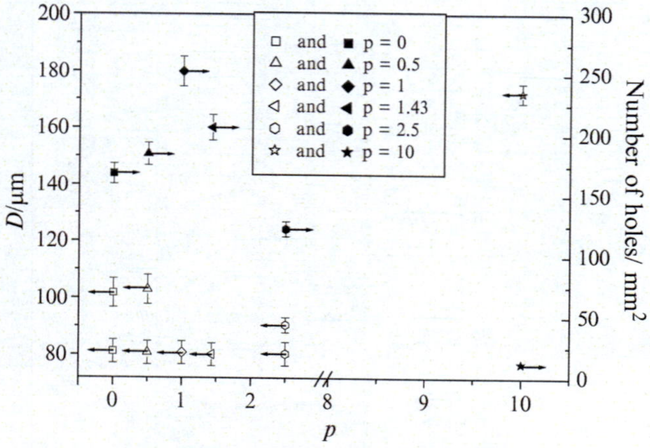

Fig. 4.22 The average diameter, *D*, and the number of holes per mm^2 surface area of copper electrodes obtained at a constant current density of 0.44 A/cm^2 and for different pause to pulse ratios. Pause duration: 10 ms (Reprinted from [30] with permission from Elsevier.)

corresponds to suppressed coalescence of hydrogen bubbles. The decrease of the number of holes and the appearing of dish-like holes can be ascribed to the decrease of quantity of evolved hydrogen and hence to the decrease of the effectiveness of solution stirring by evolved hydrogen with the increasing pause to pulse ratio.

Cauliflower-like agglomerates of copper grains can be classified into two groups. In the first group are inserted cauliflower-like agglomerates of copper grains constructing the honeycomb-like structures. These copper agglomerates are very disperse and consisted of small agglomerates of copper grains, as shown in Fig. 4.23a which shows the typical cauliflower-like agglomerates of copper grains obtained by the square-wave PC with a deposition pulse of 10 ms. From Fig. 4.23a can be seen that the size of these small copper grains agglomerates of which large cauliflower-like agglomerates formed among holes were constructed was about 4 μm. Also, the agglomerates of copper grains were surrounded by irregular channels for which the origin was of evolved hydrogen. In the second group are inserted cauliflower-like agglomerates

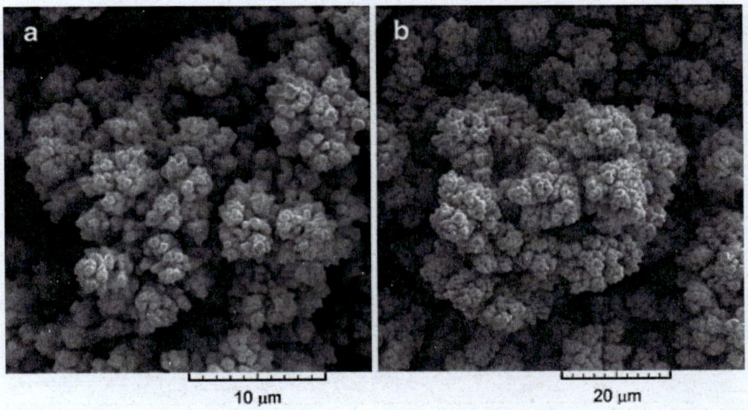

Fig. 4.23 The typical cauliflower-like agglomerates of copper grains formed by the PC regimes with a deposition pulse of: (**a**) 10 ms and (**b**) 1 ms. Pause duration: 10 ms (Reprinted from [30] with permission from Elsevier.)

of copper grains formed by the PC regime with a deposition pulse of 1 ms. Also, this type of agglomerates can be noticed in the transitional structure formed with a deposition pulse of 4 ms. The typical agglomerate of copper grains from this group is given in Fig. 4.23b which shows the one obtained with a deposition pulse of 1 ms. Copper grains agglomerates from this group are formed at the electrode surface independently from the formed hydrogen bubbles. They were larger and more compact than those formed among hydrogen bubbles at the constant current density and by the PC regimes with deposition pulses of 7, 10, and 20 ms [30].

Cross section of copper deposits obtained with deposition pulses of 4, 7, 10, and 20 ms is shown in Fig. 4.24. From Fig. 4.24, it can be seen that the interior of these structures was very porous and consisted of disperse particles surrounded by irregular channels for which the origin is of evolved hydrogen [44]. Also, it is necessary to note that dendritic character of these particles decreased with the prolonging duration of deposition pulse. As already mentioned, the porous interior of these deposits is very important for electrocatalytic purposes because the pores facilitate the transport of electroactive species through the interior of the structures, what is very desirable

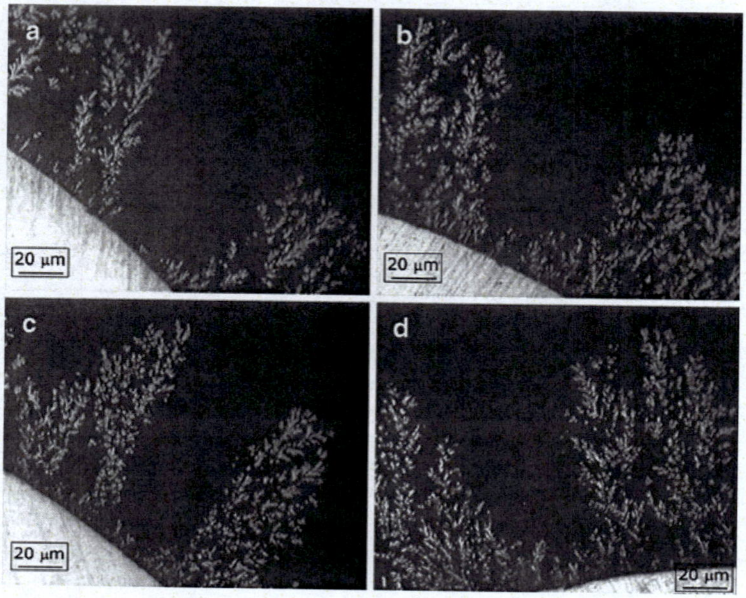

Fig. 4.24 Cross section of the copper deposits obtained by the PC regimes with deposition pulses of: (**a**) 4 ms; (**b**) 7 ms; (**c**) 10 ms; and (**d**) 20 ms. The current density amplitude: 0.44 A/cm^2. Pause duration: 10 ms (Reprinted from [44] with permission from Springer.)

for the evaluation of electrochemical reactions. For example, copper shows a high activity for nitrate ion reduction [46, 47], as well as for a reaction in which nitrate reduces to ammonia in high yield in aqueous acidic perchlorate and sulfate media [48].

4.6.1.3 Discussion of the Effect of Different Parameters of the PC Regimes on Electrodeposition of Copper in the Hydrogen Codeposition Range

From the above consideration, the existence of the strong effect of the selected parameters of the square-waves PC on hydrogen evolution reaction and hence morphology of electrodeposited copper is very clear. The change of morphology of electrodeposited copper from very branchy dendrites and dish-like holes to the honeycomb-like structures can be explained by the analysis of the effectiveness of solution stirring by evolved hydrogen in the following way: the effectiveness of stirring of the solution by hydrogen generated at the cathode surface during electrochemical deposition process increases with intensification of hydrogen evolution reaction. In the one moment, hydrogen evolution will become vigorous enough to cause the decrease of the cathode diffusion layer thickness and the increase of the limiting diffusion current density leading to the change of the hydrodynamic conditions in the near-electrode layer [16]. According to Eqs. (4.5) and (4.10), it means the decrease of the degree of diffusion control of electrodeposition process with intensification of hydrogen evolution reaction. The degree of diffusion control of electrodeposition process will additionally decrease due to the smaller values of the effective average current densities, $j_{av}*$, in relation to those obtained in the absence of hydrogen evolution [Eqs. (4.3) and (4.13)]. Then, the overpotential amplitude corresponding to copper electrodeposition will be smaller than the one specified by pulse rectifiers and this value is denoted as effective overpotential amplitude of electrodeposition process and it is presented by Eq. (4.15).

Hence, the effective overpotential amplitude, $\eta_{A,eff}$, decreases with the intensification of hydrogen evolution reaction due to the decrease of the degree of diffusion control of electrodeposition process. So, the validity of the concept of "effective overpotential" can be expanded to include metal electrodeposition in the hydrogen codeposition range by the regime of pulsating current (PC), and then this concept applied for the PC regimes can be denoted as "effective overpotential amplitude" one [44].

Anyway, the decrease of dendritic character of particles and hence the increase of dispersity of particles in the formed honeycomb-like structures (Fig. 4.24), as well as the change of the surface morphology from dendrites (Fig. 4.19a) to agglomerates of grains formed around holes (Fig. 4.23a), clearly confirm the decrease of the degree of the diffusion control of electrodeposition process with the increase of the quantity of evolved hydrogen. The different degree of diffusion control of electrodeposition process is attained by the choice of parameters of the square-waves PC. For example, using the value $j_A = 0.20$ A/cm^2, $t_c = 1$ ms, and $t_p = 10$ ms, as well as $\eta_{I,av}(H_2) = 5.5\%$, the effective average current density, j_{av}*, calculated by the use of Eq. (4.13) was very close to the value of the limiting diffusion current density for this system (i.e., for 0.15 M $CuSO_4$ in 0.50 M H_2SO_4; $j_L \approx 16.0$ mA/cm^2) [16]. Hence, with these parameters of the PC regimes, electrodeposition process was controlled by the diffusion of ions to the electrode surface, what is confirmed by the formation of very branchy dendrites and small cauliflower-like agglomerates of copper grains (Fig. 4.19a). Anyway, the nonuniformity of the electrode surface clearly pointed out that the diffusion layer of the macroelectrode was not disturbed by evolved hydrogen during this electrodeposition process.

On the other hand, the increase of the effectiveness of solution stirring, the decrease of the cathode diffusion layer thickness, and the increase of the limiting diffusion current density occur with the intensification of hydrogen evolution reaction leading to the formation of the honeycomb-like structures. The uniform distribution of morphological forms, i.e., holes and copper grains agglomerates, means the existence of the same hydrodynamic conditions over the whole electrode surface [19]. The size of grains in cauliflower-like agglomerates approached to nanosized dimensions with the intensification of hydrogen evolution reaction, as shown in Fig. 4.25 for cauliflower-like ones obtained with $t_c = 20$ ms. From Fig. 4.25, it can be clearly seen that the size of grains in these agglomerates is between 100 and 300 nm. Also, the porosity of the honeycomb-like structures was additionally increased by numerous nanopores situated between copper grains, as observed from Fig. 4.25. The increased effectiveness of the solution stirring by evolved hydrogen with the intensification of hydrogen evolution can also be noticed by the analysis of the internal structures of the honeycomb-like deposits shown in Fig. 4.24. The decrease of dendritic character of particles and hence the increase of dispersity of deposits is just a consequence of the increased effectiveness of the solution stirring by evolved hydrogen with the intensification of hydrogen evolution reaction caused by the prolongation of deposition pulse duration.

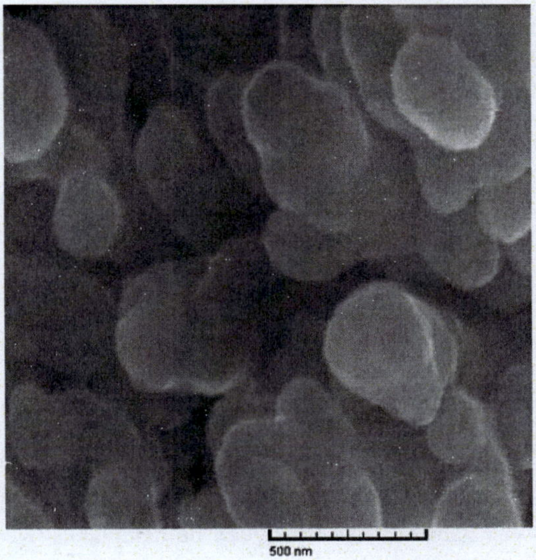

Fig. 4.25 Copper grains agglomerates obtained by the PC regime with a current density amplitude of 0.44 A/cm^2, a deposition pulse of 20 ms, and a pause duration of 10 ms (Reprinted from [44] with permission from Springer.)

4.6.2 Optimization of the Formation of the Honeycomb-Like Structure by the PC Regime

For the galvanostatic mode of operation, the specific energy consumption, w, defined as the energy spent in the process per unit mass of deposited copper, may be expressed according to [49] by Eq. (4.16):

$$w = \frac{nF \int_0^t U(t)\mathrm{d}t}{M \int_0^t \eta_{\mathrm{I,Cu}}(t)\mathrm{d}t} \tag{4.16}$$

or by Eq. (4.17) as

$$w = \frac{nFU_{\mathrm{av}}}{M\eta_{\mathrm{I,av}}(\mathrm{Cu})}, \tag{4.17}$$

where

$$U_{av} = (1/t) \int_0^t U(t)dt \tag{4.18}$$

and

$$\eta_{I,av}(Cu) = (1/t) \int_0^t \eta_{I,Cu}(t)dt. \tag{4.19}$$

According to Eq. (4.17) the specific energy consumption depends on the average cell voltage, U_{av}, and the average current efficiency for copper electrodeposition reaction, $\eta_{I,av}(Cu)$. The values of the average current efficiencies for copper electrodeposition, $\eta_{I,av}(Cu)$, were calculated using data from Table 4.2 as $\eta_{I,av}(Cu) = 100–\eta_{I,av}(H_2)$, and the obtained values are added to Table 4.2. The values of the average cell voltages, U_{av}, can be calculated by Eq. (4.18) using the experimentally determined dependences of the cell voltages, U, on time t [30]. The tendency of the decrease of the average cell voltages was observed with the shortening deposition pulse and these values were smaller than the average cell voltage obtained at constant current density of 0.44 A/cm^2 [30].

In square-wave PC regimes, the average current density and the average cell voltage depend on pause to pulse ratio in a similar way. Hence, the average cell voltage, U_{av}, in a function of p can be given by Eq. (4.20) [34, 35]:

$$U_{av} = \frac{U_A}{p+1}, \tag{4.20}$$

where U_A is the amplitude cell voltage, if there is no rest cell voltage during pause duration. In this study, the constant current density used in DC regime ($j = 0.44$ A/cm^2) was equal to the amplitude of current density in the PC regimes. Consequently, the cell voltage in the DC regime was used as the amplitude value in the PC regimes [30]. Then, the amplitude cell voltage, U_A, can be calculated as $U_A = (1/t) \int_0^t U(t)dt$, where U is the cell voltage in a time t for a constant regime of electrolysis, and the value of 1.1 V was obtained.

Anyway, the difference between the experimentally determined dependences and calculated in this way was observed and it can be ascribed to the existence of rest voltage during pause. Due to this rest voltage making the cell voltage during pause to be higher than zero, the amplitude cell voltage cannot be calculated by the experimentally measured average cell voltage using Eq. (4.20).

For that reason, assuming that the value of the peak voltage during deposition pulse will not be higher than the cell voltage in constant current mode, the average cell voltage obtained at the constant current density ($p = 0$) was used for the calculation of the specific energy consumption. Then, using the values of the average current efficiencies for copper electrodeposition reaction from Table 4.2 and the average cell voltage of 1.1 V obtained at the constant current density of 0.44 A/cm^2, the values of the specific energy consumption are calculated and added to Table 4.2.

From Table 4.2, it can be seen that the shortening of deposition pulse led to the decrease of the specific energy consumption, and the obtained values were smaller than the one obtained in the constant mode. On the other hand, the increasing deposition pulse duration favored the formation of the honeycomb-like electrodes. This enabled the optimization process of the formation of the honeycomb-like copper electrodes by the choice of the appropriate parameters of square-waves PC. As already mentioned, the honeycomb-like structures were formed with the pause to pulse ratio smaller than 1.43 (or with the deposition pulse of 7 ms and larger, and a pause duration of 10 ms). It is clear that energy saving of 15.3% was attained by the production of the honeycomb-like electrode by the square-wave PC with a deposition pulse of 7 ms ($p = 1.43$) in comparison to the one obtained at the constant current density. Simultaneously, the increase of the specific surface area of the honeycomb-like electrodes, manifested by the increase of the number of holes due to the suppressed coalescence of closely formed hydrogen bubbles, was observed by the application of these PC regimes (Fig. 4.22).

Anyway, the use of the appropriate square-waves PC enabled energy saving in the production of the honeycomb-like electrodes. In this moment, the largest problem for the commercial manufacturing of open porous structures by the PC regimes is high cost of a pulse rectifier which is much greater than a DC unit [50].

It is a highly regulated and sophisticated design that costs more to manufacture. At the first sight, energy savings attained by the application of the PC regimes and high cost of production of pulse rectifier are in contradiction, but further development of the electronic industry will probably decrease the cost of these rectifiers and will enable their larger application in electrochemical technologies.

For the technological application of open porous structures as possible electrodes in electrochemical devices, the deposit structural stability of these deposits determined by their adhesion with electrode surface is also very important. It is a well-known fact that the adhesion of deposits is closely associated with the quantity of hydrogen by which metal deposits are formed [51]. At high current densities, dendritic deposits are initiated on the cathode with simultaneous evolution of hydrogen gas. At higher current density, the deposit structure becomes more open, nonuniform, and finer, due to the increased nucleation rate and hydrogen evolution on the cathode [51]. Also, similar effects on copper deposit morphology are observed in potentiostatic regime of electrolysis. Increasing overpotential intensifies the hydrogen evolution reaction causing the change of morphology of electrodeposited copper from dendrite to the honeycomb-like structure constructed of holes formed by the attached hydrogen bubbles with cauliflower-like copper grains agglomerates among them [16]. The mechanical strength of the copper deposits decreases with the increasing current density and hence with the increasing quantity of evolved hydrogen what is proven by the measurement of adhesion of copper deposits by the peel strength test [51]. Hence, it can be concluded by this analysis that the mechanical strength of the honeycomb-like copper structures was improved by the application of PC regimes due to the decrease of the average current efficiency of hydrogen evolution by which they are formed.

Hence, the following conveniences in the production of open porous structures, denoted as 3-D foam or the honeycomb-like one, are attained since the appropriate square-waves PC parameters were selected: (a) energy saving, (b) the increase of the specific surface area of the electrodes, and (c) probably the improvement of the deposit structural stability due to the decrease of the quantity of evolved hydrogen needed for their formation. Also, it is clear that the benefits attained by the application of the pulsating overpotential (PO) regime can be succesfully transferred to technological attractive pulsating current (PC) regime.

4.7 The Regime of Reversing Current

The regime of reversing current (RC) is characterized by the cathodic current density, j_c, and the anodic current density, j_a, as well as by the duration of flow of the current in the cathodic and the anodic directions, t_c and t_a, respectively [34, 35]. The average current density, j_{av}, is given by Eq. (4.21):

$$j_{av} = \frac{j_c t_c - j_a t_a}{t_c + t_a} \tag{4.21}$$

or

$$j_{av} = \frac{j_c - j_a r}{1 + r}, \tag{4.22}$$

if

$$r = \frac{t_a}{t_c}. \tag{4.23}$$

For the RC regime in the millisecond range, the surface concentration of the depositing ions is determined by the average current density [34, 35], and the overpotential amplitude, η_A, can be presented by Eq. (4.24):

$$\eta_A = \frac{b_c}{2.3} \ln \frac{j_{av}(r+1) + j_a r}{j_0} + \frac{b_c}{2.3} \ln \frac{1}{1 - \frac{j_{av}}{j_L}}. \tag{4.24}$$

Equation (4.24) is valid in the frequency range 10–100 Hz, where the frequency is sufficiently high to produce constant concentration on the surface and sufficiently low that the effect of DC capacity can be neglected. It is noteworthy that it is valid if $j_{av} < j_L$ in the mixed controlled deposition.

With the applied cathodic current density pulses larger than the limiting diffusion current density, parallel to copper electrodeposition hydrogen evolution reaction occurs [52]. On the other hand, there was not any gas evolution during anodic pulses indicating that the overall gas evolution corresponds to hydrogen evolution. Then, Eq. (4.22) can be modified by Eq. (4.25):

$$j_{av}* = \frac{\eta_{I,c}(Cu)j_c - j_a r}{1 + r}, \tag{4.25}$$

where $j_{av}*$ is the effective average current density and $\eta_{I,c}(Cu)$ is the current efficiency for copper electrodeposition during cathodic pulses. Since $\eta_{I,c}(Cu) + \eta_{I,c}(H_2) = 1$, Eq. (4.25) can be presented as

$$j_{av}* = \frac{[1 - \eta_{I,c}(H_2)]j_c - j_a r}{1 + r}, \tag{4.26}$$

where $\eta_{I,c}(H_2)$ is the current efficiency for hydrogen evolution reaction during cathodic pulses. For $j_{av}* > j_L$, the overpotential amplitude is related to the hydrogen reduction, increasing with the current density of hydrogen evolution [43], and the modified Eq. (4.24) can be presented as

$$\eta_{A,eff} = \frac{b_c}{2.3}\ln\frac{j_{av}*(r+1) + j_a r}{j_0} + \frac{b_c}{2.3}\ln\frac{1}{1 - \frac{j_{av}*}{j_L*}}, \tag{4.27}$$

where $\eta_{A,eff}$ is the effective overpotential amplitude and j_L* is the effective limiting diffusion current density. Naturally, $j_{av}* < j_L*$.

The regime of pulsating current (PC) represents the special case of the reversing current (RC) regime ($j_a = 0$ mA/cm^2). For that reason, copper electrodeposition processes in the hydrogen codeposition range by the RC regimes with the different anodic current density values are compared with those obtained by the PC regime.

Table 4.3 The values of the average current efficiencies of hydrogen evolution, $\eta_{I,av}(H_2)$, in %, obtained for electrodeposition of copper with different anodic current densities

j_a (mA/cm^2)	0	40	240	440	640
$\eta_{I,av}(H_2)$ (%)	30.7	28.9	25.6	21.7	19.3

Reprinted from [52] with permission from Elsevier
j_a the anodic current density in mA/cm^2

4.7.1 The Effect of the Anodic Current Density on the Formation of the Honeycomb-Like Electrodes

In all PC and RC experiments, the cathodic current density (or the current density amplitude in the PC regime) of 0.44 A/cm^2, the cathodic time (or deposition pulse) of 10 ms, and the anodic time (or pause duration) of 5 ms were applied, while the analyzed anodic current densities were 0.040, 0.24, 0.44, and 0.64 A/cm^2. Copper was electrodeposited from 0.15 M $CuSO_4$ in 0.50 M H_2SO_4 at the room temperature using cylindrical copper working electrodes. The values of the average current efficiencies of hydrogen evolution calculated as $\eta_{I,av}(H_2) = (1/t)\int_0^t \eta_I(H_2)dt$ for the analyzed PC and RC regimes are presented in Table 4.3.

Morphologies of copper deposits obtained by the PC regime and RC regimes with different anodic current densities are shown in Fig. 4.26. Holes formed of detached hydrogen bubbles and agglomerates of copper grains or dendrites were obtained with all analyzed PC and RC regimes. From Fig. 4.26, it can be seen that the honeycomb-like structures are formed by the PC regime and by the RC regimes with the anodic current densities up to 440 mA/cm^2. Holes obtained by a coalescence of closely formed hydrogen bubbles are noticed by the analysis of the honeycomb-like structures formed by the PC regime (Fig. 4.27a) and the RC regime with j_a of 40 mA/cm^2. In the range of the examined anodic current densities from 0 (the PC regime) to 440 mA/cm^2, holes constructing the honeycomb-like structures (so-called "noncoalesced" one) were similar to each other, and the typical hole of this type obtained with

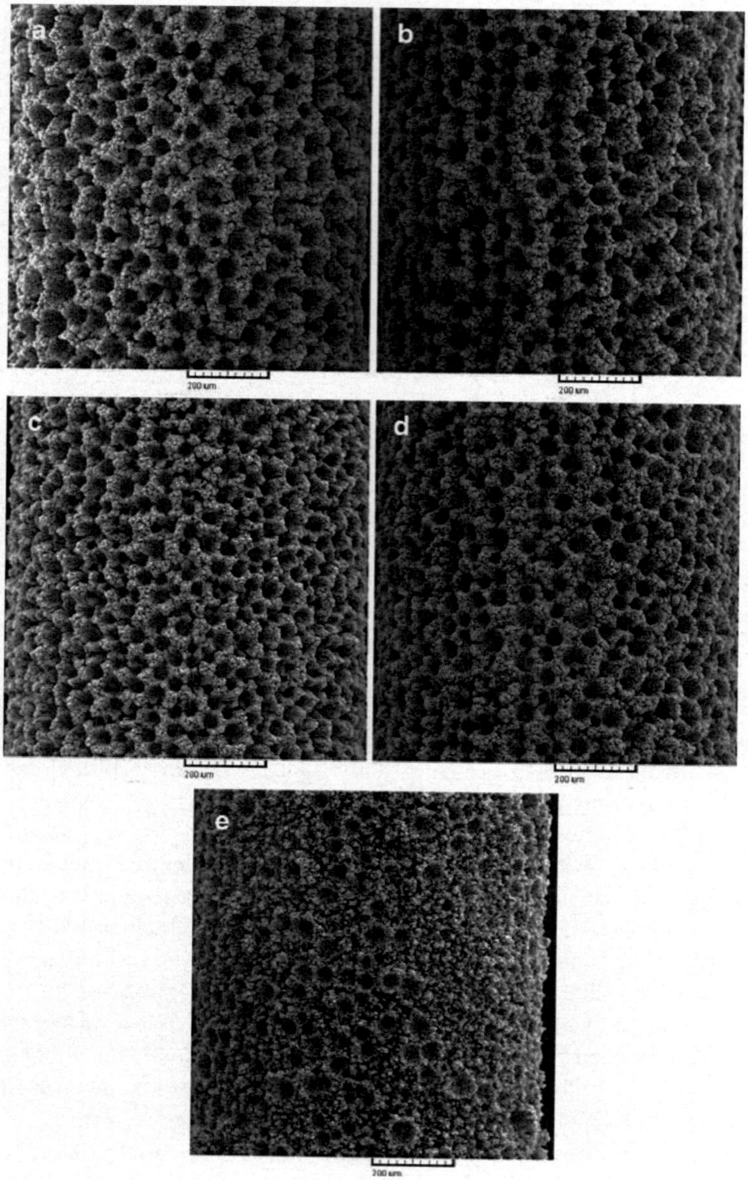

Fig. 4.26 Copper deposits obtained by: (**a**) the PC regime and by the RC regimes with the anodic current density j_a of: (**b**) 40; (**c**) 240; (**d**) 440 and (**e**) 640 mA/cm^2. The cathodic current density j_c: 440 mA/cm^2, the cathodic pulse t_c: 10 ms, and the anodic pulse, t_a: 5 ms (Reprinted from [52] with permission from Elsevier.)

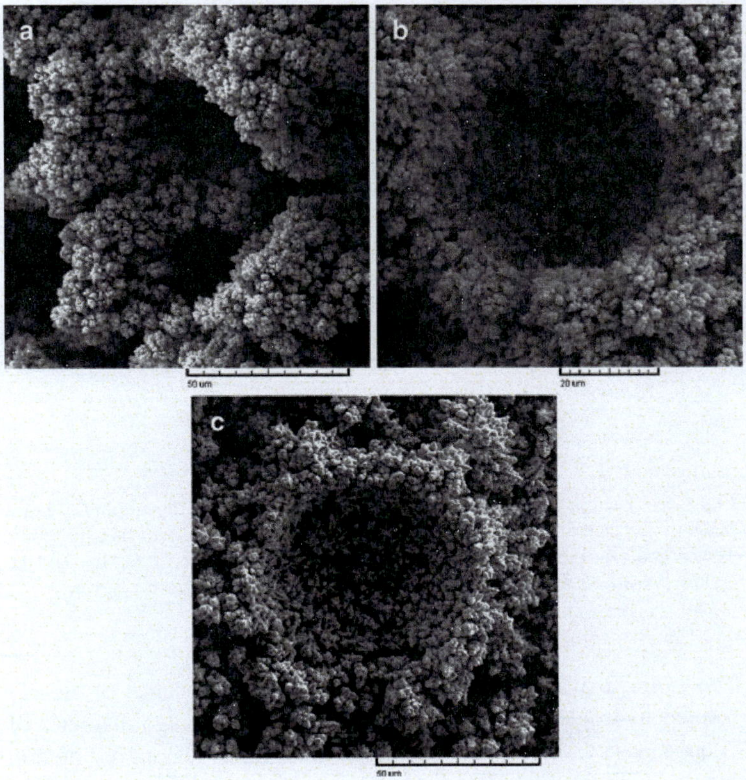

Fig. 4.27 The types of holes formed of detached hydrogen bubbles by different PC and RC regimes: (**a**) coalesced hole obtained by the PC regime; (**b**) noncoalesced hole obtained by the RC regime with $j_a = 440$ mA/cm^2 and (**c**) dish-like hole obtained by the RC regime with $j_a = 640$ mA/cm^2. The cathodic current density j_c: 440 mA/cm^2, the cathodic pulse t_c: 10 ms, and the anodic pulse, t_a: 5 ms (Reprinted from [52] with permission from Elsevier.)

j_a of 440 mA/cm^2 is shown in Fig. 4.27b. Finally, dish-like holes were dominant type of holes obtained with j_a of 640 mA/cm^2 (Fig. 4.27c).

Figure 4.28 shows the dependences of the average diameter and the number of holes formed at electrode surface in a function of the anodic current density. The average size and number of holes obtained for the PC regime ($j_a = 0$ mA/cm^2) are also included in this figure.

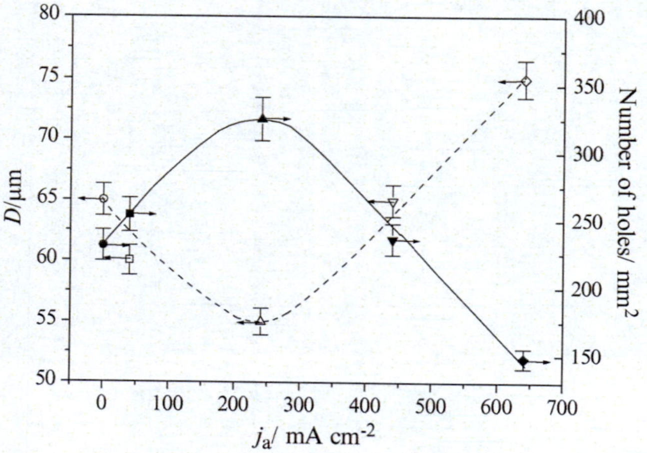

Fig. 4.28 The dependences of the average diameter, D, and the number of holes per mm^2 surface area of copper electrodes on the anodic current density value. The cathodic current density j_c: 440 mA/cm^2, the cathodic pulse t_c: 10 ms, and the anodic pulse t_a: 5 ms (Reprinted from [52] with permission from Elsevier.)

The average diameter of holes obtained by coalescence of closely formed hydrogen bubbles was larger than the average diameter of "noncoalesced" holes, and the size of these holes is not presented in Fig. 4.28. From Fig. 4.28, it can be seen that the dependence of the average diameter of holes on the anodic current density is of a parabolic shape with a minimum at $j_a = 240$ mA/cm^2. On the other hand, the dependence of the number of the holes per mm^2 surface area of the electrode on the anodic current density value had the shape of reverse parabola with the maximal number of holes formed with $j_a = 240$ mA/cm^2. Due to the existence of holes formed by coalescence process, the overall number of holes formed by both the PC regime and the RC regime with $j_a = 40$ mA/cm^2 was smaller than the one formed by the RC regime with $j_a = 240$ mA/cm^2. The decrease of the number of holes with the anodic current density larger than $j_a = 240$ mA/cm^2 can be ascribed to the strong effect of the anodic current densities on both the hydrogen evolution reaction and copper electrodeposition rate.

Morphologies of copper deposits formed around holes are shown in Fig. 4.29. The change of morphology of the electrodeposited copper from cauliflower-like agglomerates of copper grains to dendrites was observed with the increasing anodic current density values. Cauliflower-like agglomerates of copper grains obtained by the regimes of PC (Fig. 4.29a) and RC with $j_a = 40$ mA/cm^2 (Fig. 4.29b) are very disperse, and the presence of deep irregular channels of detached hydrogen bubbles [19] formed around small copper grains agglomerates was easily noticed. From Fig. 4.29a, b, it can be clearly seen that these cauliflower-like agglomerates of copper grains are very similar to each other, what is understandable due to the absence or the small value of the anodic component of the current density in comparison with the cathodic current density. With the increasing anodic current density, deep irregular channels around small copper agglomerates were lost making the honeycomb-like structures obtained with $j_a = 240$ mA/cm^2 (Fig. 4.26c) and $j_a = 440$ mA/cm^2 (Fig. 4.26d) more compact than those obtained in the PC regime (Fig. 4.26a) and the RC regime with $j_a = 40$ mA/cm^2 (Fig. 4.26b). Also, holes approached each other (or the wall width decreased) with the increase of the anodic current density (Fig. 4.29a–c). The increase of the anodic current density led to the formation of dendrites on the top of agglomerates of copper grains (Fig. 4.29d). Finally, very branchy dendrites were formed with $j_a = 640$ mA/cm^2 (Fig. 4.29e).

The increase of the compactness of the honeycomb-like structures can be explained by the strong effect of the anodic current density on both the hydrogen evolution reaction and copper electrodeposition rate as follows: in the growth process, due to the current density distribution effect, during cathodic pulses both the copper nuclei and the hydrogen bubbles are primarily formed at the top of agglomerates of grains formed around initially formed hydrogen bubbles. During anodic pulses, due to the same effect, the anodic current density (or the anodic current lines) will be concentrated at the top of these freshly formed copper nuclei causing their dissolution. The dissolution process of copper nuclei facilitates to freshly formed hydrogen bubbles to find path to coalesce with initially formed hydrogen bubbles leading to their growth with electrolysis time. With the increasing anodic current density, the dissolution of copper

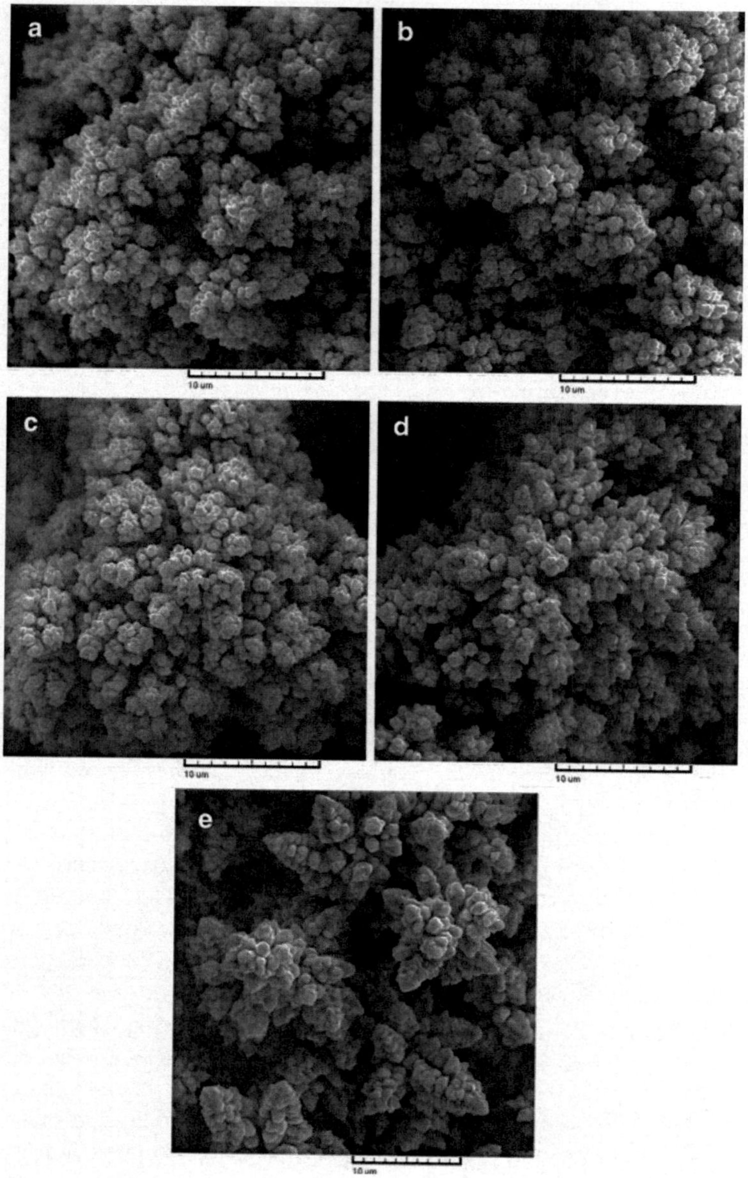

Fig. 4.29 Morphology of electrodeposited copper formed around holes by: (**a**) the PC regime and by the RC regimes with the anodic current density j_a of: (**b**) 40; (**c**) 240; (**d**) 440 and (**e**) 640 mA/cm^2. The cathodic current density j_c: 440 mA/cm^2, the cathodic pulse t_c: 10 ms, and the anodic pulse t_a: 5 ms (Reprinted from [52] with permission from Elsevier.)

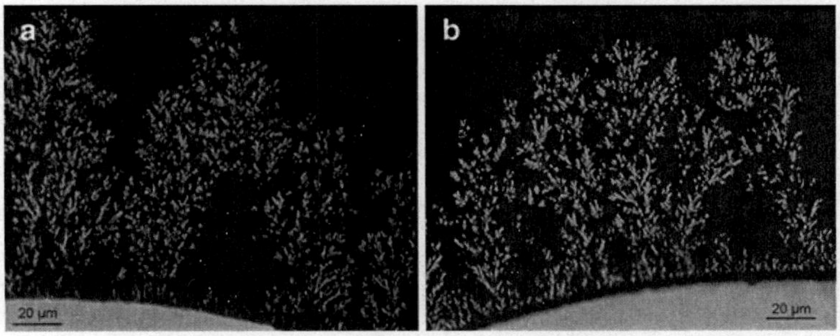

Fig. 4.30 Cross section of the honeycomb-like deposits obtained with: (**a**) $j_a = 40$ mA/cm^2 and (**b**) $j_a = 440$ mA/cm^2. The cathodic current density j_c: 440 mA/cm^2, the cathodic pulse t_c: 10 ms, and the anodic pulse t_a: 5 ms (Reprinted from [52] with permission from Elsevier.)

nuclei increases, and the number of freshly formed hydrogen bubbles which will coalesce with the initially formed hydrogen bubbles is increased. Simultaneously, the number of hydrogen bubbles which will remain trapped among freshly formed copper nuclei and constructing channel structure through the interior of the deposit is decreased. It means that these simultaneous processes will lead to a reduction of irregular channels formed around small copper grains agglomerates, and hence, up to the increase of the compactness of the honeycomb-like structures.

Anyway, the ratio of hydrogen creating channel structure to the overall quantity of evolved hydrogen decreases with the increasing anodic current density values. Also, due to the increasing dissolution effect, the size of grains constructing agglomerates decreased with the increase of the anodic current density.

The complete insight into the strong effect of the anodic current density on the macromorphology of copper deposits is obtained by the analysis of their internal structures. Figure 4.30 shows cross sections of the honeycomb-like deposits obtained with $j_a = 40$ mA/cm^2 (Fig. 4.30a) and $j_a = 440$ mA/cm^2 (Fig. 4.30b). At the first sight, a clear difference in the internal structures of these copper deposits can be seen by the analysis of Fig. 4.30. The interior of copper deposit obtained with $j_a = 40$ mA/cm^2 was constructed from fine particles (Fig. 4.30a), while a well-defined dendritic structure can be noticed in the copper deposit obtained with $j_a = 440$ mA/cm^2 (Fig. 4.30b). The second strong effect of the increasing anodic current density on the internal structure of the copper deposits was related to the shape of cavities formed in the interior of deposits. The absence of the deposit at the bottom of the cavity and its regular shape close to the electrode surface following the shape of hydrogen bubbles clearly indicate that the origin of the cavity in the deposit obtained with $j_a = 40$ mA/cm^2 is of hydrogen bubbles formed at the electrode surface in the initial stage of electrodeposition. The formed hydrogen bubbles isolate the electrode surface preventing the growth of deposit at the position of their formation. In the growth process, due to the current density distribution effect, they remained included in the interior of deposit making this deposit type to be very porous. This type of cavity is not found in the copper deposits obtained with the larger anodic current densities indicating that the increase of the number of holes formed of detached hydrogen bubbles (Fig. 4.28) can be not only ascribed to suppressed coalescence of closely formed hydrogen bubbles but also to the improved current density distribution at the growing electrode surface. The second type of the cavity had irregular shape (Fig. 4.30b) and this cavity type is formed by the overlap of closely formed branchy dendrites. From Fig. 4.30b, it can be seen that before the overlap of dendrites, copper electrodeposition occured over the whole electrode surface. The improvement of the current density distribution effect was observed with the increasing anodic current density.

The formation of dendrites instead of fine particles can be explained by the decrease of the quantity of the hydrogen generated during copper electrochemical deposition processes with the increase of the anodic current density as follows: due to the decrease of the quantity of evolved hydrogen which leads to a stirring of solution in the near-electrode layer, the thickness of the diffusion layer increases and the limiting diffusion current density decreases causing the increase of the degree of diffusion control of the electrodeposition process. From the point of view of mechanism of the electrodeposition processes in the hydrogen codeposition range, it means that the effective overpotential amplitude, $\eta_{A,eff}$ [Eq. (4.27)] will increase with the increasing anodic current density. The change of morphology of the electrodeposited copper from very disperse agglomerates of copper grains to branchy dendrites (Fig. 4.29) clearly points out the increase of the degree of diffusion control of the electrodeposition process with the increasing anodic current density.

4.7.1.1 Analysis of the Applied Regimes of RC on Both the Specific Surface Area and the Structural Stability

On the basis of the maximal number of holes and the minimal wall width (or the distance among holes), it is clear that the largest specific surface area of the honeycomb-like electrode is obtained by the RC regime with $j_a = 240$ mA/cm^2.

As already mentioned, the average current efficiency of hydrogen evolution decreased with the increasing anodic current density. In the range of the anodic current densities from 0 (the PC regime) to 440 mA/cm^2 which enabled the formation of the honeycomb-like electrodes, the decrease of $\eta_{I,av}(H_2)$ of 29.3% was reached. The honeycomb-like electrode with the maximal specific surface area (the RC regime with $j_a = 240$ mA/cm^2) was formed with 16.6% smaller the average current efficiency of hydrogen evolution than the one obtained by the PC regime. The change of $\eta_{I,av}(H_2)$ was increased when the comparison between the average current efficiencies of hydrogen evolution obtained for the honeycomb-like structures formed in the constant galvanostatic regime at the current density which corresponded to the cathodic current density of 440 mA/cm^2 and the RC regime with $j_a = 240$ mA/cm^2 was made. In the constant regime of electrolysis, at $j = 440$ mA/cm^2, $\eta_{I,av}(H_2) =$ 36.0% [30], and the decrease of $\eta_{I,av}(H_2)$ of 28.9% was obtained. It is well known fact that the structural stability of a deposit is closely associated with the quantity of hydrogen evolved at the electrode surface during the electrodeposition process [51]. The smaller the quantity of evolved hydrogen, the better structural stability of a deposit is obtained. The decrease of a quantity of evolved hydrogen spent for the formation of the honeycomb-like electrodes by the RC regime clearly indicates the improvement of the structural stability of these electrodes in relation to the one obtained by other regimes of electrolysis, such as the PC regime and the constant galvanostatic one.

Some of the advantages of the production of the honeycomb-like electrodes by the RC regime in relation to the constant galvanostatic and PC regimes are both the increase of the specific surface area and the improvement of their structural stability. It is clear that these advantages make this regime superior in relation to other regimes of electrolysis.

4.8 Comparison of Open Porous Copper Structures Obtained by Different Current Regimes of Electrolysis

Using the well-known fact that electrodeposition of metal by the PC and RC regimes in the millisecond range occurs at the average current density [34, 35], the formation of open porous copper structures by these regimes of electrolysis at the same average current density is examined. The obtained surface morphologies were compared with the one obtained in galvanostatic regime at the current density which was equal to the selected average current density. In all experiments for which results are presented here, copper electrodeposition was performed from 0.15 M $CuSO_4$ in 0.50 M H_2SO_4 at room temperature using cylindrical copper wires as working electrodes [53]. In all PC and RC experiments, the cathodic current density of 0.44 A/cm^2 and a deposition pulse of 10 ms were selected. The average current density, j_{av}, of 0.12 A/cm^2 was reached by a selection of pause duration of 26.6 ms [for the PC regime; see Eq. (4.3)], as well as by the regulation of ratios between the anodic current density j_a and the anodic pulse duration t_a [for the RC regimes; see Eq. (4.21)]. Then, the following j_a and t_a values were selected:

(a) $j_a = 0.040$ A/cm^2; $t_a = 20$ ms (denoted as RC40)
(b) $j_a = 0.20$ A/cm^2; $t_a = 10$ ms (denoted as RC200)
(c) $j_a = 0.52$ A/cm^2; $t_a = 5$ ms (denoted RC520)

Figure 4.31 shows the morphologies of electrodeposited copper obtained by the RC regimes (RC40—Fig. 4.31a; RC200—Fig. 4.31b;

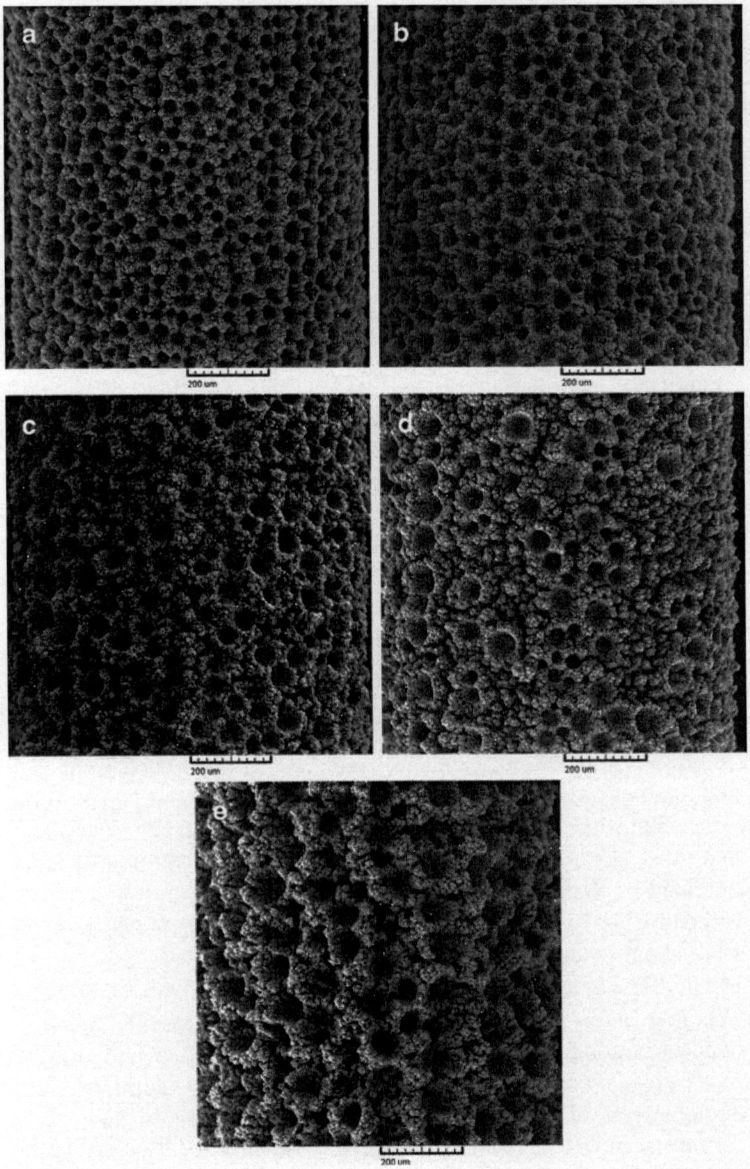

Fig. 4.31 Copper deposits obtained by: (**a**) RC40; (**b**) RC200; (**c**) RC520 regimes; (**d**) PC regime; and (**e**) the galvanostatic regime

and RC520—Fig. 4.31c), by the PC regime (PC—Fig. 4.31d); and in the constant galvanostatic regime (DC—Fig. 4.31e). As already mentioned, the average current density was the same in all analyzed PC and RC regimes, while the applied current density in the DC mode corresponded to the average current density in the PC and RC regimes. From Fig. 4.31, it can be seen that holes formed of detached hydrogen bubbles and cauliflower-like agglomerates of copper grains or dendrites were formed under these electrodeposition conditions. These copper deposits were formed by approximately the same quantity of evolved hydrogen which corresponded to $\eta_{I,av}(H_2)$ of 22.0 ± 0.8% [53]. It was understandable due to the fact that electrodeposition processes in the millisecond range at periodically changing rate occur at the average current density [34, 35].

The obtained surface morphologies can be classified into two groups. In the first group are the honeycomb-like structures obtained by the RC regimes, denoted as RC40 (Fig. 4.31a) and RC200 (Fig. 4.31b), as well as in the DC mode (Fig. 4.31e). The characteristic of the second group of the obtained copper deposits is the dominant presence of dish-like holes and independently formed cauliflower-like agglomerates of copper grains. These morphological forms were obtained by the RC520 (Fig. 4.31c) and PC (Fig. 4.31d) regimes.

From Fig. 4.31a, b, e, it can be seen a clear difference in the size and number of holes, as well as in the wall width among them in the honeycomb-like structures obtained by the RC40 and RC200 regimes and those obtained at a constant current density. The typical holes obtained by the RC40 and the DC mode are shown in Fig. 4.32a, b, respectively. The holes obtained by the RC40 and RC200 regimes were about twice smaller than those obtained at the constant current density. The average size of these holes was about 50 μm, while holes obtained in the DC mode were about 100 μm. Morphologies of deposits among holes obtained by the RC40 regime and the DC mode are shown in Fig. 4.32c, d, respectively. The cauliflower-like agglomerates of copper grains with small dendrites on them were formed by the RC40 (Fig. 4.32c) and RC200 regimes. The nucleation exclusion zones can be clearly observed around these cauliflower-like agglomerates of copper grains. On the other hand, the relatively large cauliflower-like copper grains agglomerates surrounded by deep irregular channels for which the origin is of evolved hydrogen were formed in the DC mode (Fig. 4.32d). Anyway, it is clear that the application of appropriate RC regime led to the redistribution of evolved hydrogen favoring growth of hydrogen bubbles with electrolysis time and hence decreasing the contribution of generated hydrogen to the creating of channel structure through the interior of the deposit (the current density distribution effect).

Dish-like holes were the dominant type of holes obtained by the PC and RC520 regimes. The typical dish-like hole obtained by the PC regime is given in Fig. 4.32e. The average diameter of this type of holes was about 100 μm. The cauliflower-like agglomerates of copper grains, and the mixture of cauliflower-like copper grains agglomerates and dendrites were formed by the PC and RC520 regimes, respectively. The typical independently formed cauliflower-like agglomerate of copper grains obtained by the PC regime is presented in Fig. 4.32f.

From technological point of view, the first group of copper deposits (the honeycomb-like structures) is especially important for the application as electrodes in electrochemical devices [1] and in catalysis [3]. It is shown that the size of holes and wall width among holes were decreased, while the number of holes was increased when the appropriate RC parameters were applied (RC40 and RC200). In this way, the specific surface of the honeycomb-like electrodes was considerably increased confirming superiority the regime of reversing current in relation to other current regimes of electrolysis (the PC regime and the constant galvanostatic regime).

Comparing all periodically changing regimes of electrolysis, it can be noticed that the effects attained by the application of the RC regimes on microstructural characteristics of the honeycomb-like structures were more similar to those attained by the regime of pulsating overpotential (PO) than those attained by the pulsating current (PC) regime. It was understandable due to the existence of anodic current during “off period” in the PO regimes [37].

Anyway, from the point of view of the formation of the honeycomb-like electrodes by the electrodeposition techniques, the regime of reversing current is more suitable than other current regimes of electrolysis.

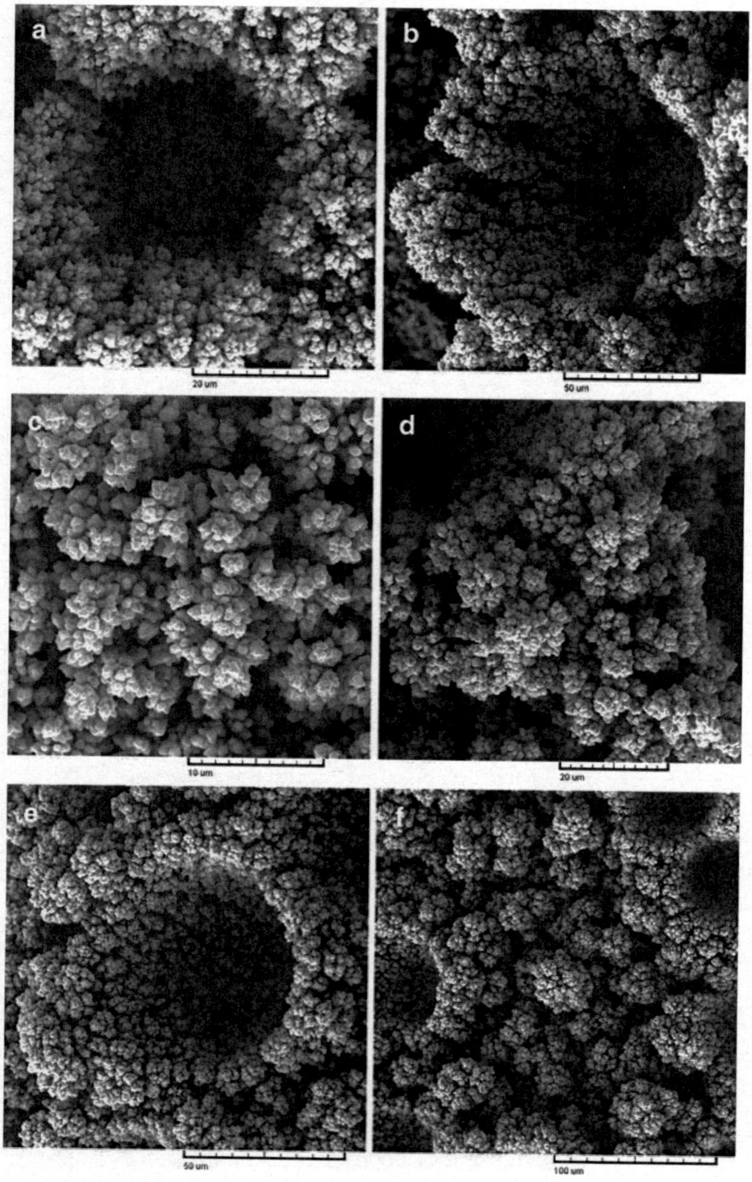

Fig. 4.32 The typical holes obtained by (**a**) RC40, (**b**) galvanostatic, and (**e**) PC regimes and morphological forms obtained among holes by (**c**) RC40, (**d**) galvanostatic, and (**f**) PC regimes

4.9 Conclusions

The comprehensive survey of the formation of open porous copper structures by both the constant and periodically changing regimes of electrolysis is presented. The main characteristics of this structure type, denoted as the honeycomb-like or the 3D foam ones, are holes or pores formed of detached hydrogen bubbles surrounded by cauliflower-like agglomerates of copper grains or dendrites. In the constant regimes of electrolysis, these structures are formed at overpotentials outside the plateau of the limiting diffusion current density (the potentiostatic regime) or current densities higher than the limiting diffusion current density (the galvanostatic regime), where parallel to copper electrodeposition hydrogen evolution reaction occurs. The number, size, and distribution of holes in the honeycomb-like structures depended on the Cu(II) ions and H_2SO_4 concentrations, temperature of electrolysis, the type of the working electrode used, and a time of electrolysis.

In periodically changing regimes of electrolysis, such as the pulsating overpotential (PO), the pulsating current (PC), and the reversing current (RC) regimes, the overpotential amplitude (in the PO regime), the current density amplitude (in the PC regime), or the cathodic current density (in the RC regime) is outside (the PO regime) or higher than the limiting diffusion current density (the PC and RC regimes). The following conveniences in the production of the honeycomb-like structures can be attained since the appropriate square-waves parameters of periodically changing regimes of electrolysis are selected (a) energy saving; (b) the increase of the specific surface area of the electrodes; and (c) probably the improvement of the deposit structural stability due to the decrease of the quantity of evolved hydrogen needed for their formation.

Acknowledgments The author is grateful to Prof. Dr. Konstantin I. Popov for helpful discussion during the preparation of this chapter.

Also, the author is grateful to Dr. Goran Branković and Dr. Ljubica Pavlović for SEM analysis of investigated systems, as well as to Dr. Vesna Maksimović for the cross section analysis of the obtained deposits.

The work was supported by the Ministry of Education and Science of the Republic of Serbia under the research project "Electrochemical synthesis and characterization of nanostructured functional materials for application in new technologies" (No. 172046).

References

1. Shin H-C, Dong J, Liu M (2003) Adv Mater 15:1610
2. Shin H-C, Liu M (2005) Adv Funct Mater 15:582
3. Yin J, Jia J, Zhu L (2008) Int J Hydrogen Energy 33:7444
4. Shin H-C, Liu M (2004) Chem Mater 16:5460
5. Li Y, Jia W-Z, Song Y-Y, Xia XH (2007) Chem Mater 19:5758
6. Guo YL, Yui H, Minamikawa H, Yang B, Masuda M, Ito K, Shimizu T (2006) Chem Mater 18:1577
7. Kazeminezhad I, Barnes AC, Holbrey JD, Seddon KR, Schwarzacher W (2007) Appl Phys A Mater Sci Process 86:373
8. Yuan JH, He FY, Sun DC, Xia XH (2004) Chem Mater 16:1841
9. Yuan JH, Wang K, Xia XH (2005) Adv Funct Mater 15:803
10. Qiu JD, Peng HZ, Liang RP, Li J, Xia XH (2007) Langmuir 23:2133
11. Wang CH, Yang C, Song YY, Gao W, Xia XH (2005) Adv Funct Mater 15:1267
12. Chen W, Xia XH (2007) Chemphyschem 8:1009
13. Meldrum FC, Seshadri R (2000) Chem Commun 1:29
14. Bartlett PN, Birkin PR, Ghanem MA, Toh C-S (2001) J Mater Chem 11:849
15. Briseno AL, Han S, Rauda IE, Zhou F, Toh C-S, Nemanick EJ, Lewis NS (2004) Langmuir 20:219
16. Nikolić ND, Popov KI, Pavlović LjJ, Pavlović MG (2006) J Electroanal Chem 588:88
17. Nikolić ND, Popov KI, Pavlović LjJ, Pavlović MG (2006) Surf Coat Technol 201:560
18. Nikolić ND, Popov KI, Pavlović LjJ, Pavlović MG (2007) J Solid State Electrochem 11:667
19. Nikolić ND, Pavlović LjJ, Pavlović MG, Popov KI (2007) Electrochim Acta 52:8096
20. Nikolić ND, Popov KI, Pavlović LjJ, Pavlović MG (2007) Sensors 7:1
21. Nikolić ND, Pavlović LjJ, Krstić SB, Pavlović MG, Popov KI (2008) Chem Eng Sci 63:2824
22. Nikolić ND, Branković G, Pavlović MG, Popov KI (2008) J Electroanal Chem 621:13
23. Nikolić ND, Popov KI (2010) Hydrogen co-deposition effects on the structure of electrodeposited copper. In: Djokić SS (ed) Electrodeposition: theory and practice, vol 48, Modern aspects of electrochemistry. Springer, New York, pp 1–70
24. Nikolić ND, Maksimović V, Pavlović MG, Popov KI (2009) J Serb Chem Soc 74:689
25. Casas JM, Alvarez F, Cifuentes L (2000) Chem Eng Sci 55:6223
26. Nikolić ND, Pavlović LjJ, Branković G, Pavlović MG, Popov KI (2008) J Serb Chem Soc 73:753
27. Nikolić ND, Pavlović LjJ, Pavlović MG, Popov KI (2007) J Serb Chem Soc 72:1369

28. Amadi A, Gabe DR, Goodenough M (1991) J Appl Electrochem 21:1114
29. Vogt H, Balzer RJ (2005) Electrochim Acta 50:2073
30. Nikolić ND, Branković G, Popov KI (2011) Mater Chem Phys 125:587
31. Kim J-H, Kim R-H, Kwon H-S (2008) Electrochem Commun 10:1148
32. Oniciu L, Muresan L (1991) J Appl Electrochem 21:565
33. Muresan L, Varvara S (2005) Leveling and brightening mechanisms in metal electrodeposition. In: Nunez M (ed) Metal electrodeposition. Nova Science, New York, pp 1–45
34. Popov KI, Maksimović MD (1989) Theory of the effect of electrodeposition at periodically changing rate on the morphology of metal deposition. In: Conway BE, Bockris JO'M, White RE (eds) Modern aspects of electrochemistry, vol 19. Plenum, New York, pp 193–250
35. Popov KI, Djokić SS, Grgur BN (2002) Fundamental aspects of electrometallurgy. Kluwer Academic/Plenum, New York
36. Nikolić ND, Branković G, Pavlović MG, Popov KI (2009) Electrochem Commun 11:421
37. Nikolić ND, Branković G, Maksimović VM, Pavlović MG, Popov KI (2010) J Solid State Electrochem 14:331
38. Popov KI, Nikolić ND, Živković PM, Branković G (2010) Electrochim Acta 55:1919
39. Popov KI, Stojilković ER, Radmilović V, Pavlović MG (1997) Powder Technol 93:55
40. Barton L, Bockris JO'M (1962) Proc Roy Soc A 268:485
41. Nikolić ND, Branković G, Maksimović VM, Pavlović MG, Popov KI (2009) J Electroanal Chem 635:111
42. Ko W-Y, Chen W-H, Tzeng S-D, Gwo S, Lin K-J (2006) Chem Mater 18:6097
43. Popov KI, Pavlović MG (1993) Electrodeposition of metal powders with controlled grain size and morphology. In: White RE, Bockris JO'M, Conway BE (eds) Modern aspects of electrochemistry, vol 24. Plenum, New York, pp 299–391
44. Nikolić ND, Branković G, Maksimović V (2012) J Solid State Electrochem 16:321
45. Nikolić ND, Branković G (2010) Electrochem Commun 12:740
46. Dima GE, de Vooys ACA, Koper MTM (2003) J Electroanal Chem 554–555:15
47. Ko W-Y, Chen W-H, Cheng C-Y, Lin K-J (2009) Sens Actuators B Chem 137:437
48. Pletcher D, Poorbedi Z (1979) Electrochim Acta 24:1253
49. Gorgievski M, Božić D, Stanković V, Bogdanović G (2009) J Hazard Mater 170:716
50. Chandrasekar MS, Pushpavanam M (2008) Electrochim Acta 53:3313
51. Sun BK, O'Keefe TJ (1998) Surf Coat Technol 106:44
52. Nikolić ND, Branković G, Maksimović VM (2011) J Electroanal Chem 661:309
53. Nikolić ND, Branković G (2012) Mater Lett 70:11

Chapter 5
Morphology, Chemical, and Phase Composition of Electrodeposited Co–Ni, Fe–Ni, and Mo–Ni–O Powders

V.D. Jović, U.Č. Lačnjevac, and B.M. Jović

5.1 Introduction

The alloy powders of the iron-group metals are of great interest for many industrial applications [1–88].

Ultrafine Co–Ni powders showed significant promise for future development of hard materials [1, 2], magnetic materials [3–5], commercial batteries [6], catalysts [7–9], catalyzing electroplates [10, 11], hydrogen absorbing alloy anodes [12], and magnetoresistive sensors [13, 14]. The powders for these applications were made by several different techniques such as mechanical alloying–ball milling [15, 16], sonochemical synthesis [17–19], chemical reduction [13, 14, 20–23], synthesis in microemulsion [24, 25], gas evaporation [26, 27], multimellow reduction [28, 29], coprecipitation [30], gas reduction [31, 32], freeze-drying [33], and spray pyrolysis [34]. Depending on the technique used for powders synthesis, dimensions of Co–Ni powder particles varied from 40 nm [33] to 700 nm [34]. Unfortunately, limited number of papers concerning Co–Ni powder

V.D. Jović (✉) • U.Č. Lačnjevac • B.M. Jović
Department of Materials Science, Institute for Multidisciplinary Research, University of Belgrade, Kneza Višeslava 1, P.O. Box 33, Belgrade 11030, Serbia
e-mail: vladajovic@imsi.rs; uros.lacnjevac@imsi.rs; bjovic@imsi.rs

Stojan S. Djokić (ed.), *Electrochemical Production of Metal Powders*, Modern Aspects of Electrochemistry 54, DOI 10.1007/978-1-4614-2380-5_5,

electrodeposition exists in the literature [35–40]. Among them four are our publications [37–41]. As shown in these papers [37–41], the morphology and composition of electrodeposited powders were found to be sensitive to the solution composition (to the ratio of Ni^{2+}/Co^{2+} ions concentration).

Fe–Ni-based alloy powders are known as promising soft magnetic materials with low coercivity and high permeability [42–66]. Most of the literature data concerning the procedure of their production deal with the mechanical alloying using ball-milling process [42–55]. It is shown that the grain size of particles is closely related to the magnetic properties of Fe–Ni alloy powders obtained by mechanical alloying, being also very sensitive to the conditions of ball-milling process (the vial rotation speed, the disc rotation speed, and the time of milling). Gas condensation method using Ar with H_2 as convection gas is used for preparing nanograined Fe–12Ni powders [56]. In this investigation it is shown that the fcc-to-bcc phase transformation might be inhibited if the grain size is sufficiently small (below 27 nm). Nanocrystalline $FeNi_3$ powders with the particles size of 40–50 nm, prepared by powder processing (calcination of the Fe: Ni = 1:3 mixture of Fe and Ni oxalates with subsequent annealing in hydrogen atmosphere at 600°C), showed higher coercive force and lower permeability, with such behavior being explained by the role of interfacial regions in the samples [57]. Hydrodynamic electrospray technique [58] and electroless plating procedure [59] were also used for the synthesis of nanosized Fe–Ni powders. Recently it is discovered that these powders could be used for the absorption of electromagnetic waves [60, 61], which is a very important issue for both commercial and military application. Fe–Ni nanofibers for this application were synthesized by calcination and subsequent hydrogen reduction of electrospun Fe, Ni nitrates, and polystyrene [61].

Among our recent papers [62, 63] the electrodeposition of Fe–Ni alloy powders was the subject of only few papers. Zhelibo et al. [64, 65] suggested a method for producing very fine Fe–Ni alloy powder by electrolysis in a two-layer electrolytic bath, using a hydrocarbon solvent from an oil refining fraction as an upper organic layer with evaporation at 180°C, and subsequent reduction annealing in a hydrogen atmosphere. The influence of the reduction annealing temperature [64] and the electrolysis temperature [65] on the formation, chemical and phase composition, structure, and magnetic properties of highly dispersed Fe–Ni alloy powders were investigated and the optimal thermal conditions for the production of powders with micron-sized particles were determined [64, 65]. The effect of complexing agents (citric and oxalic acid) on the process of Fe–Ni alloy powders was also investigated [66]. It was shown that complexing agents influence the kinetics of powders electrodeposition as well as the morphology of the Fe–Ni powders. Finer powders were produced in the presence of citric acid in comparison with those obtained in the presence of oxalic acid [66].

The subject of our recent investigations was the analysis of morphology and composition of electrodeposited Fe–Ni powders as a function of the solution composition [62], as well as the possibility of $NiFe_2O_4$ formation after additional oxidation of electrodeposited powders [63].

One of the components of the Fe–Ni system, formed by oxidation of Fe–Ni mixture at high temperatures, is a nickel ferrite, $NiFe_2O_4$, an inverse spinel in which the tetrahedral sites are occupied by Fe^{3+} ions and octahedral sites by Fe^{3+} and Ni^{2+} ions [67].

According to the literature [68, 69], the $NiFe_2O_4$ is the most suitable material for device applications in the upper microwave and lower millimeter wave ranges. At the same time, $NiFe_2O_4$ was found to be a highly reproducible humidity [70] and gas [71, 72] sensor material. Numerous techniques, such as ferrite plating [73, 74], oxidation of metallic films [75], arc plasma method [76], chemical transport [77, 78], chemical vapor deposition [79–82], the dip coating process [83], spray pyrolysis [84, 85], and pulsed wire discharge [86], have been used for the preparation of $NiFe_2O_4$ films. The main difficulty of these methods is the limit in the choice of substrate material, since it must be kept at a high temperature after deposition. In the work of Sartale et al. [87] less expensive and more popular method for the preparation of the spinel nanocrystalline $NiFe_2O_4$ films has been proposed. The procedure was based on the electrodeposition of $NiFe_2$ alloy coating from a nonaqueous ethylene glycol sulfate bath and subsequent electrochemical oxidation of the alloy in an aqueous alkaline bath at room temperature. It is shown that the air annealing of the as-deposited $NiFe_2O_4$ thin films at 500°C for 5 h improved the crystallinity and morphology of the films.

In the work of Fang et al. [88] the $NiFe_2O_4$ ultrafine powder with high crystallinity has been prepared through a reverse microemulsion route. After optimizing the composition in the starting solution, the resulting $NiFe_2O_4$ was formed at a temperature of about 550–600°C, which is much lower than that observed from the solid state reaction. Magnetic investigation indicates that samples are soft magnetic materials with low coercivity and with the saturation magnetization close to the bulk value of nickel ferrite. Most recently $NiFe_2O_4$ nanoparticles were synthesized via solid state reaction process of the $Fe_{67}Ni_{33}$ alloy nanopowder at different annealing temperatures in the air [89]. It was shown by XRD and TEM analysis that $NiFe_2O_4$ started to form at around 450°C, being well defined after annealing at 550°C with the powder particle size ranging between 15 and 50 nm.

The Mo–Ni alloys possess several useful properties: exceptional corrosion and wearing resistance [90–92], high catalytic activity for hydroprocessing of aromatic oils [93], and gas phase hydrogenation of benzene [94], as well as high hardness [95]. Their catalytic activity for hydrogen evolution has been one of the most investigated properties in the literature [96–110].

The Mo–Ni alloys can be produced by several methods, from which metallurgical ones are not convenient because of easy oxidation and high melting temperature of molybdenum. The powder metallurgy and mechanical alloying [111, 112], spark plasma sintering [113], and laser cladding [114] are mostly used for the Mo–Ni alloy preparation. Recently, coated superconductor tapes of high mechanical strength were made of the Mo–Ni alloy powder by rolling assisted biaxially textured substrates method (RABiTS™) [115].

All the above-mentioned methods are expensive in comparison with the electrodeposition of Mo–Ni alloy coatings. Although molybdenum cannot be separately deposited from aqueous solutions, it can be codeposited with the iron-group metals (Fe, Co, Ni) in the presence of appropriate complexing agents, by the type of alloy electrodeposition defined by Brenner [116] as induced codeposition.

Most of the papers concerning electrodeposition of compact Mo–Ni alloy coatings are dealing with the mechanism of their electrodeposition (mechanism of induced codeposition), and according to the literature, the most probable mechanism is the one reported by Podlaha and Landolt [117–120].

It should be emphasized here that, except our recent papers [121–123], there are no papers in the literature concerning electrodeposition of Mo–Ni alloy powders (actually powders of the system Mo–Ni–O) and their characterization.

5.2 System Co–Ni

Electrodeposition of Co–Ni powders was performed in two types of ammonium hydroxide containing electrolytes: 1 M $(NH_4)_2SO_4$ + 0.7 M NH_4OH and 1 M NH_4Cl + 0.7 M NH_4OH. In sulfate containing electrolytes cobalt and nickel sulfates were used, while in chloride containing electrolytes cobalt and nickel chlorides were used as a source of Co^{2+} and Ni^{2+} ions [37–40]. The Co–Ni alloy powders were also electrodeposited from a solution containing 0.4 M H_3BO_3 + 0.2 M Na_2SO_4 and Ni and Co sulfate salts [41]. Three different Ni^{2+}/Co^{2+} concentration ratios were used: $Ni^{2+}/Co^{2+} = 1.00$ (0.01 M $NiSO_4$/0.01 M $CoSO_4$); $Ni^{2+}/Co^{2+} = 0.50$ (0.01 M $NiSO_4$/0.02 M $CoSO_4$); and $Ni^{2+}/Co^{2+} = 0.33$ (0.01 M $NiSO_4$/0.03 M $CoSO_4$). From this solution Co–Ni alloy powders were electrodeposited at a constant current density of approximately 70 mA cm^{-2}, corresponding to the cell voltage of about 5.0 V [41].

5.2.1 Polarization Curves for Co–Ni Alloy Powders Electrodeposition from Ammonium Sulfate–Ammonium Hydroxide Electrolyte

In Fig. 5.1 are shown polarization curves corrected for IR drop (see Chap. 2), for the processes of Co, Ni, and Co–Ni alloy powders electrodeposition from ammonium sulfate–ammonium hydroxide containing supporting electrolyte (1 M $(NH_4)_2SO_4$ + 0.7 M NH_4OH). As can be seen their shape is identical to that for pure Co, Fe, and Ni powders electrodeposition, characterized with two inflection points, A and B. For Co electrodeposition, sharp increase of current occurs at

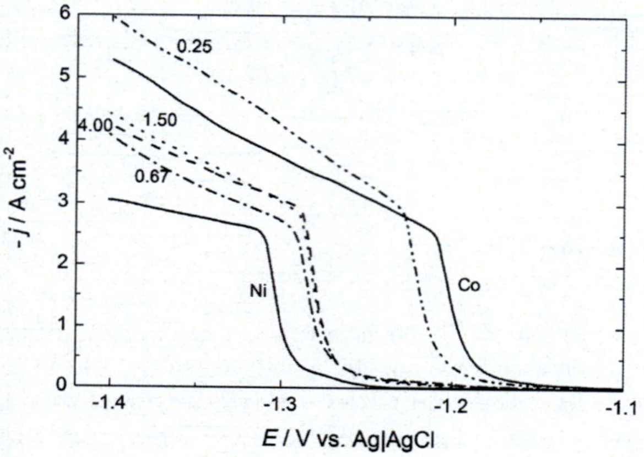

Fig. 5.1 Polarization curves for the electrodeposition of cobalt (Co), nickel (Ni), and Co–Ni alloy powders after IR drop correction recorded for different Ni^{2+}/Co^{2+} ions ratios: 4.00, 1.50, 0.67, and 0.25 (marked in the figure) (Reprinted from [40] with the permission of Elsevier.)

about −1.19 V, while for Ni electrodeposition this phenomenon is moved to more negative potentials (at about −1.30 V), indicating that the overvoltage for Ni electrodeposition in this solution is for about 0.1 V higher than that for Co electrodeposition, as it is the case for hydrogen evolution from pure supporting electrolytes (Fig. 2.24). In the case of alloy electrodeposition sharp increase of current density on the polarization curves recorded for Ni^{2+}/Co^{2+} ratios 4.00, 1.50, and 0.67 takes place at almost identical potentials, while the one for $Ni^{2+}/Co^{2+} = 0.25$ is moved to more positive potential, close to the polarization curve for pure Co.

Polarization curves corresponding only to the processes of electrodeposition of pure metals and alloy powders are shown in Fig. 5.2a, while corresponding η_j vs. E curves for alloy powders electrodeposition are presented in Fig. 5.2b.

As can be seen, significantly different polarization curves are obtained indicating different contribution of the hydrogen evolution current in all cases. It is quite interesting (Fig. 5.2a) that diffusion limiting current densities for Co and Ni are significantly different for the same cations concentration (0.1 M), $j_{l(Co)} = -0.76$ A cm^{-2} and $j_{l(Ni)} = -0.36$ A cm^{-2}. Taking into account these values, it is

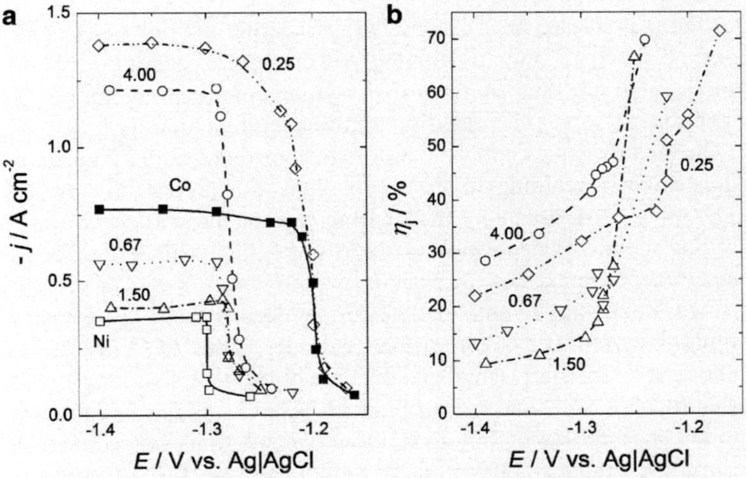

Fig. 5.2 (**a**) Polarization curves for the electrodeposition of cobalt (Co), nickel (Ni), and Co–Ni alloy powders (after subtraction of the current density for hydrogen evolution) recorded for different Ni^{2+}/Co^{2+} ions ratios: 4.00, 1.50, 0.67, and 0.25 (marked in the figure). (**b**) Corresponding current efficiency vs. potential curves for Co–Ni powders electrodeposition (Reprinted from [40] with the permission of Elsevier.)

obvious that at the same concentration of their ions the difference in their diffusion limiting current densities could be either due to different complexes that those metals may form with ammonia [124], or due to different metal electrodeposition process (difference in the contribution of hydrogen evolution current). Since both metals form the same number of complexes with ammonia (6 complexes) and their stability constants are very similar [124], it is most likely that the complex formation is not responsible for the observed difference in diffusion limiting current densities for Ni and Co. On the other side, hydrogen evolution current density in the case of Co electrodeposition is higher than that for Ni electrodeposition in the region of the diffusion limiting current, indicating that most likely this phenomenon could be ascribed to the increase of the real surface area during metal (powder) electrodeposition. It is important to note that it is quite difficult to make any convincing conclusion concerning this phenomenon, since there is no possibility to obtain the data about the increase of the real surface area of the electrode during the processes of Co and Ni powders and Co–Ni alloy powders electrodeposition. If the increase of the real surface area is more pronounced during Co electrodeposition, the diffusion limiting current density could be lower and its value could be closer to the one obtained for Ni electrodeposition. According to the morphology of Co and Ni powders [37–40], it is most likely that the increase of the real surface area during the electrodeposition of Co is much higher than that for Ni electrodeposition.

It is interesting to note that the current density plateaus (diffusion limiting current densities) for the electrodeposition of Co–Ni alloys at the Ni^{2+}/Co^{2+} ratios 4.00 and 0.25 are higher than that for pure Co electrodeposition. The reason for such behavior is the fact that the total concentration of cations in these two solutions was not 0.1 M, since the compositions of these solutions were 0.1 M $NiSO_4$ + 0.025 M $CoSO_4$ + 1 M $(NH_4)_2SO_4$ + 0.7 M NH_4OH; 0.025 M $NiSO_4$ + 0.1 M $CoSO_4$ + 1 M $(NH_4)_2SO_4$ + 0.7 M NH_4OH (total concentration of cations 0.125 M). At the same time, the current efficiency for Co–Ni powder electrodeposition is the highest in these two solutions. Such behavior is most probably the consequence of the independence of the current density for hydrogen evolution on the total concentration of cations, so that the same curve for hydrogen evolution is subtracted from the higher values of j_{tot} (see Fig. 5.2a).

5.2.2 *Polarization Curves for Co–Ni Alloy Powders Electrodeposition from Ammonium Chloride–Ammonium Hydroxide Electrolyte*

The polarization curves corrected for IR drop for the processes of pure cobalt (Co), pure nickel (Ni), and Co–Ni alloy (powders) electrodeposition from ammonium chloride–ammonium hydroxide containing supporting electrolyte are presented in Fig. 5.3 (in this case the total concentration of cations in all investigated solutions was 0.1 M). As can be seen, cobalt electrodeposition (Co) commences at about −1.1 V, while sharp increase of current density (massive Co

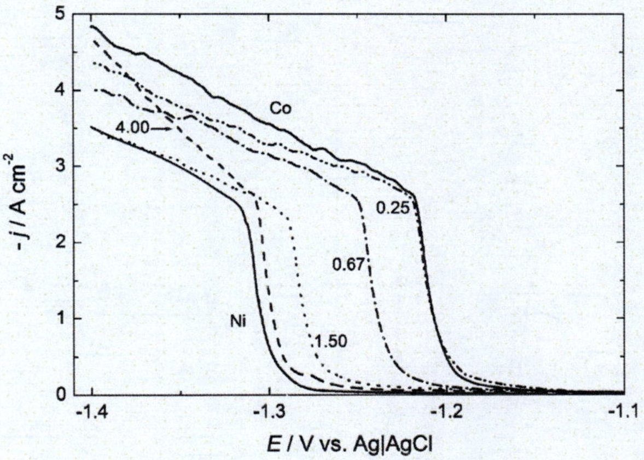

Fig. 5.3 Polarization curves for the electrodeposition of cobalt (Co), nickel (Ni), and Co–Ni alloy powders after IR drop correction recorded for different Ni^{2+}/Co^{2+} ions ratios: 4.00, 1.50, 0.67, and 0.25 (marked in the figure)

electrodeposition and hydrogen evolution) occurs at about −1.19 V. In the case of nickel electrodeposition (Ni), sharp increase of current density is moved to more negative potentials (at about −1.30 V), indicating that the overvoltage for massive Ni electrodeposition (and hydrogen evolution) in this solution is for about 0.1 V higher than that for Co electrodeposition, as it is the case for hydrogen evolution from pure supporting electrolytes (see Fig. 2.24, Chap. 2). In the case of Co–Ni alloy powders electrodeposition, polarization curves recorded for all Ni^{2+}/Co^{2+} ratios are placed between the polarization curves for pure Co and pure Ni.

Polarization curves corresponding only to the processes of electrodeposition of pure metals and alloy powders are shown in Fig. 5.4a, while corresponding η_j vs. E curves for pure metals and alloy powders electrodeposition are presented in Fig. 5.4b, respectively.

As in the case of sulfate supporting electrolyte (Fig. 5.2a), "diffusion limiting current density" (in this case not represented with the plateau) for Co is higher than that for Ni (Fig. 5.4a). At the same time, all polarization curves for Co–Ni alloy powders electrodeposition are characterized with higher values of current densities, indicating more efficient process of powders formation in the chloride

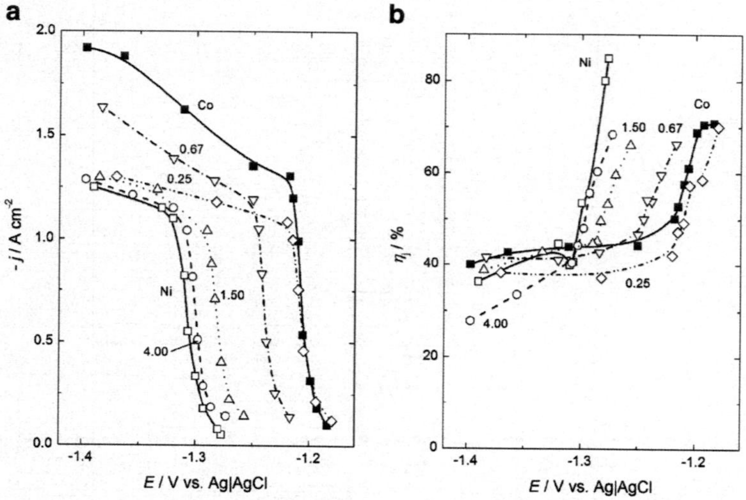

Fig. 5.4 (**a**) Polarization curves for the electrodeposition of cobalt (Co), nickel (Ni), and Co–Ni alloy powders (after subtraction of the current density for hydrogen evolution) recorded for different Ni^{2+}/Co^{2+} ions ratios: 4.00, 1.50, 0.67, and 0.25 (marked in the figure). (**b**) Corresponding current efficiency vs. potential curves for cobalt (Co), nickel (Ni), and Co–Ni powders electrodeposition (Reprinted from [39] with the permission of Elsevier.)

supporting electrolyte. This statement is supported with the η_j vs. E curves presented in Fig. 5.4b, where the current efficiency for powders electrodeposition is much higher than that in the sulfate supporting electrolyte (Fig. 5.2b).

5.2.3 Chemical Compositions of the Co–Ni Alloy Powders

All powder samples for chemical analysis, phase composition analysis, and morphology investigations were electrodeposited at the room temperature in the cylindrical glass cell of the total volume of 1 dm^3 with cone-shaped bottom of the cell to collect powder particles in it. Working electrode was a glassy carbon rod of the diameter of 5 mm with a total surface area of 7.5 cm^2 immersed in the solution and placed in the middle of the cell. Cylindrical Pt–Ti mesh placed close to the cell walls was used as a counter electrode providing excellent current distribution in the cell.

In all cases certain small amount of disperse deposit remained on the glassy carbon electrode, while only powder particles that were detached from the electrode surface and collected at the cone-shaped bottom of the cell were analyzed.

Chemical composition of all electrodeposited Co–Ni alloy powders was determined by the AAS technique after dissolving certain amount of powders in HCl. The compositions of powders electrodeposited from sulfate and chloride containing supporting electrolytes were different.

5.2.3.1 Chemical Compositions of the Co–Ni Alloy Powders Electrodeposited from Ammonium Sulfate–Ammonium Hydroxide Electrolyte

The dependence of atomic percentages of both metals as a function of Ni^{2+}/Co^{2+} ratio is presented in Fig. 5.5a. As it could be expected, the content of Ni increases, while the content of Co decreases with the increase of the Ni^{2+}/Co^{2+} ratio.

It is well known that the electrodeposition of compact deposits of Co–Ni alloys belongs to the type of anomalous codeposition according to Brenner's classification [116]. The influence of the electrolyte composition on the alloy powder composition is shown in Fig. 5.5a, b. As can be seen in Fig. 5.5a, the content of Ni increases by increasing the ratio of Ni^{2+}/Co^{2+} ions, while the content of Co decreases and both dependences are approximately linear after sharp increase (decrease) at low values of Ni^{2+}/Co^{2+}. According to Brenner's classification [116], the behavior presented in Fig. 5.5b indicates two types of powder codeposition: irregular, up to about 60 at.% of Ni (or 40 at.% of Co) in the solution and anomalous at higher amounts of Ni. The irregular codeposition is expected to occur in systems in which the static potentials of the parent metals are close together and with metals that form solid solution. Both conditions are fulfilled for the system Co–Ni [116]. The characteristic of anomalous codeposition is that it occurs only under certain conditions of concentration and operating variables for a given plating bath [116]. Hence, if the deposition of the more noble metal is so strongly

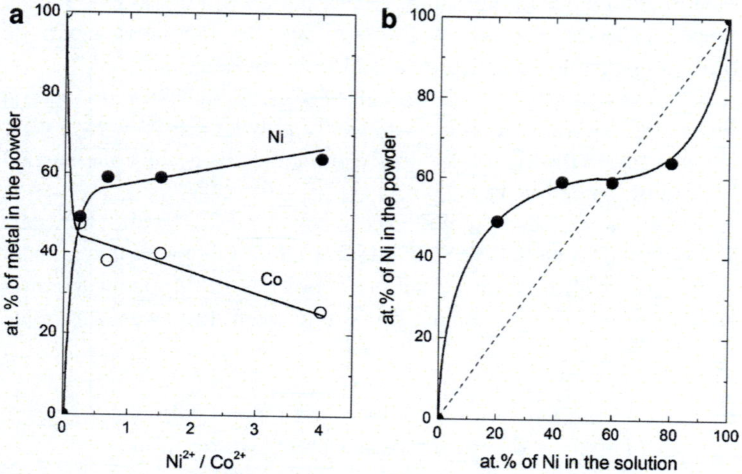

Fig. 5.5 (**a**) Composition of powders as a function of the Ni^{2+}/Co^{2+} ions ratio. (**b**) Corresponding diagram according to Brenner's classification [116] (Reprinted from [38] with the permission of Springer-Verlag.)

inhibited that no current can be recorded as the potential is driven negative, at some potential the less noble metal would start depositing causing the increase of deposition current. At some more negative potential a breakthrough in the deposition of the more noble metal may happen as a result of reduction or removal of the additive or by opening some alternative mechanism of discharge, not operative at less negative potential. Such a case is found in the deposition of Ni–Zn alloys and is explained by "the hydroxide suppression mechanism" [125, 126]. It has already been shown that from simple salt containing electrolytes codeposition of Co and Ni possesses anomalous character [116, 127–129], as well as in the electrolytes containing metal complexes [130], but in all cases compact deposits were obtained. As can be seen in Fig. 5.5b, at the conditions of powder electrodeposition where both components deposit at their diffusion limiting current densities, irregular and/or anomalous character of Co–Ni alloy deposition is achieved, depending on the solution composition.

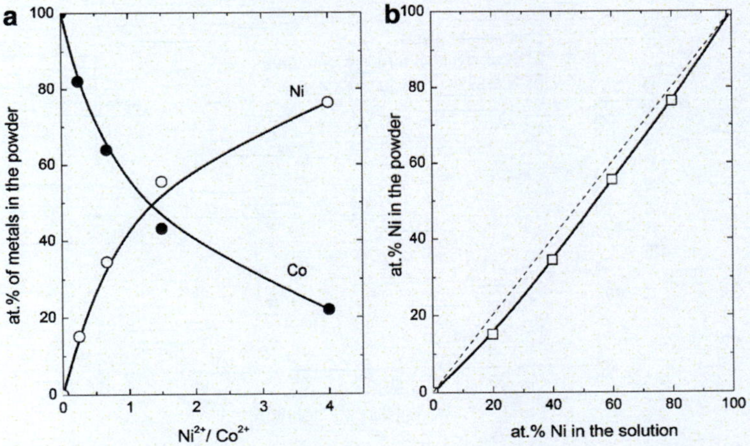

Fig. 5.6 (**a**) Composition of powders as a function of the Ni^{2+}/Co^{2+} ions ratio. (**b**) Corresponding diagram according to Brenner's classification [116] (Reprinted from [39] with the permission of Elsevier.)

5.2.3.2 Chemical Compositions of the Co–Ni Alloy Powders Electrodeposited from Ammonium Chloride–Ammonium Hydroxide Electrolyte

The same dependences as in Fig. 5.5 are presented in Fig. 5.6 for chloride containing supporting electrolyte. In this case anomalous codeposition is obtained for all Co–Ni alloy powders compositions, with the more noble metal electrodeposition (Ni) being suppressed by the presence of the less noble metal (Co) in the solution, i.e., less noble metal is more readily deposited.

5.2.3.3 Chemical Compositions of the Co–Ni Alloy Powders Electrodeposited from Borate–Sulfate Electrolyte

The polarization curves for this system showed that the current efficiency for Co–Ni powder formation is very low, about 1–2%.

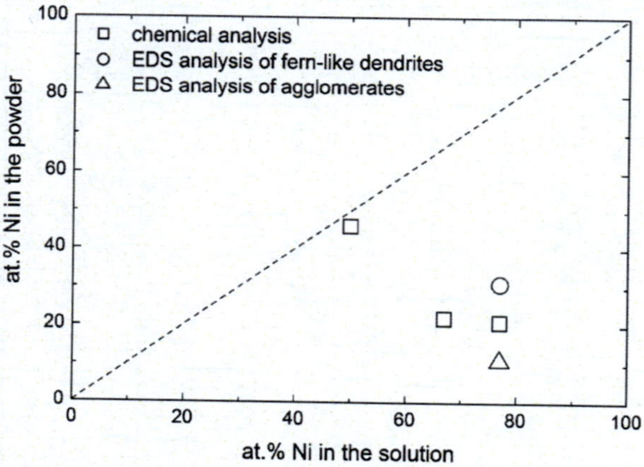

Fig. 5.7 Diagram corresponding to Brenner's classification for the Co–Ni powders electrodeposited from borate–sulfate containing electrolyte (results of chemical analysis—*open squares*, result of EDS analysis—*open triangle*—agglomerates, *open circle*—fern like dendrites)

In this case longer time for powders deposition was needed in order to obtain sufficient amount of powder to be analyzed [41].

In Fig. 5.7 is shown diagram corresponding to Brenner's classification for this system. Since the morphology of electrodeposited powders was different for different values of the Ni^{2+}/Co^{2+} ratios (see Figs. 5.18–5.20), composition of all powders was determined by the chemical analysis (□), while the powder with the highest amount of cobalt ($Ni^{2+}/Co^{2+} = 0.33$) was analyzed by the EDS analysis too. According to the results presented in Fig. 5.7, anomalous codeposition is characteristic for all investigated solutions, being more pronounced for lower values of Ni^{2+}/Co^{2+} ratios. As can be seen in Fig. 5.7, the composition of agglomerates (see Fig. 5.20a) showed lower content of Ni (△), while the composition of fern-like dendrites (see Figs. 5.18a, b, 5.19a, and 5.20b) showed higher content of Ni (○) in comparison with the results of chemical analysis (□). The average value of the EDS analysis for both morphologies was identical to the chemical composition analysis.

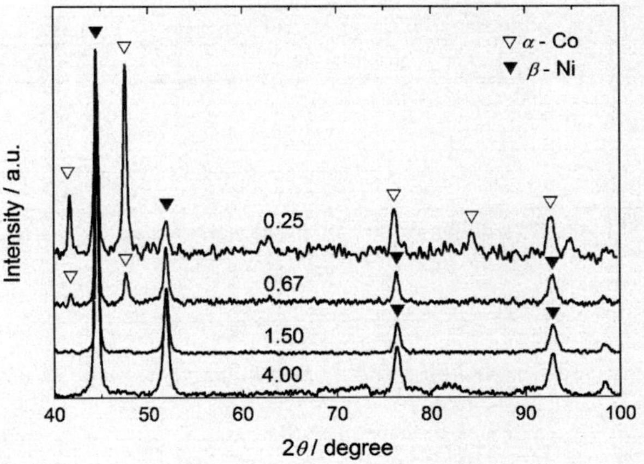

Fig. 5.8 Diffractograms of the Co–Ni powders electrodeposited from the solutions of different Ni^{2+}/Co^{2+} ions ratios (marked in the figure). The phase composition was identical for powders electrodeposited from sulfate and chloride supporting electrolytes (Reprinted from [38] with the permission of Springer-Verlag.)

5.2.4 Phase Compositions of the Co–Ni Alloy Powders

The phase composition of all alloy powder samples was determined by the X-ray technique. The X-ray diffraction patterns of all four samples electrodeposited from ammonium hydroxide containing solutions are shown in Fig. 5.8.

As can be seen, XRD patterns contain the characteristic peaks for Ni-based solid solution (Ni-rich phase, β-Ni phase) (▼) with face-centered cubic lattice (f.c.c., space group Fm3m) and h.c.p. α-Co phase (▽). It is seen in Fig. 5.8 that with the decrease of Ni^{2+}/Co^{2+} ions ratio peaks for β-Ni phase (▼) become smaller and some of them disappear, while the peaks for h.c.p. α-Co phase (▽) become more pronounced. For samples electrodeposited from the solutions containing higher concentration of Ni^{2+} ions than Co^{2+} ions ($Ni^{2+}/Co^{2+} = 4.00$ and 1.50) only the presence of β-Ni phase (▼) was detected, while at concentration ratios $Ni^{2+}/Co^{2+} < 1$ h.c.p. α-Co phase (▽) together with

Table 5.1 Crystallite size of phases detected in the Co–Ni powders

Ni^{2+}/Co^{2+}	β-Ni phase (nm)	α-Co phase (nm)
4.00	24	
1.50	28	
0.67	31	17
0.25	30	16

Reprinted from [38] with the permission of Springer-Verlag

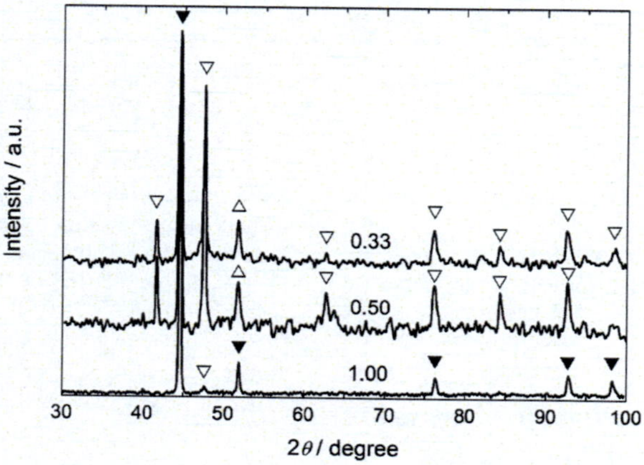

Fig. 5.9 Diffractograms of the Co–Ni powders electrodeposited from the solutions of different Ni^{2+}/Co^{2+} ions ratios (marked in the figure) from the borate–sulfate containing solutions (β-Ni phase—*inverted filled triangle*, α-Co phase—*inverted open triangle*, f.c.c. Co phase—*open triangle*)

β-Ni phase (▼) was detected in the alloy powder deposit. The dimensions of crystallites for both phases were determined from the diffractograms presented in Fig. 5.8 using Scherrer formula [131]. The results are presented in Table 5.1.

The X-ray diffraction patterns of three samples electrodeposited from borate–sulfate containing solutions are shown in Fig. 5.9.

The XRD patterns contain the characteristic peaks for Ni-based solid solution (Ni-rich phase, β-Ni phase) (▼) with face-centered cubic lattice (f.c.c., space group Fm3m) and h.c.p. α-Co phase (▽). It is seen in Fig. 5.9 that with the decrease of Ni^{2+}/Co^{2+} ions ratio peaks for β-Ni phase (▼) become smaller and some of them disappear, while the peaks for h.c.p. α-Co phase (▽) become more pronounced and f.c.c. Co phase (△) becomes visible (0.50 and 0.33). According to the literature [132], the f.c.c. Co phase could be expected at temperatures higher than 450°C, while in the case of alloy powder electrodeposition this structure has been formed at the room temperature, indicating that during the process of alloy powder electrodeposition the conditions for the formation of this phase are fulfilled and that in the case of alloy powder electrodeposition the data available in the literature for equilibrium conditions of corresponding alloy formation [132] are not fully reliable.

5.2.5 The Morphology of the Co–Ni Alloy Powders

The morphology of pure Co and pure Ni powders electrodeposited from ammonium sulfate or ammonium chloride supporting electrolytes, as well as the mechanism of powder agglomerates formation, has been discussed in Chap. 2. The mechanism of alloy powders formation is practically the same as that of pure metal powders. The morphology was found to depend mainly on the Ni^{2+}/Co^{2+} ions ratio (i.e., the composition of the alloy powders) and to some extent on the presence of sulfate or chloride anions, or borate.

5.2.5.1 The Morphology of the Co–Ni Alloy Powders Electrodeposited from the Sulfate Containing Supporting Electrolyte

The morphology of the Co–Ni alloy powders electrodeposited at the current density $j = 0.5\ \mathrm{A\ cm^{-2}}$ from the electrolytes containing different Ni^{2+}/Co^{2+} ions ratio is shown in Figs. 5.10–5.13. As can be seen, the morphology of Co–Ni alloy powder is sensitive to the Ni^{2+}/Co^{2+} ions ratio, i.e., to the composition of the alloy powder.

In the powder containing the highest amount of Ni, of about 64 wt.% Ni, electrodeposited at the $Ni^{2+}/Co^{2+} = 4.00$, more than 90% of agglomerates were typical spongy agglomerates varying in

Fig. 5.10 Typical powder agglomerate detected in the powder electrodeposited at the $Ni^{2+}/Co^{2+} = 4.00$ (Reprinted from [38] with the permission of Springer-Verlag.)

size from about 50 µm to about 200 µm, shown in Fig. 5.10. Very few cauliflower-like agglomerates of slightly different shape than those typical for pure Ni powder were also detected in the alloy powder. One such particle is marked with the arrow in the upper right corner of Fig. 5.10. It is important to note that this particle has a flat surface on one side, indicating that this type of growth belongs to the beginning of the powder formation with this particle falling off from the electrode surface before a bigger, typical spongy agglomerate is formed. Spongy agglomerates are more similar to the ones detected in pure Co powder, with the presence of large number of cavities, but contrary to the situation in the case of pure Co powder, here practically none of the cavities possess fern-like dendrites on their bottom, indicating that most of the agglomerates detached from the electrode surface before the hydrogen bubble in the cavity was liberated. It is interesting that practically no dendrites are found in this powder, although about 26 wt.% of Co is present in the powder.

As the Ni^{2+}/Co^{2+} ions ratio in the electrolyte decreases from 4.00 to 1.50, the composition of the Co–Ni alloy powder changes from about 78 at.% Ni—22 at.% Co to about 56 at.% Ni—44 at.% Co (Fig. 5.5) and this change influences the morphology of the powder. All powder agglomerates electrodeposited at $Ni^{2+}/Co^{2+} = 1.50$ are

Fig. 5.11 Typical powder agglomerates detected in the powder electrodeposited at the $Ni^{2+}/Co^{2+} = 1.50$ (Reprinted from [38] with the permission of Springer-Verlag.)

covered with the fern-like dendrites (Fig. 5.11b). Only a few agglomerates are bigger (about 200 μm) and the number of cavities observed on them is significantly reduced (Fig. 5.11a) in comparison with the one obtained at the ratio $Ni^{2+}/Co^{2+} = 4.00$. As can be seen in Fig. 5.11a, significant number of dendritic agglomerates is detected in this powder deposit (around the agglomerate). Very few spongy-like agglomerates of the size of about 500 μm are also present in the powder and they are also covered with the fern-like dendrites, indicating that in the presence of higher amount of Co in the powder dendritic growth prevails. With further decrease of the Ni^{2+}/Co^{2+} ions ratio (0.67) in the electrolyte, the composition of the Co–Ni alloy powder changes towards lower content of Ni (Fig. 5.5) and the morphology of powder agglomerates is slightly different. In this case the number of agglomerates increases in comparison with the ratio $Ni^{2+}/Co^{2+} = 1.50$, with the shape of agglomerates being different, as can be seen in Fig. 5.12a–c. The common characteristic of all agglomerates obtained at $Ni^{2+}/Co^{2+} = 0.67$ is that they are covered with the fern-like dendrites. At the same time, higher amount of dendritic particles (around the agglomerate shown in Fig. 5.12b) is formed during Co–Ni powder electrodeposition. The size of dendrite particles vary from about 5 μm to about 50 μm.

In the Co–Ni alloy powder deposit with the highest percentage of Co (85 at.%), Fig. 5.13, different types of agglomerates could be detected (1) spongy-like agglomerates of the size of about 500 μm

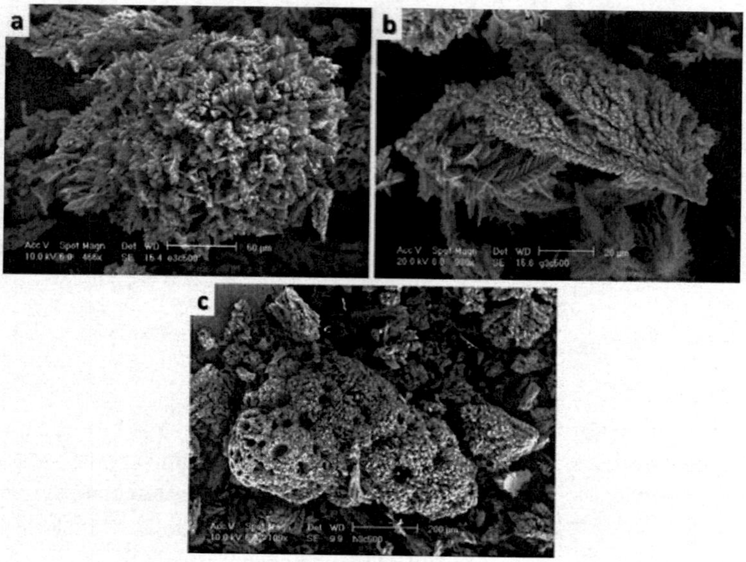

Fig. 5.12 Typical powder agglomerates detected in the powder electrodeposited at the $Ni^{2+}/Co^{2+} = 0.67$ (Reprinted from [38] with the permission of Springer-Verlag.)

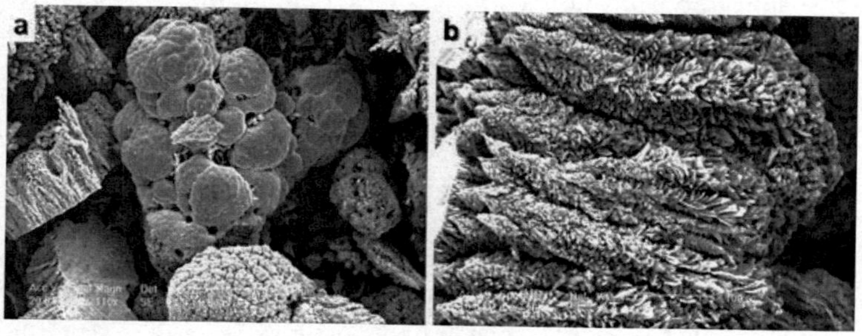

Fig. 5.13 Typical powder agglomerates detected in the powder electrodeposited at the $Ni^{2+}/Co^{2+} = 0.25$ (Reprinted from [38] with the permission of Springer-Verlag.)

covered with the fern-like dendrites, also found in the powder obtained at $Ni^{2+}/Co^{2+} = 0.67$ (Fig. 5.12c); (2) compact agglomerates, characteristic for pure Co powder deposit (Fig. 5.13a); and (3) agglomerates composed of closely packed fern-like dendrites about 200 μm long (Fig. 5.13b). By the EDS analysis of the compact agglomerate, shown in Fig. 5.13a, it is found that it does not represent pure Co particle, but the Co–Ni alloy of almost identical composition as the rest of the powder, indicating that the formation of alloy agglomerates follows a growth mechanism for Co powder formation.

Considering all detected types of agglomerates in electrodeposited Co–Ni alloy powders of different compositions, one can conclude that only one powder (with 78 at.% of Ni) contains very small amount of cauliflower-like agglomerates typical for pure Ni powder, but slightly different, while the rest of them contain some of the agglomerates detected in pure Co powder. The difference between the agglomerates detected in pure Co powder and in Co–Ni alloy powder (except for the agglomerates obtained from the electrolyte with the highest Ni^{2+}/Co^{2+} ions ratio, Fig. 5.11) is the presence of fern-like dendrites on the surface of these agglomerates. Such a behavior indicates that for these three alloy compositions the presence of Ni^{2+} and Co^{2+} ions promotes dendrite formation on the surface of all agglomerates detected in Co–Ni alloy powders. Hence, it appears that although some similarities between the type of agglomerates detected in pure Co and Co–Ni alloy powders exists, it is practically not possible to strictly correlate composition of alloy powders with their morphology.

To explain the difference in the morphology as a function of alloy composition (Ni^{2+}/Co^{2+} ions ratio), one should refer to the polarization curves presented in Fig. 5.2a. As can be seen in the figure, at the applied current density of -0.5 A cm^{-2} only one alloy, electrodeposited at the $Ni^{2+}/Co^{2+} = 1.50$, deposits at its diffusion limiting current density. The one electrodeposited at the $Ni^{2+}/Co^{2+} = 0.67$ deposits at a slightly lower current density than its diffusion limiting current density, while for the alloys electrodeposited at $Ni^{2+}/Co^{2+} = 4.00$ and 0.25 the applied current density lies in the region of mixed control of deposition, being much less than the diffusion limiting current density for those electrolyte compositions. Bearing in mind that the currents presented in this figure represent only 20% of the total current and that massive hydrogen evolution takes place together with the process of alloy electrodeposition, it is quite difficult to find good explanation for the phenomena observed. Different theories concerning powder formation during metal electrodeposition exist in the literature [133] with one of them claiming that powder formation takes place at the diffusion limiting current at potentials of hydrogen evolution, which is necessary process for powder formation. In the case of Co–Ni alloy powder electrodeposition hydrogen evolution commences already at the beginning of the electrodeposition process, being catalyzed by the electrodeposition. Hence, any comparison between these two processes is not realistic, but it is obvious that hydrogen evolution plays an important role in the process of powder electrodeposition.

5.2.5.2 The Morphology of the Co–Ni Alloy Powders Electrodeposited from the Chloride Containing Supporting Electrolyte

In Figs. 5.14–5.17 are shown SEMs of Co–Ni alloy powders obtained from a solution containing different Ni^{2+}/Co^{2+} ions ratios. At the Ni^{2+}/Co^{2+} ions ratio of 4.00, Fig. 5.14, two types of agglomerates are detected: cauliflower ones (a), varying in size between 60 and 100 μm, and bigger (varying in size between 200 and 400 μm), spongy ones (b), similar to the morphology of pure Ni powder. In the case of alloy, these two types of agglomerates are not equally distributed, but most of the agglomerates (over 80%) are spongy-like agglomerates.

With increasing the content of Co^{2+} ions in the solution, as well as in the powder (Ni^{2+}/Co^{2+} ions ratio of 1.50), very few cauliflower-like agglomerates can be detected in the powder deposit (Fig. 5.15a), being of the size of about 100 μm, while most of them are spongy-like ones of a bigger size of about 200 μm (Fig. 5.15b). It is characteristic for spongy agglomerates that two types of cavities could be detected on all powder agglomerates: narrow (cylindrical) cavities with fern-like dendrites formed at their bottom (Fig. 5.15c), indicating that before agglomerates detached from the electrode surface conditions of planar diffusion in the cavity were established causing formation of fern-like dendrites and wider, cone-shaped cavities without fern-like dendrites on their bottom (Fig. 5.15a, b). It is most likely that the

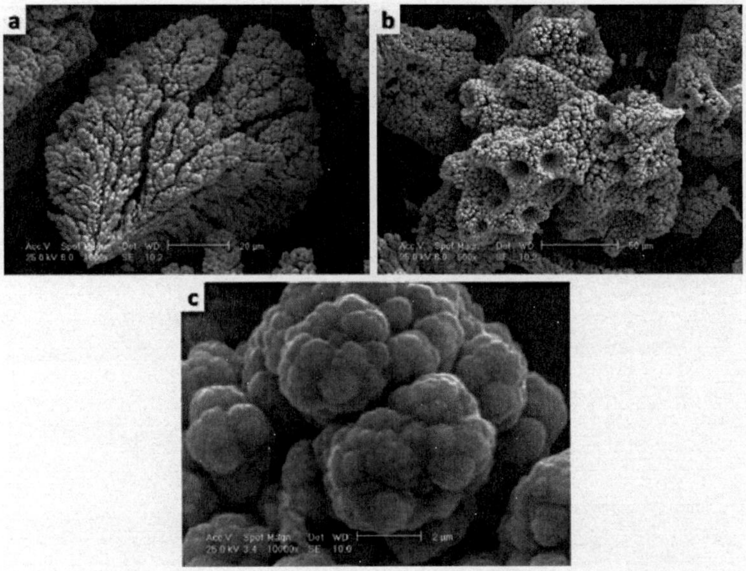

Fig. 5.14 Typical agglomerates detected in the powder electrodeposited at the $Ni^{2+}/Co^{2+} = 4.00$ (Reprinted from [39] with the permission of Elsevier.)

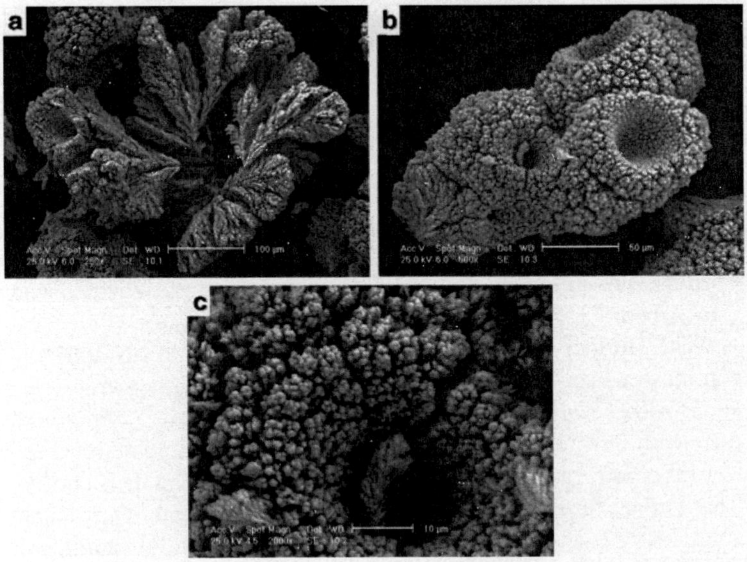

Fig. 5.15 Typical powder agglomerates detected in the powder electrodeposited at the $Ni^{2+}/Co^{2+} = 1.50$ (Reprinted from [39] with the permission of Elsevier.)

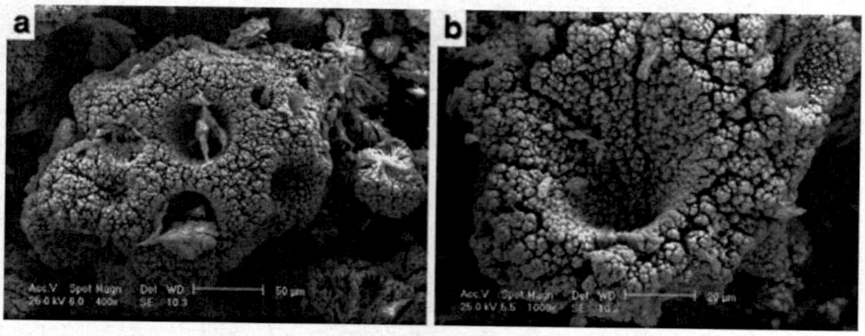

Fig. 5.16 Typical powder agglomerates detected in the powder electrodeposited at the $Ni^{2+}/Co^{2+} = 0.67$ (Reprinted from [39] with the permission of Elsevier.)

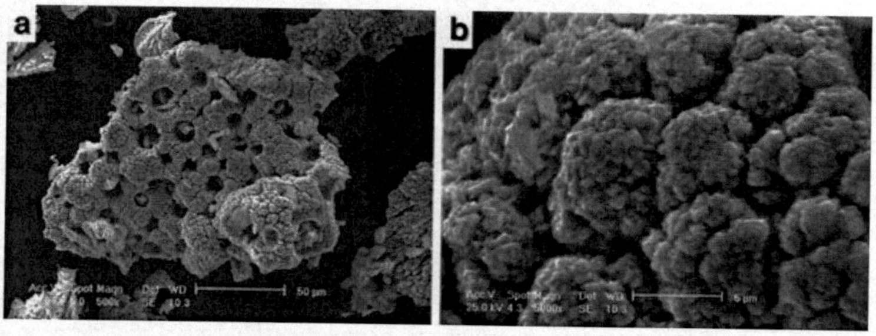

Fig. 5.17 Typical powder agglomerates detected in the powder electrodeposited at the $Ni^{2+}/Co^{2+} = 0.25$ (Reprinted from [39] with the permission of Elsevier.)

absence of fern-like dendrites at the bottom of cone-shaped cavities is the consequence of the fact that because of such a shape, with wide opening, no condition for planar diffusion in these cavities was established.

With further decrease of Ni^{2+}/Co^{2+} ions ratio to 0.67 no cauliflower-like agglomerates are detected. All agglomerates are spongy-like, again with two types of cavities: narrow (cylindrical) ones with fern-like dendrites on their bottom (Fig. 5.16a) and cone-shaped ones (Fig. 5.16b) without fern-like dendrites on their bottom. The top surface of all agglomerates obtained from these three solutions is practically identical, being characterized by cauliflower nature, as shown in Fig. 5.14c.

At the lowest Ni^{2+}/Co^{2+} ions ratio of 0.25 the same types of spongy agglomerates as ones obtained from the solution containing Ni^{2+}/Co^{2+} ions ratio of 0.67 are obtained, Fig. 5.17a. In this case two types of spongy agglomerates, concerning the top surface of the agglomerates, are detected: agglomerates with flat noodle-like endings, Fig. 5.14c, and agglomerates with the presence of well-defined crystals on the cauliflower endings, as shown in Fig. 5.17b. It is most likely that the detachment of agglomerates shown in Fig. 5.17b occurred later than the detachment of agglomerates shown in Fig. 5.14c, allowing the formation of a second layer of their growth to take place. Hence, in comparison with pure metal powders, agglomerates of alloy powders are characterized by the presence of two types of cavities: cylindrical ones with the fern-like dendrites on their bottom, indicating that hydrogen bubbles are liberated before the powder agglomerates detachment, and cone-shaped ones without fern-like dendrites on their bottom (because of the cavity shape planar diffusion cannot be established in the cavity), detected at smaller powder agglomerates. The appearance of such cavity is most probably due to different (higher) rate of hydrogen bubble formation in this solution. Comparing results obtained in chloride containing electrolyte with the results obtained in sulfate containing electrolyte, the following remarks could be made (1) polarization curves for hydrogen evolution in supporting electrolytes are different, with the one recorded in the chloride containing supporting electrolyte being characterized with higher current densities, indicating faster hydrogen evolution in the presence of smaller anions (see Fig. 2.24); (2) correct polarization curves for pure metals and alloy powders electrodeposition are also characterized with higher current densities, as well as higher current efficiencies for powder electrodeposition, indicating faster and easier powder formation in chloride containing electrolyte; (3) relatively similar morphologies of powder agglomerates are detected in both solutions, with the distribution of agglomerates being different in all cases. The characteristic feature of powder agglomerates obtained from chloride containing electrolytes is the presence of cone-shaped cavities.

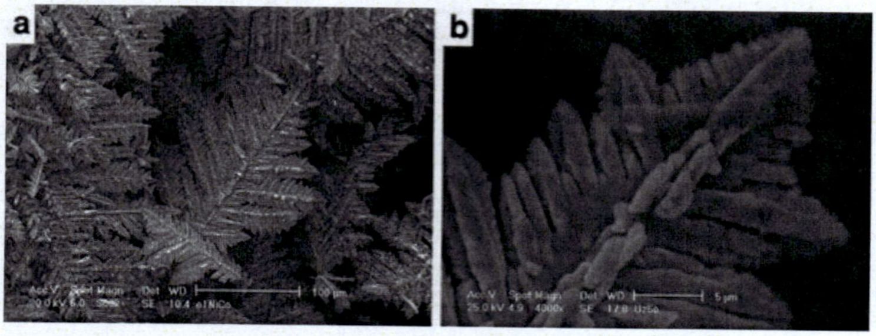

Fig. 5.18 Typical powder particles detected in the powder electrodeposited at the $Ni^{2+}/Co^{2+} = 1.00$

5.2.5.3 The Morphology of the Co–Ni Alloy Powders Electrodeposited from the Borate–Sulfate Containing Supporting Electrolyte [41]

At the highest Ni^{2+}/Co^{2+} ions ratio in the solution all powder agglomerates possess 2D fern-like shape, varying in the size from about 100 μm to about 200 μm, as can be seen in Fig. 5.18. They are all practically two dimensional (Fig. 5.18a), although at higher magnification (Fig. 5.18b) it can be seen that a third dimension of their growth also exists, but it seems that this is practically a two-dimensional layer-by-layer growth following original dendrite shape. It can also be seen that on the main branch of such dendrite small dendritic agglomerates, growing in the direction normal to the layer-by-layer growth, could be detected, Fig. 5.18b.

With the decrease of the Ni^{2+}/Co^{2+} ions ratio in the solution (0.50), i.e., the increase of the Co content in the alloy powder, among 2D fern-like agglomerates of a smaller size than in a previous case (varying in size between 50 and 100 μm—Fig. 5.19a), agglomerates of the size of about 100 μm composed of a densely packed dendritic agglomerates were detected, as can be seen in Fig. 5.19b. It is interesting to note that dendrites covering the surface of such agglomerates are not 2D fern-like type. As can be seen in Fig. 5.19c, they are branching in four directions, but they are more compact and branches are much smaller, as are the dendrites (about 10 μm long and about 2 μm wide).

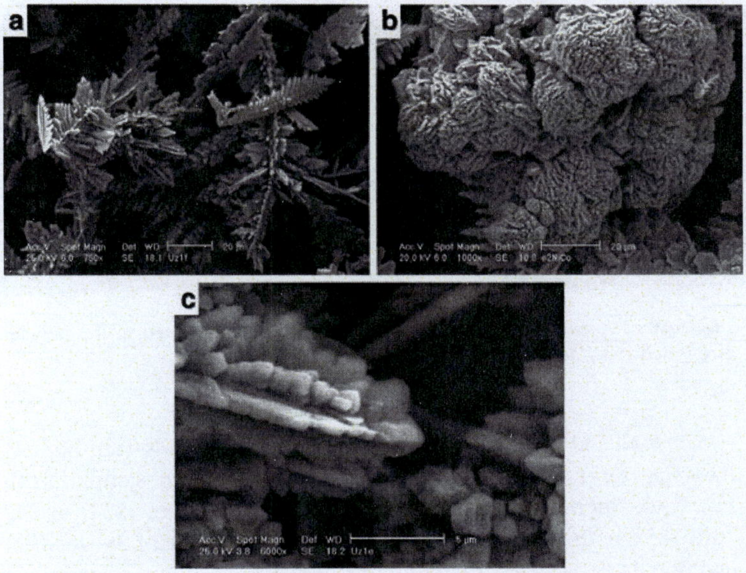

Fig. 5.19 Typical powder particles and agglomerates detected in the powder electrodeposited at the $Ni^{2+}/Co^{2+} = 0.50$

With further decrease of the Ni^{2+}/Co^{2+} ions ratio in the solution (0.33), compact agglomerates, typical for pure Co powder, could be clearly detected in the powder deposit, as can be seen in Fig. 5.20a, together with the 2D fern-like dendrites. By the EDS analysis of compact agglomerates it was confirmed that they do not represent pure Co particle, but the Co–Ni alloy, indicating that similar shapes of agglomerates characterize pure Co and Co–Ni alloy powder deposit. The presence of densely packed 3D dendritic agglomerates can also be seen in this powder, as shown in Fig. 5.20b. Again, these dendrites are more compact and smaller than those recorded on powder agglomerates electrodeposited at the Ni^{2+}/Co^{2+} ions ratio 0.50.

It is noteworthy that with the increase of the h.c.p. α-Co (100) phase in the powder deposit (Fig. 5.9) and the decrease of the f.c.c. β-Ni (111) phase, the shape of dendrite agglomerates changes from typical 2D fern-like dendrites to 3D dendrites. This is in accordance with the statement [134] that in the presence of the (111) orientation

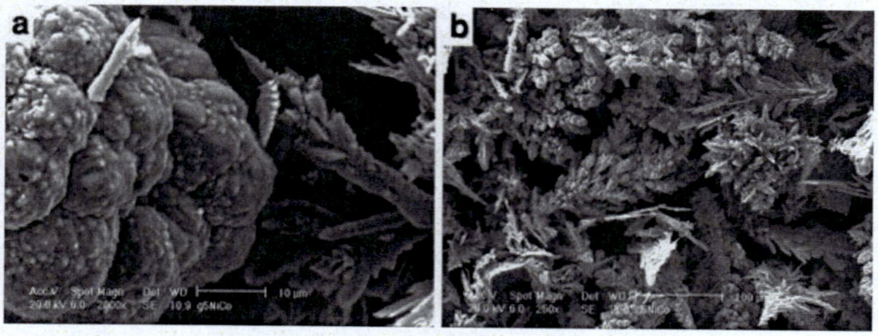

Fig. 5.20 Typical powder particles and agglomerates detected in the powder electrodeposited at the $Ni^{2+}/Co^{2+} = 0.33$

(f.c.c. β-Ni (111) phase) 2D dendrite growth prevails (the highest intensity for f.c.c. β-Ni (111) phase is detected in sample 1.00, Fig. 5.9). Such dendrites are denoted in the literature [134] as 2D {100}60°, with the angle of 60° between the main tree of the dendrite and the branches. To explain the influence of other orientations (h.c.p. α-Co (100) etc.) on the growth of dendrites in the investigated powder and appearance of 3D dendrites in the powder deposit, additional experiments and more detailed analysis are needed.

It should be emphasized here that the morphology of Co–Ni powders presented in the paper of El-Halim et al. [35] is completely different than the one presented in this chapter. This difference is most probably the consequence of the fact that in our work only powder agglomerates detached from the electrode surface were analyzed by the SEM, while in the paper of El-Halim et al. [35] disperse deposit still on the electrode surface was investigated.

5.3 System Fe–Ni

Electrodeposition of Fe–Ni powders was performed in ammonium chloride–sodium citrate containing electrolyte of the composition 1 M NH_4Cl + 0.2 M $Na_3C_6H_5O_7$. Iron(III) and iron(II) chloride and nickel chloride were used as a sources of Fe^{3+} or Fe^{2+} and Ni^{2+} ions. The total concentration of cations was kept at 0.1 M. Four different Ni/Fe concentration ratios were investigated: Ni/Fe = 9/1; Ni/Fe = 3/1; and Ni/Fe = 1/1; Ni/Fe = 1/3. In the case of iron(III) ions pH of the solution was adjusted to 4.5, while in the case of iron(II) ions pH of the solution was adjusted to 4.0. All solutions were made from analytical grade purity chemicals and distilled water by the following procedure: $Na_3C_6H_5O_7$ was first dissolved and the pH was adjusted to slightly higher value than desired by HCl; in the next step metal (Fe(III), or Fe (II) or Ni) salts were dissolved; finally, ammonium chloride was added and pH adjusted to the exact value. Concerning the stability of solutions it is well known that during investigation some Fe(II) become oxidized into Fe(III). This should be particularly pronounced during the deposition of powders for 1 or 2 h. Taking into account that in all cases Fe(II) or Fe(III) made very stable complexes with citrate anions, we did not experience problems in the case of polarization measurements (polarization curves were practically the same after 3–4 measurements), but for any case before each experiment fresh solution was made and used for investigation as well as for powder electrodeposition.

It should be mentioned here that an attempt was made to deposit Fe–Ni powders from the electrolytes of the same composition but of low pH 2. The powder agglomerates were successfully produced on the cathode, but immediately after detaching from the cathode surface they started dissolving with gas evolution in all investigated electrolytes. Only in the case of very short time of electrolysis and removal of remained powder (which had not yet been dissolved) it was possible to obtain small amount of powder for further analysis. That was the reason why all experiments were performed in the solutions of pH 4.0 or 4.5, since in this solution powders were stable after deposition.

From all solutions Fe–Ni alloy powders were electrodeposited at a constant current density corresponding to the slightly lower value (j_{pd}) than the inflection point B [marked with (●)] on the polarization curves (see Fig. 5.21).

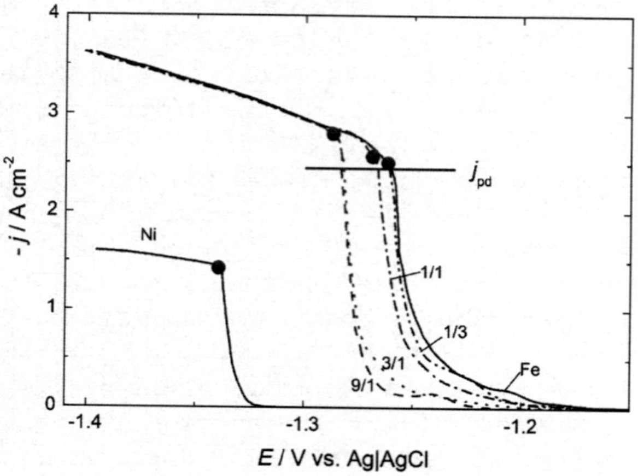

Fig. 5.21 Polarization curves for the electrodeposition of iron (Fe), nickel (Ni), and Fe–Ni alloy powders after IR drop correction recorded for different Ni/Fe ions ratios 9/1, 3/1, 1/1, and 1/3 (marked in the figure), in the solution of Fe(II) and Ni(II) species (Reprinted from [61] with the permission of Elsevier.)

5.3.1 Polarization Curves for Fe–Ni Alloy Powders Electrodeposition from Ammonium Chloride–Sodium Citrate Electrolyte

The polarization curves corrected for IR drop for the processes of Fe, Ni, and Fe–Ni alloy powders electrodeposition from ammonium chloride–sodium citrate containing supporting electrolyte in the presence of Fe(II) and Ni(II) species are shown in Fig. 5.21. In the case of Fe(II) salts polarization curve for iron electrodeposition (Fe) was placed at more positive potentials than that for nickel (Ni) as it is expected from the values of their reversible potentials. The polarization curves for Fe–Ni alloy powders electrodeposition are placed in between and all of them were placed at more positive potentials than expected from the Ni/Fe ratio, indicating anomalous codeposition.

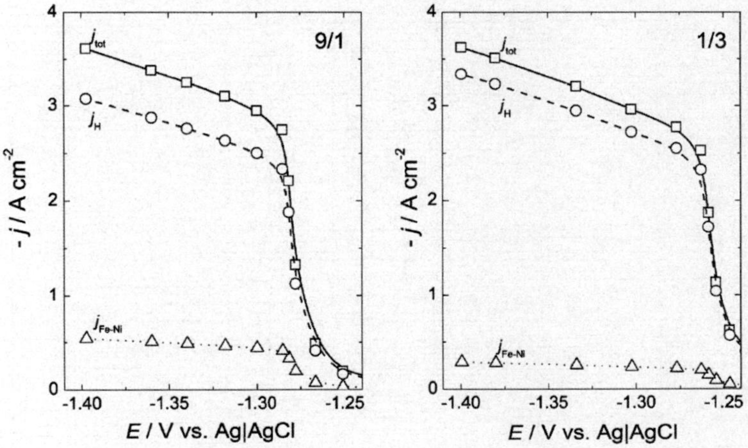

Fig. 5.22 Polarization curve for the Fe–Ni alloy powders electrodeposition (j_{tot}) (*open square*), polarization curve for hydrogen evolution (j_H) (*open circle*), and polarization curve for Fe–Ni powder electrodeposition after subtraction of hydrogen evolution current (j_{Fe-Ni}) (*open triangle*). The values of Ni/Fe ions ratios are marked in the figure (Reprinted from [61] with the permission of Elsevier.)

It was found that for the Fe(III) salts electrolytes the current efficiency was very low, 1–2% (the polarization curves for powder deposition (j_{tot}) and for hydrogen evolution (j_H) practically overlapped), and it was necessary to deposit powders at least for 2 h to obtain the amount of powder that could be used for the morphology and composition analysis (SEM, EDS). In the case of Fe(II) salt electrolytes, current efficiency at the potentials more negative than the second inflection point (● in Fig. 5.21) varied between 8% and 15% depending on the Ni/Fe ratio, as shown in Fig. 5.22. The average values for the diffusion limiting current densities for alloy powder electrodeposition were $j_{Fe-Ni} = -0.26$ A cm^{-2} for the ratio 1/3 and $j_{Fe-Ni} = -0.49$ A cm^{-2} for the ratio 9/1.

With the increase of iron concentration (as well as with the increase of the amount of iron in the powder—decrease of Ni/Fe ratio) the current efficiency for powder electrodeposition decreased, which is in accordance with the data obtained for compact Fe–Ni alloy deposits [135].

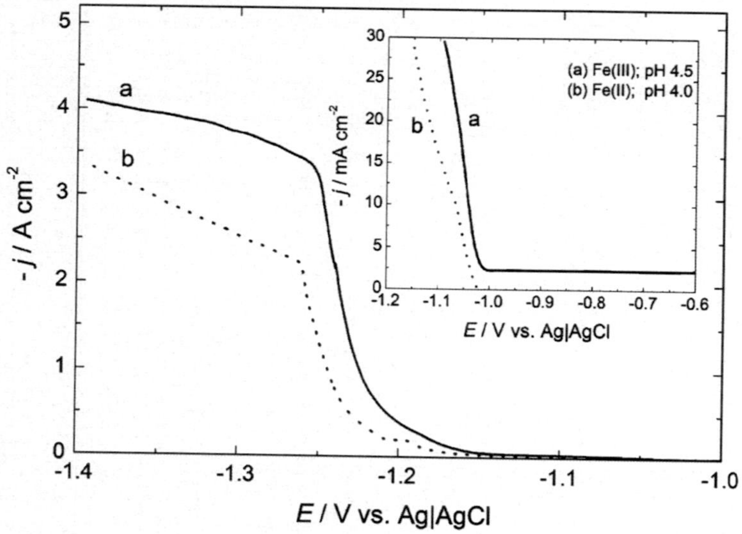

Fig. 5.23 The polarization curves for the electrodeposition of Fe powders from the electrolyte containing 0.1 M $FeCl_3$ + 0.2 M $Na_3C_6H_5O_7$ + 1 M NH_4Cl (a) and 0.1 M $FeCl_2$ + 0.2 M $Na_3C_6H_5O_7$ + 1 M NH_4Cl (b) (Reprinted from [61] with the permission of Elsevier.)

The polarization curves were recorded starting from the potential of −0.6 V vs. Ag|AgCl. By comparing the polarization curves for Fe powder deposition from Fe(III) (curve a) and Fe(II) (curve b) salts in ammonium chloride–sodium citrate electrolyte, shown in Fig. 5.23, certain cathodic current density (~−2.5 mA cm^{-2}) has been detected already at the starting potential for curve a (inset of Fig. 5.23). This cathodic current density remained constant down to the potential of about −1.0 V and it started rising at the same potential value at which the current density recorded in the presence of Fe(II) salt started to rise.

A low current efficiency for the Fe–Ni alloy powder deposition from the solution containing Fe(III) salt is the consequence of the first step in the overall reaction being reduction of Fe(III) species into Fe(II) species, which takes place at all potentials more negative than −0.2 V.

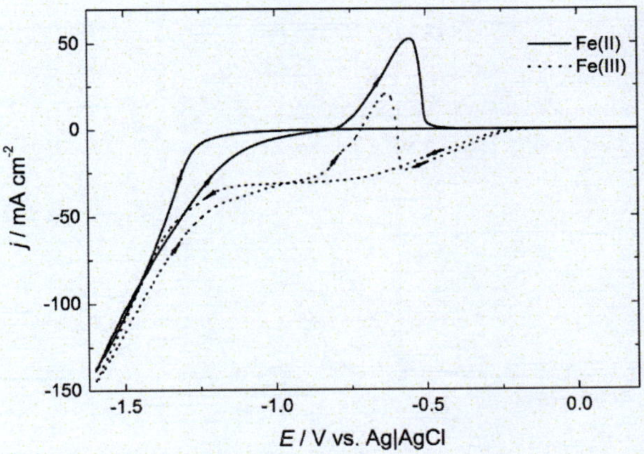

Fig. 5.24 The cyclic voltammograms recorded at the sweep rate of 10 mV s^{-1} and RPM = 1,000 onto Au electrode in the electrolytes containing 0.1 M Fe(III) (*dotted line*) and 0.1 M Fe(II) (*solid line*) salts (Reprinted from [61] with the permission of Elsevier.)

5.3.1.1 Cyclic Voltammograms of Fe Electrodeposition from Fe(III) and Fe(II) Salts

To investigate this phenomenon more closely, RDE electrode was used. The cyclic voltammograms at the sweep rate of 10 mV s^{-1} and RPM = 1,000 were recorded in both solutions in the potential limit from −1.6 to 0.6 V. The results are shown in Fig. 5.24 (from 0.2 to 0.6 V there was no indication of any reaction).

In the conditions of convective diffusion the deposition of Fe starts at about −1.3 V in both solutions, with the peaks of Fe dissolution being placed between −0.8 V and −0.5 V. In the presence of Fe(III) salt well-defined cathodic wave, typical for the diffusion controlled process, has been detected in the potential range between −0.2 V and −1.2 V. Hence, it is obvious that before the commencement of Fe deposition from Fe(III) salt, the reduction of Fe(III) into Fe(II) occurs. The reduction process of Fe(III) into Fe(II) has been investigated at different rotation rates. As can be seen in Fig. 5.25a, the current density plateau for this process increases with increasing rotation rate. By plotting the values of the current density plateaus recorded

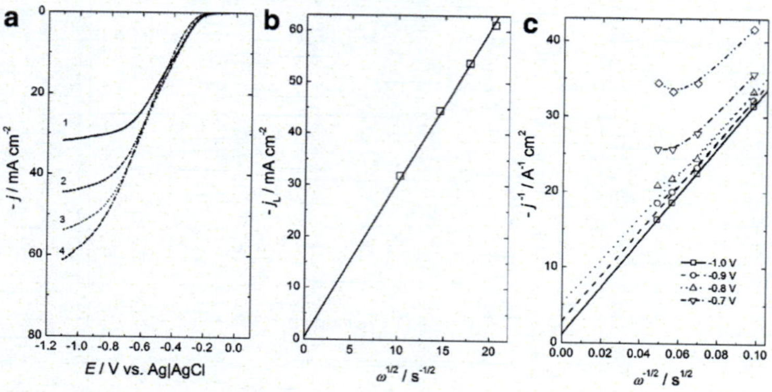

Fig. 5.25 (**a**) Cathodic cyclic voltammograms for Fe(III) reduction ($v = 10$ mV s^{-1}) recorded at different RPM's: 1—RPM = 1,000; 2—RPM = 2,000; 3—RPM = 3,000 and 4—RPM = 4,000. (**b**) j_L vs. $\omega^{1/2}$ recorded at the potential of −1.0 V. (**c**) j^{-1} vs. $\omega^{-1/2}$ recorded at different potentials (marked in the figure). Electrolyte: 0.1 M $FeCl_3$ + 0.2 M $Na_3C_6H_5O_7$ + 1 M NH_4Cl (Reprinted from [61] with the permission of Elsevier.)

at the potential of −1.0 V (j_L) as a function of the $\omega^{1/2}$, well-defined linear relation has been obtained, Levich's equation [136], as shown in Fig. 5.25b. The diffusion coefficient of the species undergoing reduction during this process, calculated from the slope of this dependence, amounts to ~3.7×10^{-5} cm^2 s^{-1}. In comparison with the diffusion coefficient for $Fe(CN)_6^{3-}$ ions (0.6×10^{-5} cm^2 s^{-1}) [136] higher value has been obtained, indicating that this process is not a simple diffusion-controlled one electron exchange from Fe(III) to Fe (II), since in both cases the reduction of complexes ($Fe(CN)_6^{3-}$ and $FeC_6H_5O_7$) is considered. Taking into account that the concentration of reacting Fe(III) species (actually $FeC_6H_5O_7$ complex) is high (0.1 M), a mixed activation–diffusion control could be expected. Considering plots presented in Fig. 5.25c (j^{-1} vs. $\omega^{-1/2}$ recorded at different potentials marked in the figure), it is obvious that the reduction of Fe(III) species into Fe(II) species is not under pure diffusion control. As can be seen the intercept on the j^{-1} axis exists at potentials of −1.0 V, −0.9 V, and −0.8 V, while at more positive potentials this dependence deviates from linearity, indicating the presence of activation-controlled current density [136]. It should be emphasized here

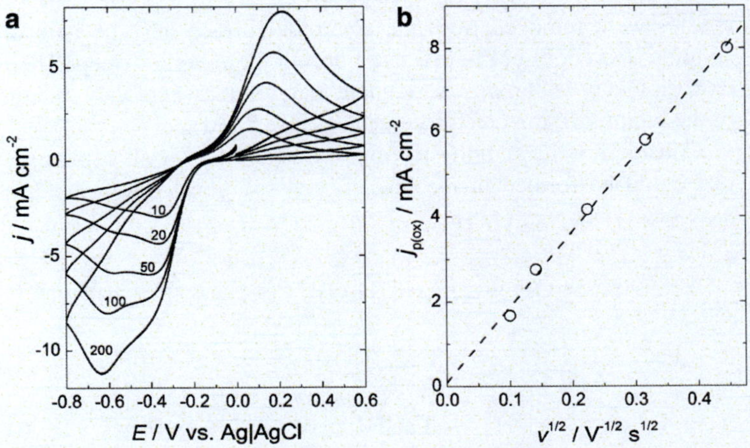

Fig. 5.26 (**a**) The cyclic voltammograms of Fe(III) species reduction and Fe(II) species oxidation recorded at different sweep rates (marked in the figure in mV s^{-1}) in the stagnant electrolyte (RPM = 0) onto Au electrode in the electrolyte containing 0.1 M $FeCl_3$ + 0.2 M $Na_3C_6H_5O_7$ + 1 M NH_4Cl. (**b**) $j_{p(ox)}$ vs. $\nu^{1/2}$ dependence obtained from the anodic current density peaks of oxidation shown in (**a**) (Reprinted from [61] with the permission of Elsevier.)

that in the case of convection (RPM > 0) no counter peak for the oxidation of these species could be detected at potentials more positive than 0.0 V. In the case of stagnant electrolyte (RPM = 0) the reduction and oxidation peaks of these species are clearly seen on the cyclic voltammograms, as shown in Fig. 5.26a.

The presence of two cathodic peaks and one anodic peak confirms the statement that this process is not a simple diffusion-controlled one electron exchange from Fe(III) to Fe(II), as well as not pure diffusion controlled process (activation control is also involved [136]). By plotting current density of oxidation peaks ($j_{p(ox)}$) versus $\nu^{1/2}$, well-defined linear dependence is also obtained (Fig. 5.26b), indicating diffusion-controlled oxidation [136], but the value of the diffusion coefficient calculated from the slope of this dependence (assuming that the concentration of reduced species is 0.1 M) was one order of magnitude lower than expected. Hence, it could be concluded that Fe(III) species reduce in the potential range between −0.2 V and −1.0 V and that under the convective diffusion reduced species were removed from the electrode surface into the bulk of solution and could not be oxidized during the reverse sweep. Their oxidation is possible only in stagnant electrolyte, where they remain in the vicinity of the electrode surface (Fig. 5.26a).

Considering the stability (formation) constants for all complexes that could be formed in the investigated solutions, following data have been obtained: [124]

$$Fe^{3+} + HC_6H_5O_7{}^{2-} \rightarrow [FeHC_6H_5O_7]^{+} \quad K_1 = 3.17 \times 10^{12} \tag{5.1}$$

$$Fe^{3+} + C_6H_5O_7{}^{3-} \rightarrow [FeC_6H_5O_7] \quad K_1 = 1.0 \times 10^{25} \tag{5.2}$$

$$Fe^{3+} + OH^{-} \rightarrow [FeOH]^{2+} \quad K_1 = 7.4 \times 10^{11} \tag{5.3}$$

$$Fe^{3+} + 2OH^{-} \rightarrow [Fe(OH)_2]^{+} \quad K_2 = 1.5 \times 10^{21} \tag{5.4}$$

$$Fe^{3+} + 3OH^{-} \rightarrow [Fe(OH)_3] \quad K_3 = 4.7 \times 10^{29} \tag{5.5}$$

$$Fe^{2+} + HC_6H_5O_7{}^{2-} \rightarrow [FeHC_6H_5O_7] \quad K_1 = 1.2 \times 10^{3} \tag{5.6}$$

$$Fe^{2+} + C_6H_5O_7{}^{3-} \rightarrow [FeC_6H_5O_7]^{-} \quad K_1 = 3.16 \times 10^{15} \tag{5.7}$$

$$Fe^{2+} + OH^{-} \rightarrow [FeOH]^{+} \quad K_1 = 3.6 \times 10^{5} \tag{5.8}$$

$$Fe^{2+} + 2OH^{-} \rightarrow [Fe(OH)_2] \quad K_2 = 5.9 \times 10^{9} \tag{5.9}$$

$$Fe^{2+} + 3OH^{-} \rightarrow [Fe(OH)_3]^{-} \quad K_3 = 4.7 \times 10^{9} \tag{5.10}$$

$$Fe^{2+} + 4OH^{-} \rightarrow [Fe(OH)_4]^{2-} \quad K_4 = 3.8 \times 10^{8} \tag{5.11}$$

As can be seen, 11 different species could be formed, five with Fe^{3+} ions and six with Fe^{2+} ions. Taking into account their stability constants and concentration of complexing agents, it appears that for Fe(III) salt the most stable ones are $[FeC_6H_5O_7]$ and $[Fe(OH)_3]$, while for Fe(II) salt the most stable ones are $[FeC_6H_5O_7]^{-}$ and $[Fe(OH)_2]$. The possibility of $Fe(OH)_3$ formation could be excluded since at pH 4.5 concentration of OH^- ions is 3.16×10^{-10} M and the dominant species in the Fe(III) salt electrolyte is complex $[FeC_6H_5O_7]$.

Let us reconsider the process of Fe powder deposition from the electrolyte containing Fe(III) salt. The reduction of $[FeC_6H_5O_7]$ starts at about −0.2 V producing Fe(II) species. Taking into account the presence of two cathodic peaks during this reduction process, it could not be specified which species were formed, pure Fe^{2+} ions or complex $[FeC_6H_5O_7]^-$ (the most stable complex with the Fe^{2+} ions), or both. It is most likely that both species were formed and that Fe deposition occurs by further reduction of these species with simultaneous hydrogen evolution which should be pronounced at potentials more negative than −1.3 V. Since during the Fe deposition pH in the vicinity of the electrode surface is much higher due to massive hydrogen evolution, it is possible that Fe^{2+} ions react with OH^- ions forming $[Fe(OH)_2]$ which precipitates and prevents deposition of pure Fe powder. On the other side, in the solution containing Fe(II) salt the possibility of $[Fe(OH)_2]$ formation could also be excluded since at pH 4.0 concentration of OH^- ions is 1.0×10^{-10} M and the dominant species is complex $[FeC_6H_5O_7]^-$. Accordingly, the reduction of this complex starts at about −1.3 V producing Fe deposit directly, without the possibility of eventual $[Fe(OH)_2]$ formation, and in this solution Fe powder has been successfully deposited.

In the case of the solutions for Fe–Ni alloy powder electrodeposition, among the species presented above for Fe, the following species could be formed with Ni in the investigated solutions: [124]

$$Ni^{2+} + HC_6H_5O_7{}^{2-} \rightarrow [NiHC_6H_5O_7] \quad K_1 = 1.3 \times 10^5 \tag{5.12}$$

$$Ni^{2+} + C_6H_5O_7{}^{3-} \rightarrow [NiC_6H_5O_7]^- \quad K_1 = 2.0 \times 10^{14} \tag{5.13}$$

$$Ni^{2+} + OH^- \rightarrow [NiOH]^+ \quad K_1 = 9.3 \times 10^{14} \tag{5.14}$$

$$Ni^{2+} + 2OH^- \rightarrow [Ni(OH)_2] \quad K_2 = 3.5 \times 10^8 \tag{5.15}$$

$$Ni^{2+} + 3OH^- \rightarrow [Ni(OH)_3]^- \quad K_3 = 2.0 \times 10^{11} \tag{5.16}$$

As can be seen, again the most stable one is $[NiC_6H_5O_7]^-$ and the same conclusion concerning the formation of Ni-hydroxide species in the investigated solutions as the one given above for Fe could be applied. Hence, Ni electrodeposition occurs by the reduction of citrate complex $[NiC_6H_5O_7]^-$ without the possibility for Ni-hydroxide species formation.

Among the above-mentioned complexes of Fe and Ni, ammonium complexes could be formed with Fe(II) and Ni ions in ammonium chloride solution, chloride complex could be formed with Fe(II) and Fe(III) and sulfate complexes could be formed with Fe(III) and Ni, but they all have several orders lower stability constants in comparison with citrates and cannot influence the reaction mechanism.

5.3.2 *Chemical Compositions of the Fe–Ni Alloy Powders*

All powder samples for chemical analysis, phase composition analysis, and morphology investigations were electrodeposited under the same conditions as for Co–Ni alloy powders, except that the current density for powder electrodeposition was the value presented in Fig. 5.21 with j_{pd}.

Chemical composition of all electrodeposited Fe–Ni alloy powders was determined by the AAS technique after dissolving certain amount of powders in HCl. Also, chemical composition of these alloy powders was determined from the average values of EDS analysis. All powder samples were analyzed by EDS in such a way that at least 8 powder agglomerates (maximum number 20) on the SEM micrographs were chosen and EDS analysis was performed at 3–10 different positions on each particle. In some cases the analysis was performed at a point of 1 μm^2, while in some cases rectangle surface from 120 μm^2 (10 × 12 μm) to 9,600 μm^2 (80 × 120 μm) has been analyzed. In some cases the composition of the powder agglomerates was uniform all over the agglomerates surface, while in some cases at. % of oxygen was found to vary significantly on the surface of one particle, or on specific part of the analyzed particle. This behavior was

Table 5.2 The average composition of the Fe–Ni powders electrodeposited from Fe(III) and Fe(II) salts containing electrolytes, obtained by the EDS analysis (in at.%)

	Fe(III) solution			Fe(II) solution			Ni solution	
Ni/Fe	O	Fe	Ni	O	Fe	Ni	O	Ni
Ni							20	80
9/1	22	34	44	5	21	74		
3/1	34	41	25	8	35	57		
1/1	33	55	12	12	58	30		
1/3	12	83	5	30	62	8		
Fe				20	80			

Reprinted from [61] with the permission of Elsevier

independent of the shape of the investigated particle (flat surface, cone-shaped cavities, different crystals on the particle surface), as well as of the composition of the powders. Taking into account that the EDS analysis strongly depends on the position and angle of the beam, for powder agglomerates like these presented in this work the EDS analysis results should not be considered as quantitative ones. Hence, in Table 5.2 are presented average values (obtained by the analysis of significant number of agglomerates for each powder) for the EDS analysis of all powder samples. From the presented results it is most likely that the presence of oxygen could be the consequence of washing and drying procedure. It could also be possible that the oxygen is a result of eventual [$Fe(OH)_2$] formation during the powder deposition, as explained in Sect. 5.3.1.1 for pure Fe powder deposition. In such a case it would be reasonable to expect higher percentage of oxygen in the powders containing higher amount of Fe for Fe–Ni alloy powders deposited from the electrolyte containing Fe(III) salt, while for powders deposited from the electrolyte containing Fe(II) salt percentage of oxygen should be relatively constant for all investigated compositions.

As can be seen in Table 5.2, this was not the case. Hence, it appeared that during the deposition of Fe–Ni alloy powders the behavior of pure metals cannot be considered as relevant for the properties of alloy powders, since they form solid solution-type alloys which behave in a different way than pure metals. It should be stated that the results of the chemical and EDS analysis (after subtraction of oxygen percentages) were in good agreement.

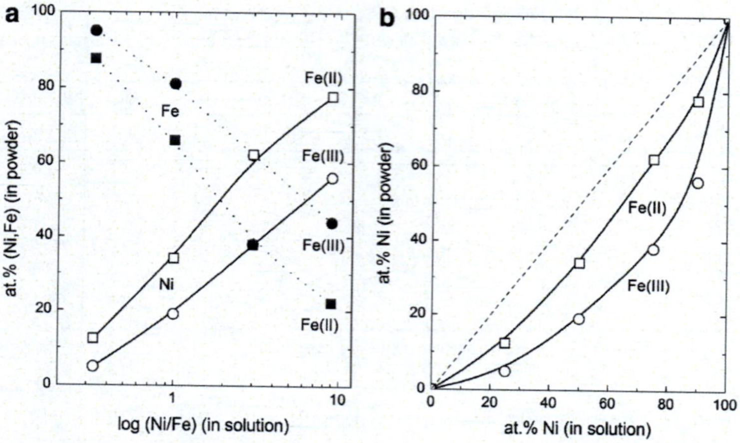

Fig. 5.27 (**a**) Dependences of at.% of Fe (*filled circle and square*) and Ni (*open circle and square*) in the powders as a function of the Ni/Fe ratio: electrolyte containing Fe(III) (*filled and open circle*) and Fe(II) (*filled and open square*) salts. (**b**) Typical Brenner's presentation of anomalous codeposition for the powders electrodeposited from the electrolytes containing Fe(III) (*open circle*) and Fe(II) (*open square*) salts [116] (Reprinted from [61] with the permission of Elsevier.)

The composition analysis was performed for powders electrodeposited from the solutions containing Fe(III) and Fe(II) salts.

The dependence of at. % of both metals as a function of the log (Ni/Fe) ratio is presented in Fig. 5.27a. As it could be expected the content of Ni increases, while the content of Fe decreases with the increase of the Ni/Fe ratio. According to Fig. 5.27a almost linear dependence of the percentage of both metals in the powder as a function of the logarithm of their concentration ratios indicates that their content in the powder depends exponentially on the Ni/Fe ratio. Considering Fig. 5.27b it could be concluded that the anomalous character is more pronounced in the electrolytes containing Fe(III) salt. This is in accordance with the statement that the formation of some amount of [$Fe(OH)_2$] is possible in this electrolyte and assuming "hydroxide suppression" mechanism [125, 126] the anomalous character of Fe and Ni deposition should be more pronounced in this electrolyte.

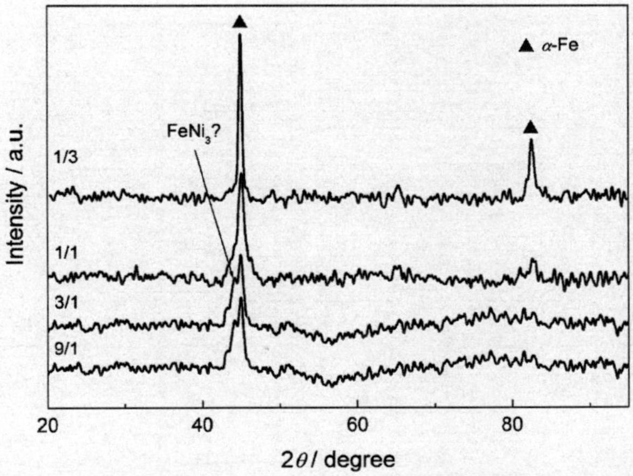

Fig. 5.28 Diffractograms of Fe–Ni alloy powders electrodeposited at different Ni/Fe ions ratios (marked in the figure) (Reprinted from [62] with the permission of Springer-Verlag.)

5.3.3 Phase Compositions of the Fe–Ni Alloy Powders

The diffractograms of Fe–Ni alloy powders electrodeposited at different Ni/Fe ions ratios from the solution of Fe(II) species are presented in Fig. 5.28. It should be stated that the dimensions of crystallites were 7 nm for the Ni/Fe ratio 9/1 and that they increased with the increase of iron content in the powder to 20 nm (for Ni/Fe = 1/3). Because of very small dimensions of crystallites, only phases with the highest intensity were determined with high certainty and these were peaks of the α-Fe phase (▲). Taking into account that the peaks of FeNi (37-0474), α-Fe (06-0696), and Ni (45-1027) practically overlap, a wide peak at about 44.5° for Ni/Fe = 9/1, 3/1, and 1/1 could be ascribed to any of these phases. The appearance of a small peak at about 44.3° for Ni/Fe = 9/1 and 3/1 indicates the presence of $FeNi_3$ {111} (38-0419), since pagoda-like crystals, typical for this phase, have been detected on the surface of powder agglomerates obtained from these solutions (see Sect. 5.3.4).

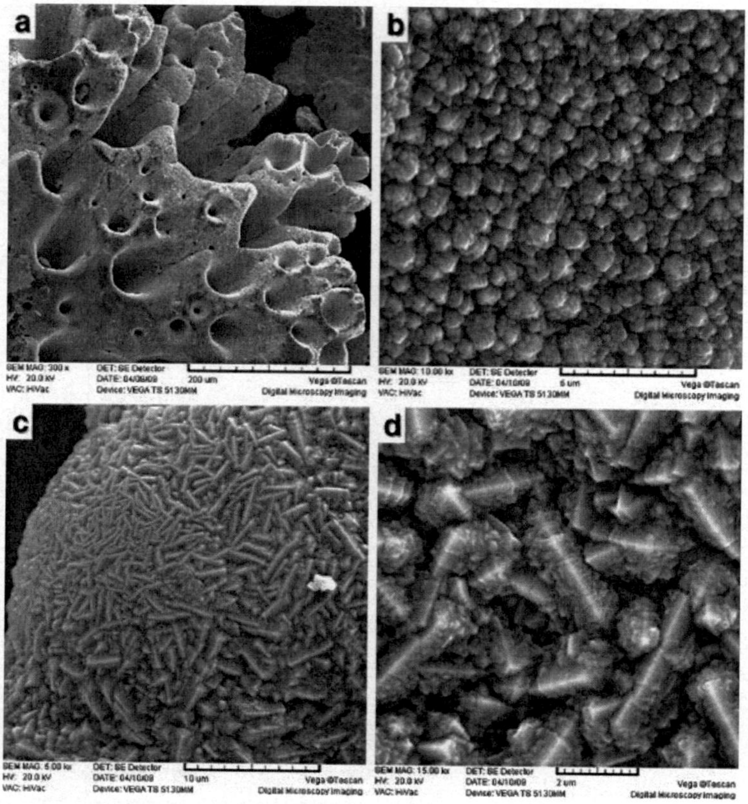

Fig. 5.29 Morphology of the Fe–Ni alloy powder electrodeposited at the Ni/Fe = 9/1

5.3.4 *Morphology of Electrodeposited Fe–Ni Powders*

5.3.4.1 The Fe–Ni Alloy Powders Obtained from the Solutions Containing Fe(II) Species

The morphology of the Fe–Ni alloy powders electrodeposited at different Ni/Fe ratios are presented in Figs. 5.29–5.32. A common characteristic of all investigated Fe–Ni powders is the formation of

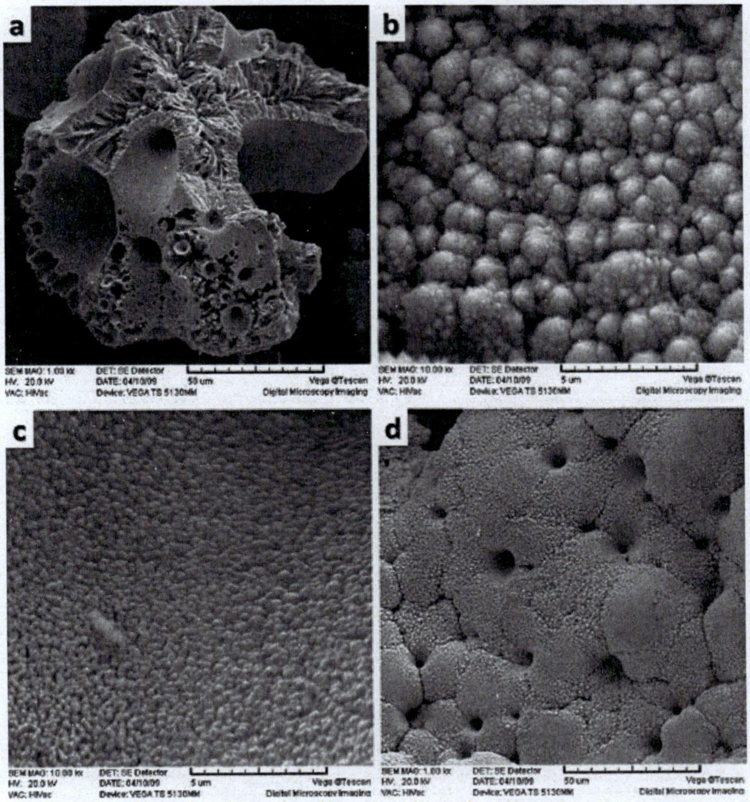

Fig. 5.30 Morphology of the Fe–Ni alloy powder electrodeposited at the Ni/Fe = 3/1

big agglomerates (up to 500 μm). Their surfaces are covered with large cone-shaped cavities, as well as with much smaller cylindrical cavities.

In Fig. 5.29 are presented typical agglomerates for Fe–Ni alloy powders electrodeposited at the Ni/Fe ratio 9/1. Large cone-shaped cavities and much smaller cylindrical cavities, corresponding to the places were hydrogen bubbles were formed, are clearly seen in Fig. 5.29a. Nodular surfaces are covered with two types of crystals: pagoda-like crystals (Fig. 5.29b) and crystals of the shape of

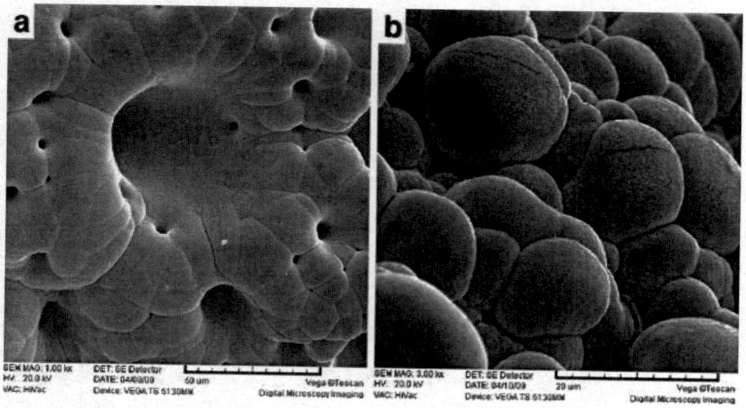

Fig. 5.31 Morphology of the Fe–Ni alloy powder electrodeposited at the Ni/Fe = 1/1 (Reprinted from [62] with the permission of Springer-Verlag.)

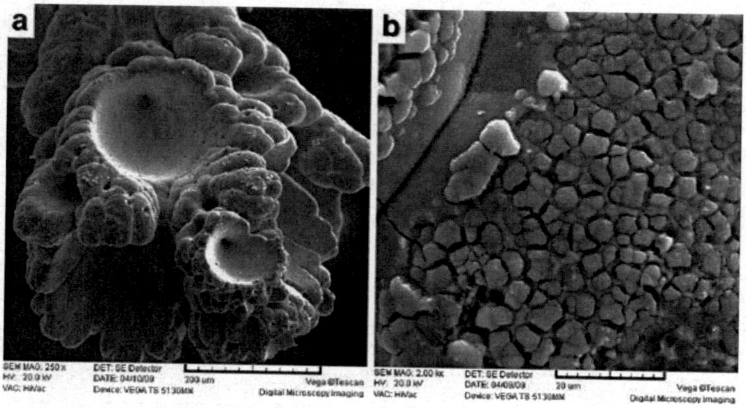

Fig. 5.32 Morphology of the Fe–Ni alloy powder electrodeposited at the Ni/Fe = 1/3 (Reprinted from [62] with the permission of Springer-Verlag.)

elongated prism (Fig. 5.29c, d), indicating possible presence of different phases.

Large cone-shaped cavities and much smaller cylindrical cavities are also present in the powder electrodeposited at Ni/Fe = 3/1

(Fig. 5.30a). Nodular surfaces (Fig. 5.30d) as well as the inside surface of cone-shaped cavities (Fig. 5.30c) are covered with spherical crystals (Fig. 5.30b).

Much smoother surface of both types of cavities and the nodular endings were obtained from the solution with Ni/Fe = 1/1 (Fig. 5.31).

In the case of the highest content of iron in the powder (Ni/Fe = 1/3) similar shape of cavities (Fig. 5.32a) was obtained as in Fig. 5.31a, while the surface of nodular endings was not smooth.

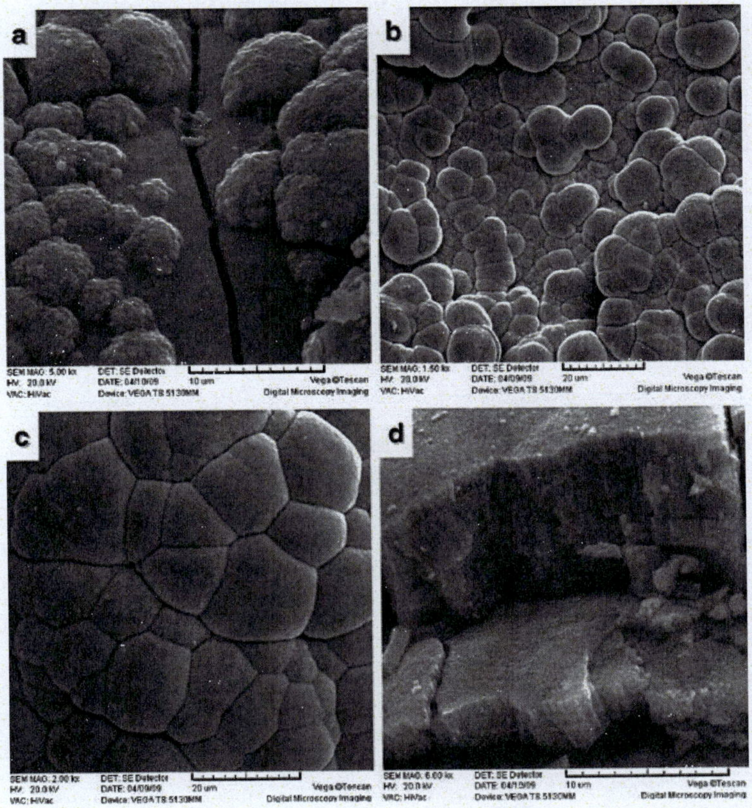

Fig. 5.33 Typical phases of the Fe–Ni powder growth

On some parts of the agglomerates surface new crystals were separated with the cracks, as shown in Fig. 5.32b.

Considering Figs. 5.29–5.32 it could be seen that on the compact surface of all agglomerates smaller spherical (rounded) grains were formed (Figs. 5.30b–d and 5.31b). The formation of such spherical grains is typical for the electrodeposition of copper from acidic sulfate electrolytes under the conditions of activation–diffusion controlled process [137]. Such mechanism has been explained for the electrodeposition of pure cobalt. The development of spherical grains in the case of Fe–Ni electrodeposition is presented in Fig. 5.33. As can be seen in Fig. 5.33a, the second generation of spherical grains grow in all directions. On the compact surface spherical grains are well defined (Fig. 5.33b) and since they are close to each other they start to overlap. During the further growth the overlapping of the diffusion zones formed over each spherical grain becomes more pronounced producing practically flat surface of the agglomerates (Fig. 5.33c). It is interesting to note that the growth of compact agglomerates of spherical grains occurs through the formation of layer-by-layer stick-like densely packed crystals of submicron dimensions, as shown in Fig. 5.33d. Presented figures are in excellent agreement with already given explanation for the electrodeposition of powders.

5.3.4.2 The Fe–Ni Alloy Powders Obtained from the Solutions Containing Fe(III) Species

The main characteristic of these powders is the presence of high number of both types of cavities on the surfaces of big agglomerates (up to 500 μm), as shown in Fig. 5.34. Taking into account that the current efficiency for the electrodeposition of these powders is only 1–2%, massive hydrogen evolution must produce larger number of cavities than in previous cases. Also, one of the characteristics of these powders is the presence of several different types of crystals on the surface of agglomerates around the cavities, which are presented in Figs. 5.35–5.38. On the agglomerates surfaces electrodeposited from the solutions with Ni/Fe = 9/1 and 3/1 mostly spherical agglomerates could be detected, as shown in Fig. 5.35a. At the

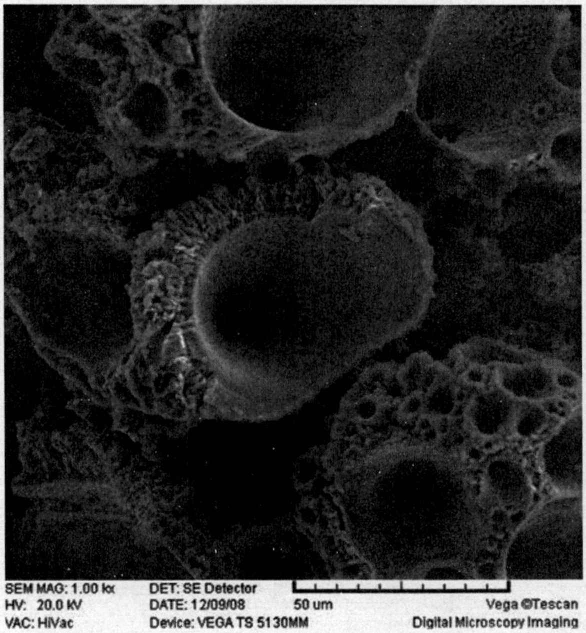

Fig. 5.34 Typical distribution of cavities on the surface of the Fe–Ni alloy powder agglomerates electrodeposited from the solution containing Fe(III) species

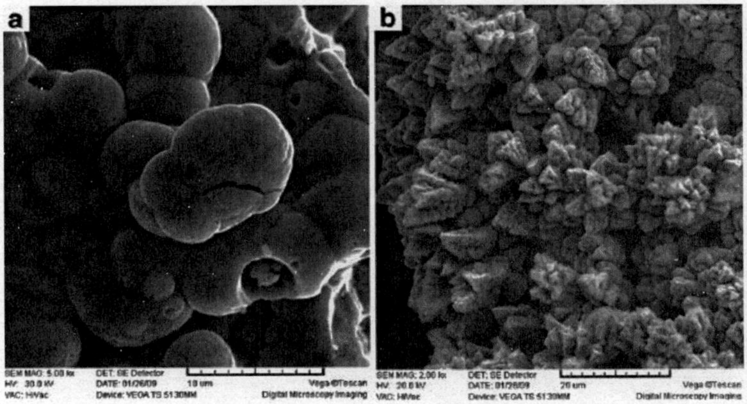

Fig. 5.35 Crystals formed on the surface of the Fe–Ni alloy powder agglomerates electrodeposited from the solution with Ni/Fe = 9/1 (Reprinted from [61] with the permission of Elsevier.)

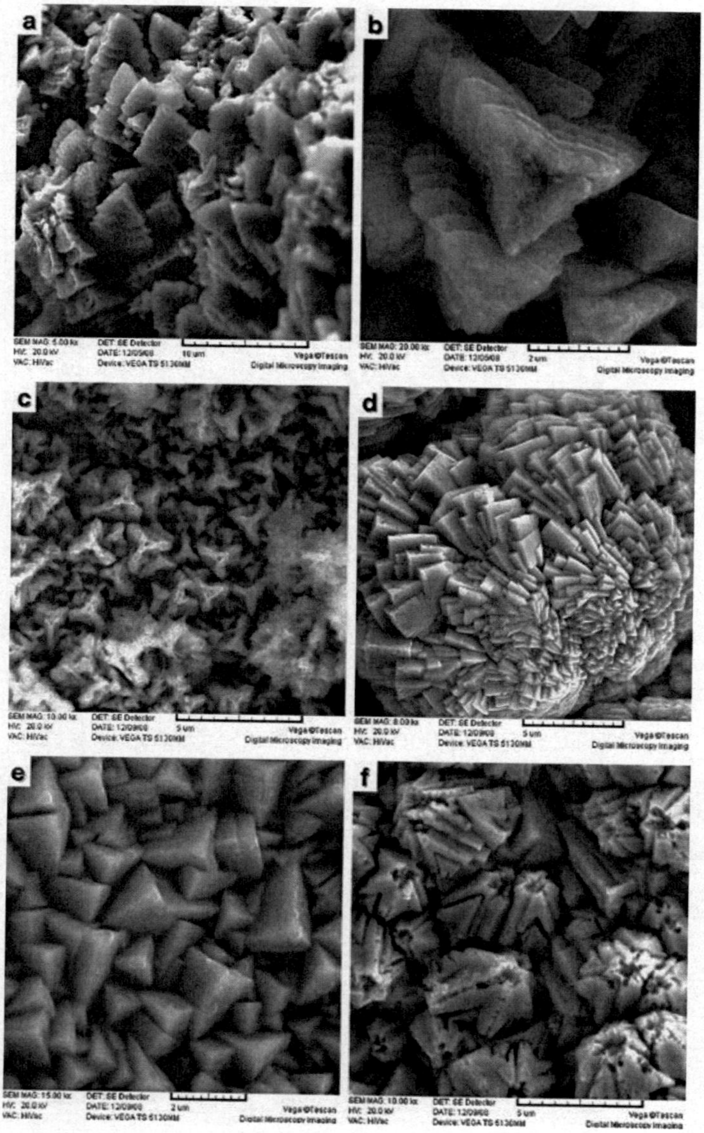

Fig. 5.36 Crystals formed on the surface of the Fe–Ni alloy powder agglomerates electrodeposited from the solution with Ni/Fe < 1

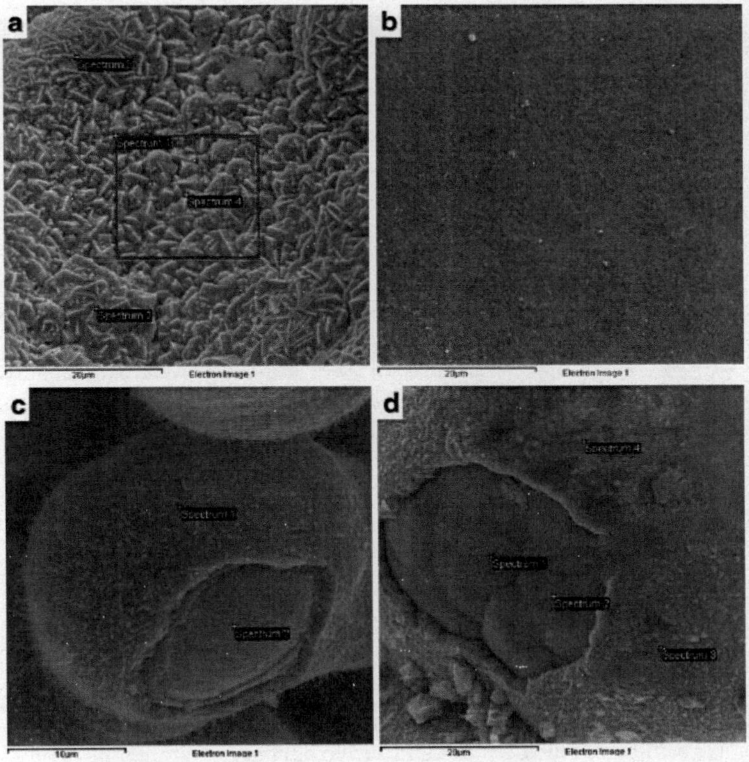

Fig. 5.37 Typical surface morphologies for all electrodeposited Fe–Ni alloy powders after annealing in air at 400°C for 3 h: (**a**) Ni/Fe = 9/1; (**b**) Ni/Fe = 3/1; (**c**) Ni/Fe = 1/1; and (**d**) Ni/Fe = 1/3 (Reprinted from [62] with the permission of Springer-Verlag.)

same time, pagoda-like crystals (b) have been detected on a certain, not negligible, number of agglomerates. Particularly interesting are crystals of the shape of pagoda. Such crystals were detected in the Fe–Ni powder synthesized in high yield by a simple and facile hydrothermal method without the presence of surfactants [137]. According to this investigation [137], $FeNi_3$ {111} single crystals were formed during the described procedure. The products obtained at 120°C were a mixture of $FeNi_3$ and Fe–Ni hydroxides composed of

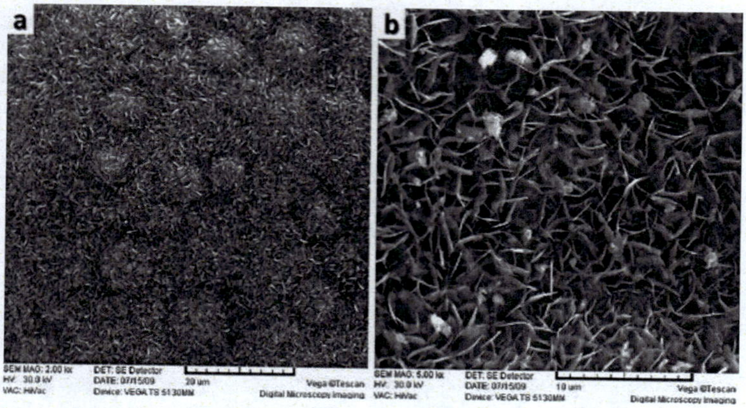

Fig. 5.38 (**a**) Typical surface morphologies for the Fe–Ni alloy powders electrodeposited at Ni/Fe ratios 9/1 and 3/1 after annealing in air at 600°C for 3 h. (**b**) Typical surface morphologies for the Fe–Ni alloy powders electrodeposited at Ni/Fe ratios 1/1 and 1/3 after annealing in air at 600°C for 3 h (Reprinted from [62] with the permission of Springer-Verlag.)

monodispersed microspheres (probably spheres presented in Fig. 5.35a). With the increase of temperature to 140°C these microspheres became micropagodas very similar to those shown in Fig. 5.35b. At higher temperatures (180°C) these agglomerates transform into perfect 3D $FeNi_3$ dendritic superstructures in certain directions [137]. Hence, comparing the crystals presented in Fig. 5.35b with those obtained by hydrothermal method [137] it seems reasonable to ascribe them to the $FeNi_3$ {111} single crystals (there is also an indication for the existence of $FeNi_3$ phase in electrodeposited powder—see XRD analysis presented in Fig. 5.28).

As the amount of iron increases in the powders (above 50 at.%), the number of different crystals detected on the agglomerates surfaces also increases. In the powders electrodeposited from the solutions with Ni/Fe = 1/3 very small amount of agglomerates is covered with spherical agglomerates, while most of them are characterized with the presence of several different shapes of crystals. These crystals are presented in Fig. 5.36a–f. It is quite difficult to explain the reasons for their appearance, since so many different phases do not exist in the Fe–Ni system and the only reasonable explanations could be that these crystals represent some superstructures formed under massive hydrogen evolution. Their shapes could be defined as follows: (a) Christmas tree-like crystals; (b) triangle-like crystals growing layer by layer; (c) propeller-like crystals; (d) plate-like crystals; (e) tetrahedral crystals, and (f) traversed polyhedron.

5.3.5 *Formation of $NiFe_2O_4$ by the Oxidation of Fe–Ni Alloy Powders*

To produce $NiFe_2O_4$, all electrodeposited powders were annealed in air, at first at the temperature of 400°C for 3 h. After SEM, EDS, and XRD analysis, powders were additionally annealed at 600°C for 3 h and analyzed by SEM, EDS, and XRD. Finally, since it was found that big agglomerates had not been completely oxidized during annealing at 600°C, powders were ground in mortar, annealed at 700°C for 3 h, and analyzed by XRD.

Typical morphologies of the surface of annealed (recrystallized) powders agglomerates (EDS) for all investigated electrolyte compositions after the first annealing procedure are shown in Fig. 5.37. The appearance of different types of crystals on powder agglomerates surface for different Ni/Fe ions ratios is characteristic for these powders. To find out what is the composition of the obtained crystals, the EDS analysis of all samples was performed and the results at different spectrum positions are given in Table 5.3. A common characteristic of the EDS analysis is the decrease of Ni content on the surface of analyzed samples with the decrease of Ni/Fe ratio (42 at.% Ni for the 9/1 ratio, Fig. 5.37a, and 27 at.% Ni for the 3/1 ratio, Fig. 5.37b), i.e., the decrease of Ni content in the alloy powders, while for the Ni/Fe ratios 1/1 and 1/3 the content of Ni on the powder agglomerates surfaces is practically zero. It is interesting to note that on the surface of samples with the Ni/Fe ratios 1/1 and 1/3 some parts of the agglomerates surfaces have been peeled off (Fig. 5.37c, d) and that the composition of the surface layer is completely different than that of the bulk of particle (Table 5.3).

Table 5.3 Results of the EDS analysis of powder samples electrodeposited from different electrolytes, obtained at different positions (spectra) at powder agglomerates presented in Fig. 5.37

Ni/Fe	Spectrum number	at.% O	at.% Fe	at.% Ni
9/1	1	39	25	36
	2	24	23	53
	3	31	19	50
	4	41	29	30
3/1	1	57	19	24
1/1	1	48	52	0
	2	42	35	23
1/3	1	29	67	4
	2	51	43	6
	3	58	42	0
	4	51	49	0

Reprinted from [62] with the permission of Springer-Verlag

According to the EDS analysis of these two samples, it appears that the surface layer is composed of iron oxide (most likely Fe_2O_3), since the content of Ni is zero. For powder samples with lower content of Fe (Ni/Fe = 9/1 and 3/1, Fig. 5.37a, b) layered structure was observed (no peeling off of the surface layer was detected) and, accordingly, only the composition of the surface was analyzed. Taking into account that for such surfaces EDS results cannot be considered as quantitative, a convincing conclusion about the properties of these powder agglomerates surfaces could not be made. Hence, it appears from the EDS analysis of the powder agglomerates surfaces that for samples with higher content of Fe (Ni/Fe = 1/1 and 1/3, Fig. 5.37c, d) the oxidation of all powders agglomerates was not complete and that only surfaces were completely oxidized due to iron oxide diffusion to the surface of the agglomerates during the oxidation process. Taking into account the dimensions of the powder agglomerates (agglomerates up to about 400 μm), it is quite possible that the oxidation process took place mostly on the agglomerates surface. It is possible that the same conclusion could be valid for samples rich in Ni, but on the SEM micrographs peeling off of the surface layer was not detected.

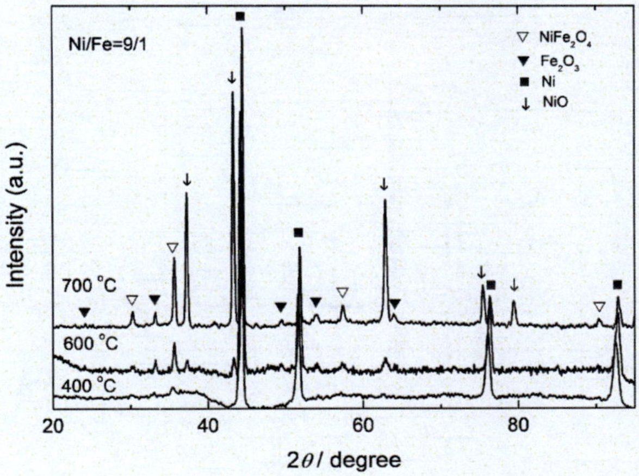

Fig. 5.39 Diffractograms for the Fe–Ni alloy powders electrodeposited at Ni/Fe = 9/1 after annealing in air at 400°C, 600°C, and 700°C for 3 h (Reprinted from [62] with the permission of Springer-Verlag.)

During the second annealing procedure (additional annealing at 600°C for 3 h in air) all powder samples were characterized with the presence of needle-like crystals on the surface of agglomerates (most likely corresponding to the Fe_2O_3 phase), with their dimensions being bigger for samples with higher content of Fe. This is shown in Fig. 5.38a for Ni/Fe ratios 9/1 and 3/1 and in Fig. 5.38b for Ni/Fe ratios 1/1 and 1/3. In comparison with the samples annealed at 400°C the EDS analysis showed higher content of Fe and O for samples obtained at Ni/Fe ratios 9/1 and 3/1, while for samples obtained at Ni/Fe ratios 1/1 and 1/3 identical results were obtained as those for the annealing at 400°C (see Table 5.3).

XRD results recorded after annealing at 400°C, 600°C, and 700°C are shown in Figs. 5.39–5.42. These results are in good agreement with the EDS analysis, showing that all powder agglomerates were oxidized during annealing, forming NiO, Fe_2O_3, and $NiFe_2O_4$ phases. As can be seen, the intensity of some peaks increases, while the intensity of some peaks decreases with increasing annealing temperature.

The intensity of Ni peaks (■) was found to decrease with increasing annealing temperature for Ni/Fe ratios 9/1, 3/1, and 1/1, with

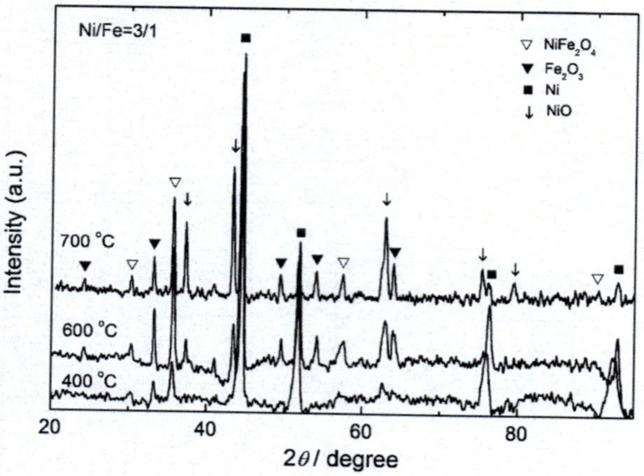

Fig. 5.40 Diffractograms for the Fe–Ni alloy powders electrodeposited at Ni/Fe = 3/1 after annealing in air at 400°C, 600°C, and 700°C for 3 h (Reprinted from [62] with the permission of Springer-Verlag.)

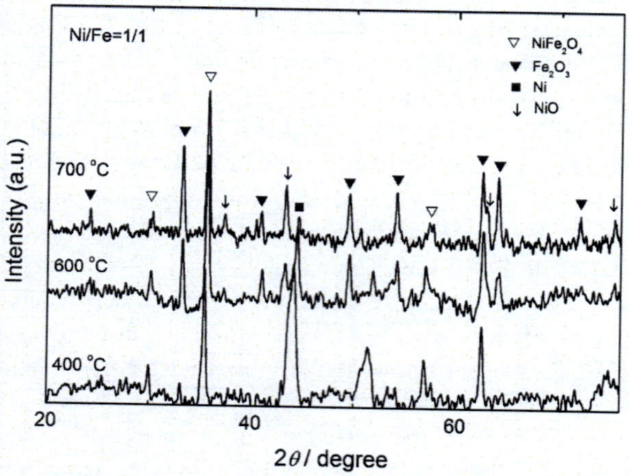

Fig. 5.41 Diffractograms for the Fe–Ni alloy powders electrodeposited at Ni/Fe = 1/1 after annealing in air at 400°C, 600°C, and 700°C for 3 h (Reprinted from [62] with the permission of Springer-Verlag.)

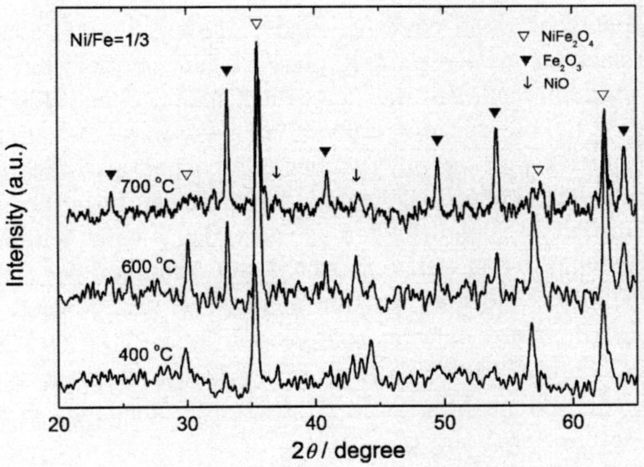

Fig. 5.42 Diffractograms for the Fe–Ni alloy powders electrodeposited at Ni/Fe = 1/3 after annealing in air at 400°C, 600°C, and 700°C for 3 h (Reprinted from [62] with the permission of Springer-Verlag.)

Ni being most likely oxidized into NiO and/or $NiFe_2O_4$ phases. The Ni phase practically disappeared in powders electrodeposited at the Ni/Fe ratio 1/3 annealed at all temperatures. Its presence on the diffractograms recorded for the annealing temperatures 400°C and 600°C in powders electrodeposited at the Ni/Fe ratios 9/1, 3/1, and 1/1 indicates that in these powders Ni cannot be completely oxidized at given annealing temperatures. The disappearance of the Ni phase in powder electrodeposited at Ni/Fe = 1/1 after annealing at 700°C confirms that during additional oxidation whole amount of Ni has been oxidized.

The intensity of the peak for NiO phase (↓) increased with increasing annealing temperature for samples rich in Ni (Ni/Fe = 9/1 and 3/1), while for other two samples NiO peaks (↓) of small intensities were detected. As can be seen in Fig. 5.42, this phase practically disappeared in the powder electrodeposited at Ni/Fe ratio 1/3 after annealing at 700°C.

The intensity of the peaks for Fe_2O_3 phase (▼) also increased with increasing annealing temperature in all samples. In samples with high content of Ni this phase has not been detected on the diffractograms recorded after annealing at 400°C only. After annealing at 700°C this phase became dominant in Fe-rich samples (Ni/Fe = 1/1 and 1/3).

Similar conclusion could be made for the $NiFe_2O_4$ phase (▽). In all samples peaks of this phase could be detected on the diffractograms recorded after annealing at 600°C and 700°C. It is interesting to note that the peaks of $NiFe_2O_4$ phase (▽) possess the highest intensity in the powder electrodeposited at Ni/Fe ratio 1/3 after annealing at 600°C, indicating that these are the best conditions for its formation. Such behavior is in accordance with the findings of Ceylan et al. [89] that well-defined $NiFe_2O_4$ nanoparticles crystallize at 550°C after solid state reaction in $Fe_{67}Ni_{33}$ nanopowder. Hence, it appears that for a powder composition 89 at.% Fe—11 at.% Ni (powder obtained for Ni/Fe ratio 1/3), which is different than that for $Fe_{67}Ni_{33}$, dominant phase is $NiFe_2O_4$ after annealing of electrodeposited powder at 600°C.

5.4 System Mo–Ni–O

The Mo–Ni–O powders were electrodeposited from two supporting electrolytes, as in the case of Co–Ni system: 1 M NH_4Cl + 0.7 M NH_4OH and 1 M $(NH_4)_2SO_4$ + 0.7 M NH_4OH, with the pH of the solutions being 9.0.

5.4.1 *Polarization Curves*

5.4.1.1 Polarization Curves in Chloride Containing Electrolytes

Three different electrolytes were used for the investigations: 0.1 M $NiCl_2$ + 1 M NH_4Cl + 0.7 M NH_4OH + x M Na_2MoO_4, with x being 0.05 M, 0.1 M, and 0.3 M, respectively. In such a way the Ni/Mo ions concentration ratio was 1/0.5, 1/1, and 1/3. All powders for microstructure, composition, and phase composition analysis were electrodeposited at the limiting current density (position of the inflection point B on the polarization curves, see Fig. 5.43). In all cases small amount of the rough compact deposit (necessary for powder formation) remained on the glassy carbon surface and only agglomerates detached from the cathode surface were analyzed.

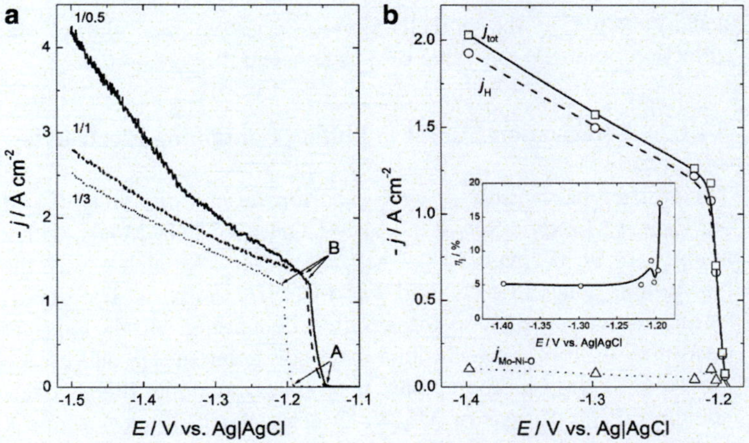

Fig. 5.43 (**a**) Polarization curves in chloride containing electrolytes for different Ni/Mo ratios (marked in the figure). (**b**) Polarization curve for powder electrodeposition and hydrogen evolution (j_{tot}), polarization curve for hydrogen evolution (j_H), and polarization curve for powder electrodeposition after subtraction of the current density for hydrogen evolution ($j_{Mo-Ni-O}$). *Inset*: current efficiency for Mo–Ni–O powder electrodeposition (η_j) as a function of potential (Reprinted from [121] with the permission of Elsevier.)

The polarization curves are presented in Fig. 5.43. As can be seen, the polarization curves characterized by two inflection points (Fig. 5.43a), as in all previous cases, were obtained. It is important to note that the potential of the beginning of alloy deposition (A) becomes more negative with the increase of molybdate ions concentration (with the decrease of Ni/Mo ratio), as it could be expected, since the potential of the Mo deposition is much more negative than that of Ni [116]. At the same time, a deposition of Mo can only take place in the presence of Ni (induced codeposition [116]). Taking into account that the concentration of Ni^{2+} ions was constant, it is quite reasonable that the value of current density of the inflection point B does not change with changing Ni/Mo ions concentration ratio (being about -1.2 A cm^{-2}).

It should also be mentioned that the current efficiency for alloy deposition in all cases was very low, about 5%. This could be seen in the inset of Fig. 5.43b for the Ni/Mo ratio 1/3, where the η_j vs. E changes from 17% to 5% in the region of sharp increase of current density, while at the potentials more negative than the second inflection point (B) the value of η_j is constant, being about 5%.

5.4.1.2 Polarization Curves in Sulfate Containing Electrolytes

Three different electrolytes were used for the investigations: 0.1 M $NiSO_4$ + 1 M $(NH_4)_2SO_4$ + 0.7 M NH_4OH + x M Na_2MoO_4, with x being 0.03 M, 0.1 M, and 0.3 M, respectively. In such a way the Ni/Mo ions concentration ratio was 1/0.3, 1/1, and 1/3. The polarization curves recorded in the solution containing sulfate ions are presented in Fig. 5.44. In this case all polarization curves are moved to slightly more cathodic potentials, indicating higher overvoltage for the process of Mo–Ni–O alloy electrodeposition in the presence of sulfate ions (Fig. 5.44a). At the same time, the total current densities in the range of potentials more negative than the inflection point (B) are smaller and the slopes of their changes are smaller in comparison with those recorded for chloride electrolyte (Fig. 5.43a). As in the case of chloride electrolyte, the potential of the beginning of alloy deposition (A) becomes more negative with the increase of molybdate ions concentration (with the decrease of Ni/Mo ratio). As shown in Fig. 5.44b, the current efficiency for alloy deposition is sensitive to the Ni/Mo ions ratio: for Ni/Mo = 1/0.3, $\eta_j = 10\%$, for Ni/Mo= 1/1, $\eta_j = 7\%$, and for Ni/Mo = 1/3, $\eta_j = 5\%$ [all η_j values correspond to the potentials more negative than the second inflection point (B)].

5.4.2 *Morphology of As-Deposited Mo–Ni–O Powders*

5.4.2.1 Morphology of Powders Electrodeposited from Chloride Containing Electrolytes

The morphologies of powder agglomerates electrodeposited onto glassy carbon electrode from electrolytes with different Ni/Mo ions concentration ratios (1/0.5, 1/1, and 1/3, samples 1, 2, and 3, respectively) are shown in Figs. 5.45 and 5.46, respectively. For the highest

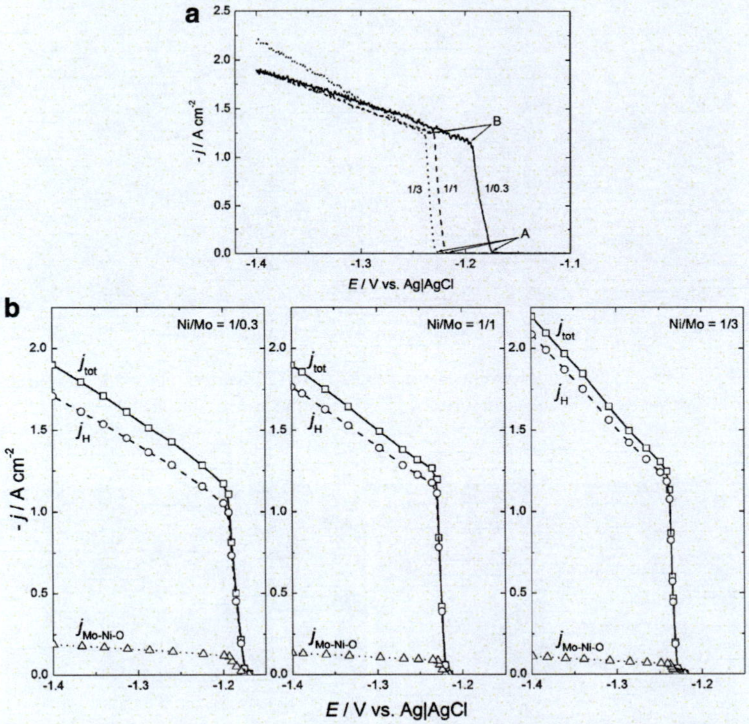

Fig. 5.44 (**a**) Polarization curves in sulfate containing electrolytes for different Ni/Mo ratios (marked in the figure). (**b**) Polarization curves for powder electrodeposition and hydrogen evolution (j_{tot}), polarization curves for hydrogen evolution (j_H), and polarization curves for powder electrodeposition after subtraction of the current density for hydrogen evolution ($j_{Mo-Ni-O}$) (Reprinted from [122] with the permission of Elsevier.)

Ni/Mo ratio of 1/0.5 (Fig. 5.46) the size of the agglomerates varies in the range of about 2–50 μm. At the same time, sharp edges could be detected on top of the surfaces of all agglomerates, indicating crystallinity of electrodeposited powder (Fig. 5.45b).

With the decrease of the Ni/Mo ratio (1/1), a cauliflower-type agglomerate, Fig. 5.46a, b, characterized with spherical edges (c) and the presence of cracks, was obtained. Its size is much higher than that of the powder particles electrodeposited at the 1/0.5 ratio, varying in

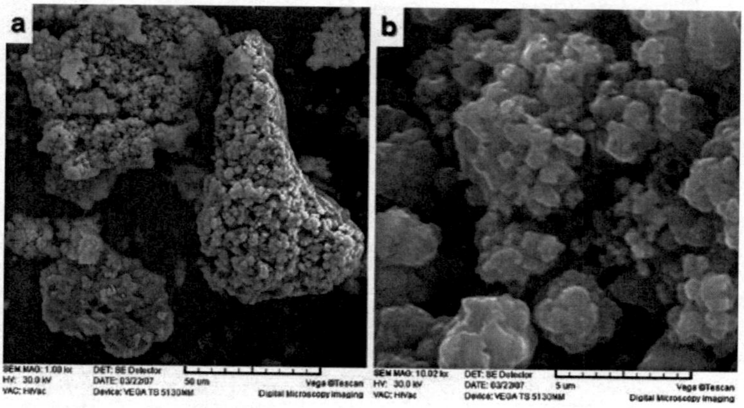

Fig. 5.45 Typical agglomerates for the Mo–Ni–O powders electrodeposited at the Ni/Mo = 1/0.5 (Reprinted from [121] with the permission of Elsevier.)

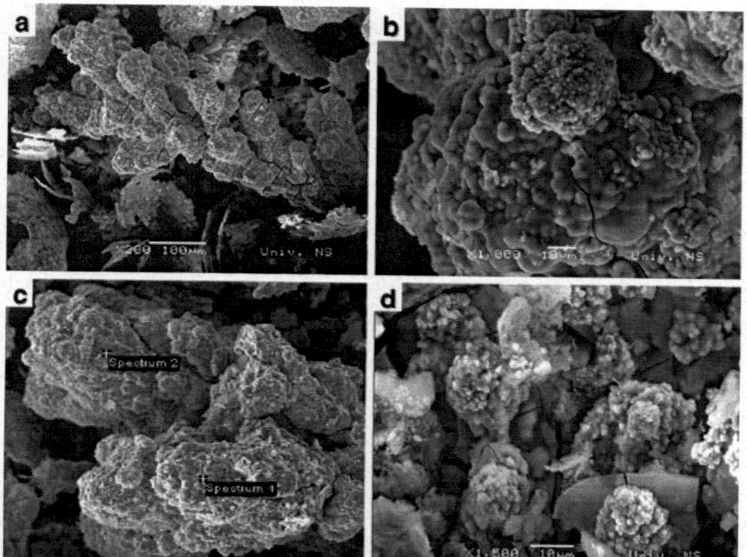

Fig. 5.46 (**a**, **b**, **c**) Typical particles for the Mo–Ni–O powders electrodeposited at the Ni/Mo = 1/1. (**d**) Typical particles for the Mo–Ni–O powders electrodeposited at the Ni/Mo = 1/3 (Reprinted from [121] with the permission of Elsevier.)

the range of about 50–500 μm. In the case of Ni/Mo ratio of 1/3, Fig. 5.46d, flat and thin parts of the powder, corresponding most likely to the compact deposit, were covered with spherical agglomerates, while the cracks are more pronounced. The appearance of a large number of cracks is the result of high tensile stresses present in the powder and the hydrogen evolution [138]. According to the presented results, it is obvious that the crystallinity of the particles becomes less pronounced with the increase of molybdate ions in the electrolyte, influencing at the same time the morphology of as-deposited powder particles.

5.4.2.2 The EDS Analysis of the As-Deposited Powders

All powder samples were analyzed by EDS in such a way that one powder particle on the SEM micrograph is chosen and EDS analysis was performed at two (or sometimes three) different positions on this particle, as shown in Fig. 5.46c. The approximate compositions of all alloy powders, evaluated by the EDS microanalysis, revealed that the composition depends not only on the Ni/Mo concentration ratio but also on the position at which the EDS analysis was performed. The results of the EDS analysis indicate the presence of significant amount of oxygen in the powders (up to about 70 at.%), as well as the nonhomogeneous composition of the as-deposited powders. At the same time the composition does not follow the Ni/Mo ions concentration ratio, although the theory of induced codeposition [116–120] predicts the increase of Mo amount in the deposit with increasing concentration of molybdate ions in the electrolyte. Such a behavior indicates that in the case of powder electrodeposition from ammonium chloride electrolyte, most likely, molybdenum is not deposited only as pure metal, but also as MoO_x species, as found in the literature for compact deposits [139–141]. At the same time, taking into account nonhomogeneous distribution of molybdenum, nickel, and oxygen in the electrodeposited powders, any conclusion concerning composition of the alloy obtained by the EDS analysis cannot be considered as reliable for this particular case.

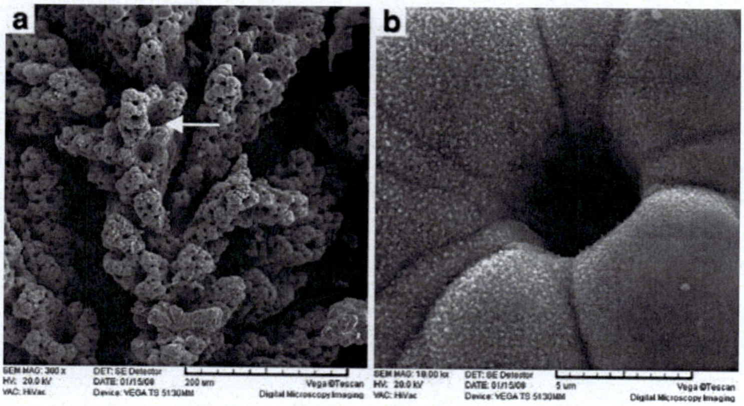

Fig. 5.47 (**a**) Typical spongy particles electrodeposited at the Ni/Mo = 1/0.3. (**b**) Cavity marked with *arrow* in (**a**) at higher magnification (Reprinted from [122] with the permission of Elsevier.)

5.4.2.3 Morphology of Powders Electrodeposited from Sulfate Containing Electrolytes

The morphology of powder particles electrodeposited from electrolytes with different Ni/Mo ions concentration ratios (1/0.3, 1/1, and 1/3) are shown in Figs. 5.47–5.49. Different morphology of the powder particles was obtained.

For the Ni/Mo ratio 1/0.3 typical spongy particles were detected, Fig. 5.47a. The characteristic of these particles is very flat surface around the holes, as shown in Fig. 5.47b.

With the decrease of the Ni/Mo ratio (1/1), two types of particles were obtained, as shown in Fig. 5.48. Both are spherical (rounded), one with flat edges (a) and another one with rough endings on top of the surface (b). Taking into account our previous investigations of alloy powders morphology [37–40], this is not unexpected. Since the growth of powder particles is characterized by the presence of two or more growth zones, it appears that spherical particles with flat edges have been detached from the electrode surface before the beginning of a growth of new zone, while the particles with rough endings have been detached after the beginning of the growth of a new crystals.

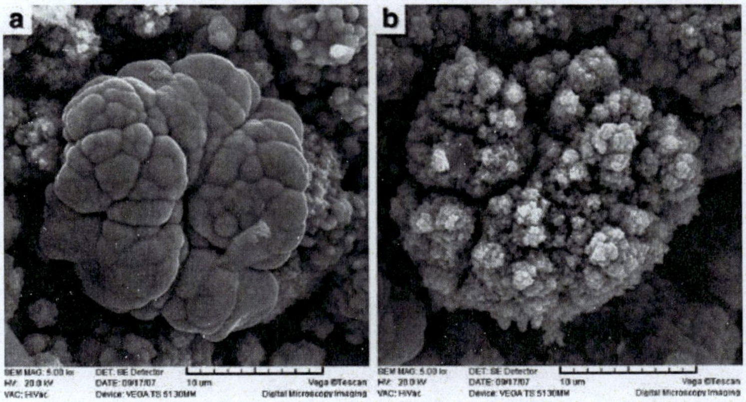

Fig. 5.48 Typical particles electrodeposited at the Ni/Mo = 1/1 (Reprinted from [122] with the permission of Elsevier.)

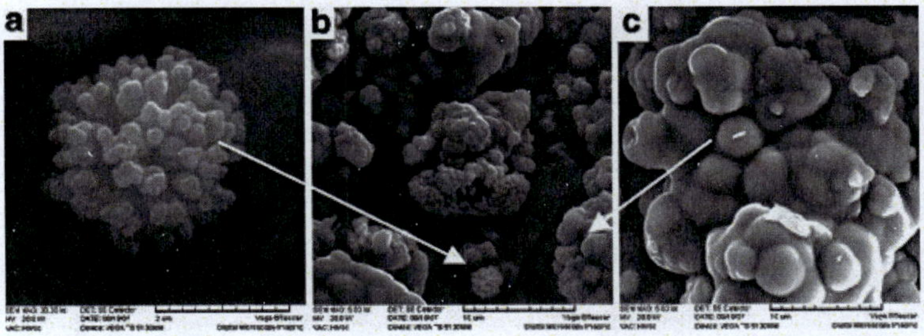

Fig. 5.49 (a) Agglomerate with dendritic surface. (b) Typical powder electrodeposited at the Ni/Mo = 1/3. (c) Agglomerate with flat surface (Reprinted from [122] with the permission of Elsevier.)

It should be noted (not shown in this figure) that some particles were characterized by the presence of cracks.

In the case of Ni/Mo ratio of 1/3 (powder with the highest amount of molybdenum) completely different morphology was detected, as shown in Fig. 5.49. The powder is composed of flat and thin parts with pronounced cracks, corresponding most likely to the compact deposit, and spherical agglomerates growing on top of a flat deposit (b). The surfaces of these agglomerates are either dendritic (a) or flat (c).

5.4.3 Composition and Phase Composition Analysis of As-Deposited and Recrystallized Powders Obtained from Chloride Electrolyte

The phase composition analysis of the powders was performed using XRD. In the case of as-deposited powders practically no peaks were detected on the diffractograms (except one broad peak observed near $2\theta = 40°$, see Fig. 5.51a—quasiamorph). It seems that, as in the case of compact deposits [121, 142], the crystallites were extremely small, i.e., of the order of nanometers and the powders were quasiamorphous. It was obvious that the Ni–Mo alloy powders should be previously recrystallized to determine their phase composition.

Two powder samples, electrodeposited from the solutions with Ni/Mo = 1/1 (sample 2) and Ni/Mo = 1/3 (sample 3), were analyzed by DSC and TGA to determine recrystallization temperature (Fig. 5.50) [121]. For both samples a multistep process with the sharp exothermic maximum on the DSC curves indicates that the recrystallization occurs at 543°C and should be performed at this or higher temperature. It should be noted that on the DSC curve for sample 2 additional small exothermic maximum, appearing at lower temperature of about 420°C, could be ascribed to the recrystallization of another phase present in the powder. At the same time, the TGA analysis revealed the weight loss of the samples of about 15%, corresponding most probably to the evaporation of electrolyte left in the pores of as-deposited powders.

Annealing (recrystallization) was performed in N_2 atmosphere at 600°C for 2 h. XRD analysis, shown in Fig. 5.51, revealed structural transformation and formation of $NiMoO_4$ and MoO_3 phases. For the recrystallized powder sample 3, Fig. 5.51a (recryst.), with the exception of few small peaks, all peaks correspond to the $NiMoO_4$ phase (▼), indicating that it is possible to obtain single phase Mo–Ni–O powder by electrochemical deposition and subsequent annealing in nitrogen. With the increase of Ni/Mo ratio (samples 1 and 2, Fig. 5.51b), among the $NiMoO_4$ phase (▼), MoO_3 phase (▽) was also detected in the powders with the intensity of its peaks increasing with the increase of the Ni/Mo ratio. From the intensity of the MoO_3 peaks it could be estimated that in sample 1 about 20% of MoO_3 is

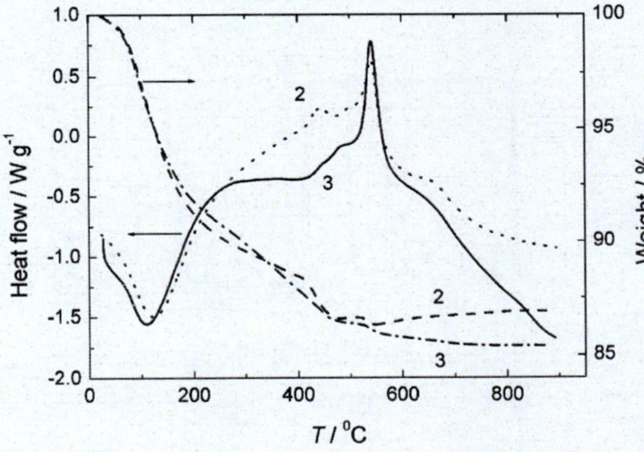

Fig. 5.50 The results of the DSC–TGA analysis for the samples 2 and 3 (marked in the figure) (Reprinted from [121] with the permission of Elsevier.)

present, while in sample 2 about 10% of MoO_3 is present. It is most probable that the peak at about $2\theta = 27°$ corresponds to the $NiMoO_4$—K_β reflection (✦) [121].

5.4.3.1 The Morphology and EDS Analysis of Recrystallized Powders

The morphology of the recrystallized powders (SEM) is shown in Fig. 5.52a–c for samples 1, 2, and 3 respectively (see Table 5.3). In all cases regular crystals were obtained with smooth surfaces and well-defined crystal planes. The size of particles is much smaller than the size of as-deposited particles, varying between about 2 μm and about 10 μm, indicating that during the annealing procedure agglomerates formed by electrodeposition separate into much smaller crystals [121].

The EDS analysis was performed on the single crystal grains, mainly with an electron beam covering the area from about 6 μm^2 to about 30 μm^2 (depending on the dimensions of the crystal grains). In some cases the point analysis (as in the case of as-deposited

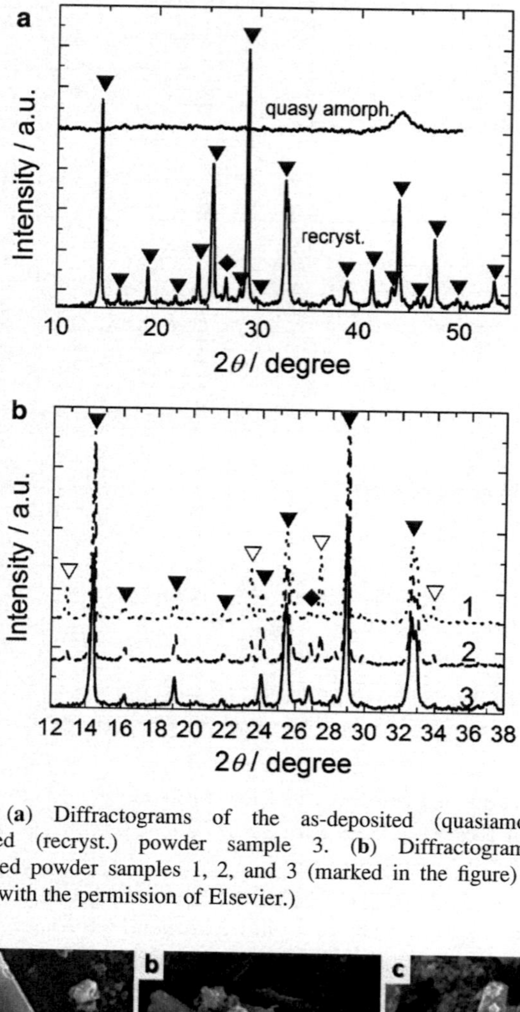

Fig. 5.51 (**a**) Diffractograms of the as-deposited (quasiamorph.) and recrystallized (recryst.) powder sample 3. (**b**) Diffractograms of the trecrystallized powder samples 1, 2, and 3 (marked in the figure) (Reprinted from [121] with the permission of Elsevier.)

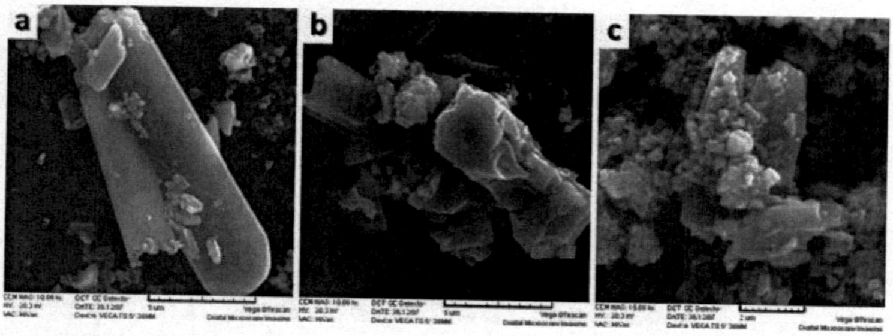

Fig. 5.52 Typical crystals for samples 1—(**a**), 2—(**b**), and 3—(**c**), detected after annealing of corresponding Mo–Ni–O powders at 600°C for 2 h (Reprinted from [121] with the permission of Elsevier.)

Table 5.4 The average composition of the recrystallized powders, determined by the EDS analysis

Sample No.	Ni/Mo	at.% O_2	at.% Ni	at.% Mo
1	1/0.5	79.49	7.54	12.97
2	1/1	80.19	8.02	11.79
3	1/3	76.99	10.28	12.73

Reprinted from [121] with the permission of Elsevier

powders, see Fig. 5.46c) was also performed. The results of the EDS analysis are presented in Table 5.4.

It is interesting to note that the average amount of Ni increases with increasing the concentration of molybdate ions in the solution (decrease of the Ni/Mo ratio), while at the same time the amount of Mo practically does not change with the decrease of the Ni/Mo ratio. Such a behavior is not in accordance with the described model, as well as with the theory of the induced codeposition [117–120]. This could be either due to the low precision of the EDS analysis, or the fact that the model and the theory were developed for the compact deposits (not applicable in the case of powder electrodeposition), or different mechanism of induced codeposition of Ni and Mo, or the consequence of solid state reaction occurring during annealing, or all of them together.

Ternary phase diagram for the system Mo–Ni–O [142, 143] is shown in Fig. 5.53. As can be seen, recrystallized samples 1, 2, and 3 are placed (according to the EDS analysis) in the region where three compounds could be expected, $NiMoO_4$, MoO_3, and O_2. Since XRD analysis cannot detect oxygen, the presence of MoO_3 and $NiMoO_4$ should be expected. At the same time the composition of all samples (EDS—Table 5.4) is not proportional to the ratio expected only for $NiMoO_4$ (1:1:4). Excess of molybdenum and oxygen clearly indicates the presence of MoO_3 and oxygen, particularly in samples 1 and 2. In the case of sample 3 only $NiMoO_4$ is detected by the XRD, most probably due to the composition of the recrystallized sample 3, which is very close to the line corresponding to $NiMoO_4$, so that the small amount of MoO_3 (less than 10%) cannot be detected. In this case the ratio expected only for $NiMoO_4$ (1:1:4) is also not achieved since some amount of oxygen (according to the ternary phase diagram [142, 143]) is present in the powder.

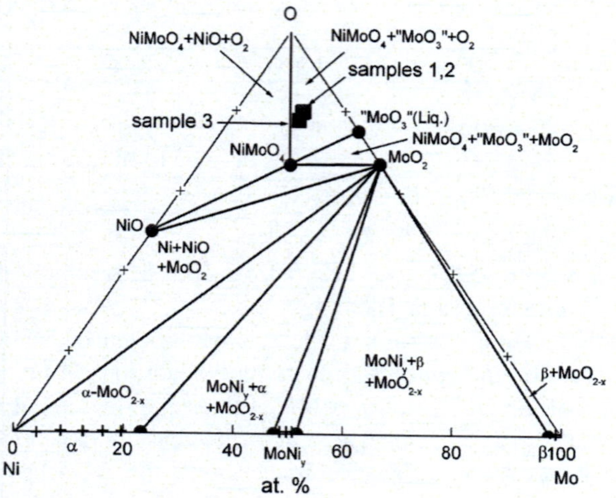

Fig. 5.53 Ternary phase diagram of the system Ni–Mo–O with the position of the recrystallized samples 1, 2, and 3 (marked in the figure—*filled square*) (Reprinted from [121] with the permission of Elsevier.)

According to the DSC analysis (Fig. 5.50), a small exothermic maximum appearing at lower temperature of about 420°C for sample 2 might correspond to the recrystallization of the MoO_3. It is most likely that during the annealing at 600°C in N_2 atmosphere for 2 h, not only recrystallization but also solid state reaction occurs, producing MoO_3 and $NiMoO_4$ phases. This assumption needed further clarification by other techniques (high temperature XRD, TEM) and is the subject of the investigations presented in Sect. 5.4.6.

5.4.4 *Morphology, Composition and Phase Composition Analysis of Recrystallized Powders Obtained from Sulfate Electrolyte*

To define temperatures of recrystallization and solid state reaction in the investigated Mo–Ni–O powders, stepwise annealing for 2 h in nitrogen atmosphere was performed (1) at 300°C, (2) at 400°C, (3) at

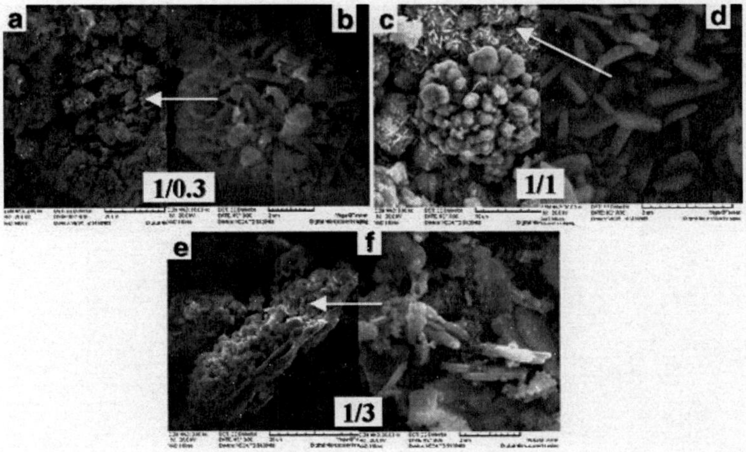

Fig. 5.54 SEM micrographs of the powders annealed at 300°C for the Ni/Mo ions ratios marked in the figure. Certain features of the powders are marked with *arrows* and shown at higher magnification in pictures (**b**) for the ratio 1/0.3, (**d**) for the ratio 1/1 and (**f**) for the ratio 1/3 (Reprinted from [122] with the permission of Elsevier.)

500°C, and (4) at 600°C. After each annealing procedure their morphology, composition, and phase composition were analyzed.

5.4.4.1 Analysis of Powders After Annealing at 300°C and 400°C

After the annealing at 300°C only small parts of the powders particles were recrystallized. As shown in Fig. 5.54, small crystals could be detected on the particles of all powders. It should be emphasized that the morphology of powder electrodeposited at Ni/Mo ratio 1/3 became completely different after this treatment. No flat and thin parts characterized by cracks (as it was the case for the as-deposited powder), corresponding most probably to the quasiamorphous (nanocrystalline) part of the powder deposit, could be detected. Hence, it appears that mostly this part of the powder crystallized.

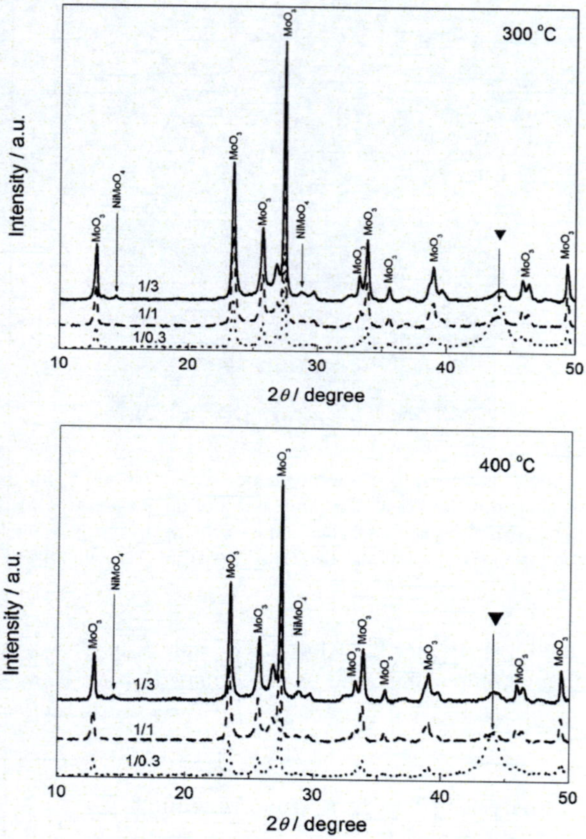

Fig. 5.55 Diffractograms for the powders annealed at 300°C and 400°C electrodeposited at different Ni/Mo ions ratios (marked in the figure) (Reprinted from [122] with the permission of Elsevier.)

Almost identical morphology was detected for the powders annealed at 400°C [122].

The corresponding diffractograms are shown in Fig. 5.55. As can be seen, already after annealing at 300°C MoO_3 phase was detected in powders electrodeposited at the Ni/Mo ions ratios 1/0.3 and 1/1, while for the powder electrodeposited at the Ni/Mo ions ratio 1/3 both MoO_3 and $NiMoO_4$ phases were detected, with the peaks for the

$NiMoO_4$ phase being very small, indicating that very small amount of this phase was present in the corresponding powder. A broad peak at about $2\theta = 44°$ (▼), particularly pronounced for the powder electrodeposited for Ni/Mo = 1/0.3, could be, most likely, ascribed to a still nanocrystalline $NiMoO_4$ phase, but there is a possibility that a small amount of the $MoNi_4$ phase is present in the powder, since the peaks of these two phases overlap at $2\theta = 44$–$45°$. With the increase of the annealing temperature to 400°C practically the same results were obtained, indicating that no additional recrystallization occurred [122].

5.4.4.2 Analysis of Powders After Annealing at 500°C

The annealing at 500°C produced certain change in the powders morphology. As can be seen in Fig. 5.56, the powder containing the smallest amount of molybdenum [1/0.3(a) and 1/0.3(b)] retained mostly the same, spongy-type morphology. At the same time, agglomerates with flat surfaces are still visible [1/0.3(b)], while the bigger crystals and higher number of crystals could be detected on their surfaces. It is characteristic that the surface of agglomerates for all samples is covered with well-defined crystals of the shape of elongated stubs, corresponding to the α-MoO_3 phase [144].

After annealing the powders at 500°C (Fig. 5.57) more pronounced peaks for the $NiMoO_4$ phase appeared on all diffractograms, reflecting the increase of this phase in all powders. A broad peak at about $2\theta = 44°$ became sharp, particularly for the powders electrodeposited at Ni/Mo ratios 1/1 and 1/3, corresponding to the $NiMoO_4$ phase with bigger dimension of crystallites (see Table 5.5).

5.4.4.3 Analysis of Powders After Annealing at 600°C

As can be seen in Fig. 5.58, the annealing at 600°C produced significant change in the morphology of powders. All powders transformed into well-defined crystals of much smaller dimensions than the dimensions of agglomerates, confirming the statement that a complete recrystallization of the Mo–Ni–O alloy powders occurred.

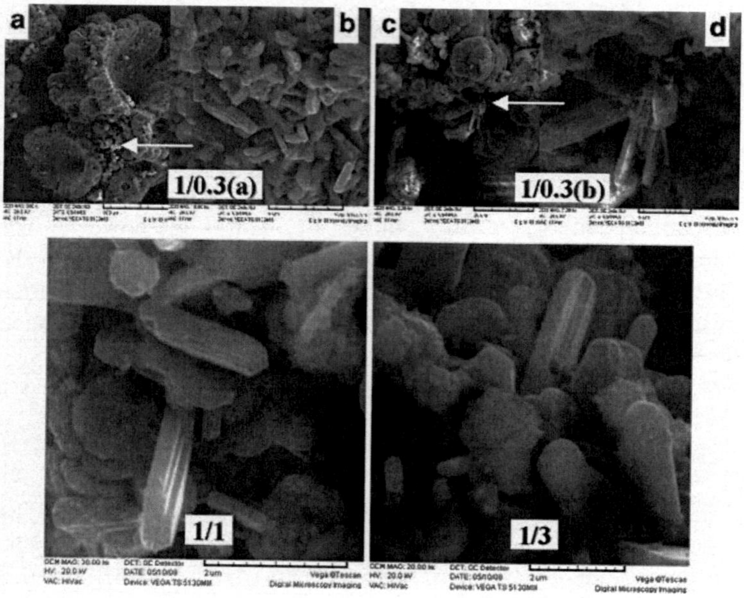

Fig. 5.56 SEM micrographs of the powders annealed at 500°C for the Ni/Mo ions ratios marked in the figure. Certain features of the powders are marked with *arrows* and shown at higher magnification in pictures (**b**) and (**d**) (Reprinted from [122] with the permission of Elsevier.)

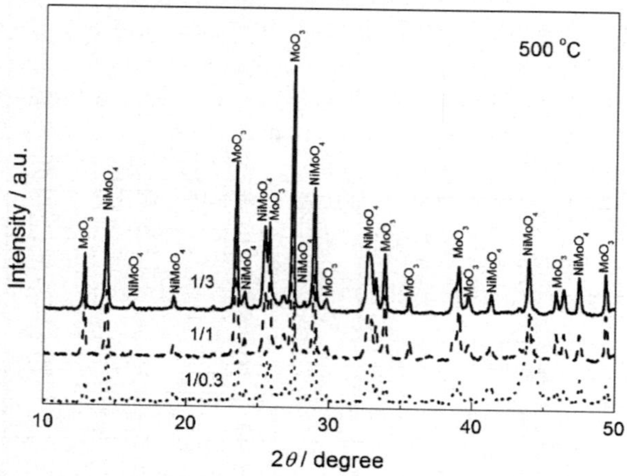

Fig. 5.57 Diffractograms for the powders annealed at 500°C electrodeposited at different Ni/Mo ions ratios (marked in the figure) (Reprinted from [122] with the permission of Elsevier.)

Table 5.5 The dimensions of crystallites of different phases of the system Ni–Mo–O as a function of the recrystallization temperature (T_r)

Phase	Ni/Mo	T_r (°C)	D_{110} (nm)[a]
MoO_3	1/0.3	300	31.37
		400	34.15
		500	43.47
	1/1	300	29.86
		400	29.57
		500	67.77
		600	95.51
	1/3	300	34.97
		400	35.49
		500	45.48
		600	82.25
Phase	**Ni/Mo**	**T_r (°C)**	**D_{220} (nm)[b]**
$NiMoO_4$	1/0.3	500	44.55
		600	70.04
	1/1	500	67.07
		600	81.28
	1/3	500	45.56
		600	75.94

Reprinted from [122] with the permission of Elsevier
[a]Reflections of the 110 (JCPDS 05-0508)
[b]Reflections of the 220 (JCPDS 33-0948)

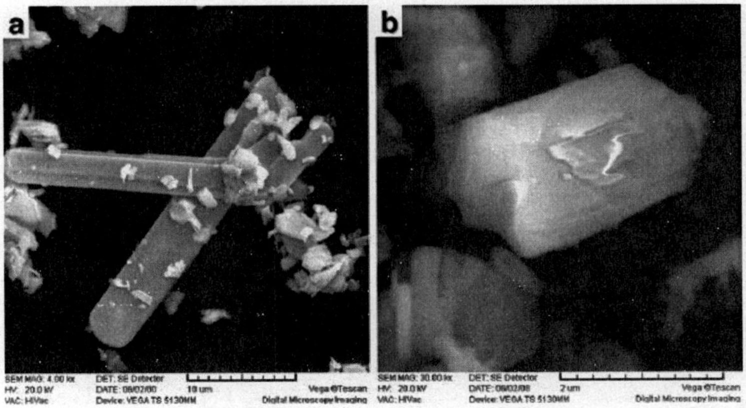

Fig. 5.58 SEM micrographs of the typical powder particles detected for all samples after annealing at 600°C (Reprinted from [122] with the permission of Elsevier.)

The most characteristic is the shape of elongated stubs (Fig. 5.58a) of bigger dimensions in comparison with the dimensions of other crystals (Fig. 5.58b).

The diffractograms recorded after annealing at 600°C are shown in Fig. 5.59a, b. As can be seen, significant change occurred for the powder electrodeposited at Ni/Mo = 1/0.3. The peaks for the MoO_3 phase disappeared, while the peaks for the $NiMoO_4$ phase became more pronounced. Such a behavior could be ascribed to the solid state reaction, with the $NiMoO_4$ phase being formed on the account of the MoO_3 phase. It is most likely that the solid state reaction took place in all powders, but only for the powder with the lowest amount of molybdenum (32 mass %, Table 5.6) all MoO_3 phase was consumed in the formation of the $NiMoO_4$ phase. It seems that the sharp peak recorded on the DSC curve at 543°C [121] corresponds to both complete recrystallization and solid state reaction of the Ni–Mo–O alloy powders.

The size of crystallites for both phases present in the Ni–Mo–O alloy powder deposits was determined by the analysis of certain peaks on the diffractograms shown in Figs. 5.55, 5.57, and 5.59a. Results are presented in Table 5.5. As expected, the size of MoO_3 and $NiMoO_4$ phase crystallites increases with increasing the annealing temperature of powders [122].

To find out whether the $MoNi_4$ phase had been formed after annealing the powder electrodeposited at the Ni/Mo = 1/0.3 at 600°C, additional XRD analysis of this sample was performed up to higher values of 2θ. The corresponding diffractogram is shown in Fig. 5.59b. As can be seen almost all peaks for this phase are present at the diffractogram, indicating that a small amount of the $MoNi_4$ phase has been formed. According to the shape (broad) and the height of the peaks, it is most likely that the crystallites were still small and that higher temperature should be applied for the complete recrystallization of this particular phase [122, 144–151].

According to the positions of the peaks on the diffractograms, for both MoO_3 and $NiMoO_4$ phases only α modifications were detected.

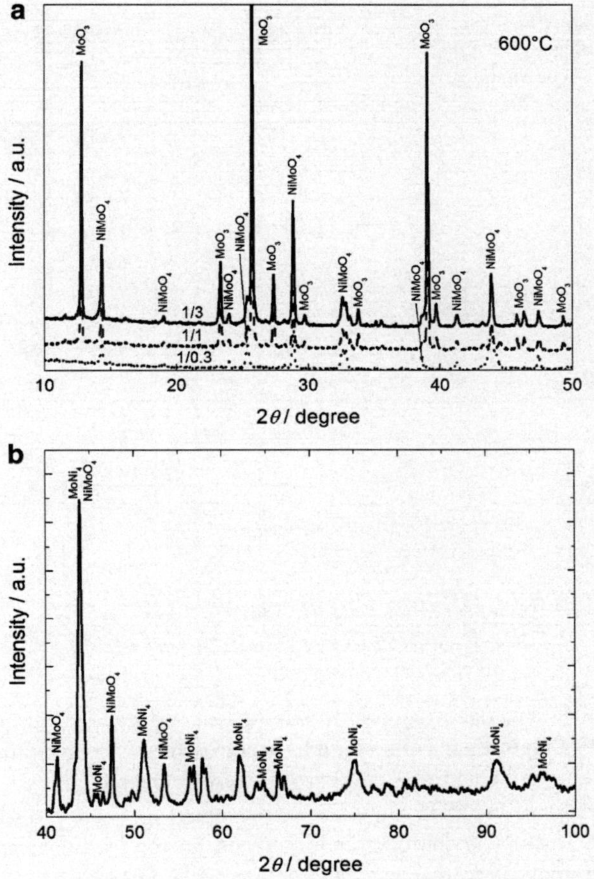

Fig. 5.59 (**a**) Diffractograms for the powders annealed at 600°C electrodeposited at different Ni/Mo ions ratios (marked in the figure). (**b**) Diffractogram for the powder annealed at 600°C electrodeposited at Ni/Mo =1/0.3, recorded at higher values of 2θ (from 40° to 100°) (Reprinted from [122] with the permission of Elsevier.)

Table 5.6 Chemical and EDS analysis of the Ni–Mo–O powders

	Chemical analysis			EDS analysis		
	Mass %		Spectrum	At.%		
Ni/Mo	Ni	Mo		O	Ni	Mo
1/0.3	68	32	1	66.69	15.80	17.52
			2	68.94	14.94	16.12
			3	61.64	13.34	16.84
1/1	50	50	1[a]	73.36	0.14	26.50
			2	69.84	13.11	17.05
			3	71.04	12.49	16.47
			4[a]	72.93	0.94	26.13
1/3	34	66	1[a]	73.72	0.00	5.46
			2[a]	72.31	0.00	25.32
			3[a]	75.59	0.00	24.41
			4	69.54	12.65	17.81
			5	65.02	13.52	14.55

Reprinted from [122] with the permission of Elsevier
[a]Particles of MoO_3

5.4.5 *The EDS and Chemical Analysis of the Powders Recrystallized at 600°C*

The EDS analysis was performed on the crystal grains (certain particles bigger than 1 μm) with an electron beam covering the area of 1 μm^2 (shown in Fig. 5.60 for all three powder compositions). The results of the EDS analysis are presented in Table 5.6 together with the results of the chemical analysis. In the chemical analysis only the mass percentages of Ni and Mo were determined.

As can be seen, the chemical composition of electrodeposited powders is identical to the ratios of Ni/Mo ions in the solution. It is characteristic that in powders electrodeposited at Ni/Mo ratio 1/0.3 all powder particles (agglomerates) are of the same, not well defined crystal shape, while at higher magnification their crystals of about 1–2 μm are well-defined (Fig. 5.60). In the case of powders electrodeposited at Ni/Mo ratios 1/1 and 1/3, two types of powder particles were detected. One, characteristic for the α-MoO_3, having a shape of very thin elongated stabs [144] reaching up to 20 μm in length, and another one of a rounded shape, belonging mainly to the

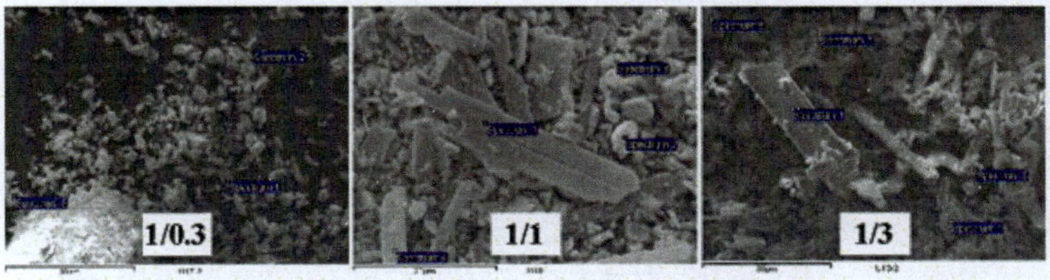

Fig. 5.60 SEM and EDS analysis of electrodeposited Mo–Ni–O powders after annealing at 600°C (Reprinted from [122] with the permission of Elsevier.)

α-$NiMoO_4$ phase [148–151]. This is confirmed by the results of the EDS analysis presented in Table 5.6. Taking into account the precision of the EDS analysis, it is quite difficult to obtain exact at. % ratio, particularly for small particles ($NiMoO_4$), but for bigger particles (like MoO_3) almost ideal Mo–O ratio has been detected on most of the analyzed samples (Fig. 5.60, Table 5.6).

As shown earlier, according to the ternary phase diagram for the system Ni–Mo–O [142, 143], the $NiMoO_4$ and MoO_3 phases should be seen at a given compositions of recrystallized powders. It is interesting to note that the previous [121] and present results [122] differ, since the presence of only one phase ($NiMoO_4$) in a previous work has been detected in the sample with the highest amount of molybdenum (Ni/Mo = 1/3), while in a present work this phase has been detected in the sample with the lowest amount of molybdenum (Ni/Mo = 1/0.3). Although previous powders were electrodeposited from the chloride containing solution, it is most likely that this difference is a consequence of the different procedures of powders annealing (recrystallization). Since in this investigation the existence of the phase transition has been confirmed, it seems more realistic that this process was more pronounced in the sample with the lowest amount of molybdenum in the powder, where the whole amount of MoO_3 phase has been consumed in the phase transition of MoO_3 phase into $NiMoO_4$ phase. Hence, single $NiMoO_4$ phase (with small amount of $MoNi_4$ phase) has been detected in this powder.

Taking into account that the investigated powders had to be annealed (recrystallized) to perform their phase composition analysis, the question arises: are these phases present in the as-deposited powders? The only way to answer this question was to perform TEM analysis of as-deposited powders.

5.4.6 The Results of TEM Analysis of As-Deposited Powders Obtained from Sulfate Electrolyte

The TEM analysis was performed on two Mo–Ni–O powders: Ni-rich (Ni/Mo = 1/0.3) and Mo-rich (Ni/Mo = 1/3) powders. The common characteristic of both powders is the presence of amorphous and crystalline particles.

5.4.6.1 The TEM Analysis of Electrodeposited Ni-Rich Mo–Ni–O Powders

The specimen is composed of large amorphous clusters containing nanosized crystalline particles. Fig. 5.61 shows a TEM image of an amorphous particle (a) and the corresponding SAED pattern (b).

Diffuse diffraction rings clearly indicate amorphous character of this powder particle. Weak spots in the diffraction pattern correspond to the {211} reflections of fine $MoNi_4$ nanoparticles (JCPDF # 65-1533) [152], shown in the close-up in Fig. 5.61c. In the crystalline parts of the Ni-rich powder the following phases were detected: NiO, MoO_3, and $MoNi_4$, in the form of idiomorphic crystallites measuring up to 70 nm in diameter. Figure 5.62 shows a cluster of well-crystallized nanoparticles. The SAED pattern (shown in Fig. 5.62b) from these clusters best corresponds to the monoclinic MoO_3 phase (JCPDF # 89-1554) [153], or one of its structural variants, depending on the position in the sample. Typical morphology of euhedral MoO_3 crystals is shown in Fig. 5.62a and on the close-up in Fig. 5.62c. Most of the crystals are tabular on {200}. The strongest SAED diffraction rings in the pattern correspond to: $D_1 = \{200\}$, $D_2 = \{211\}$, $D_3 = \{220\}$, and $D_4 = \{022\}$. In the Ni-rich areas of the sample we observe crystalline $MoNi_4$ and NiO. $MoNi_4$ appears in the form of irregularly intergrown grains measuring up to 10 nm in average.

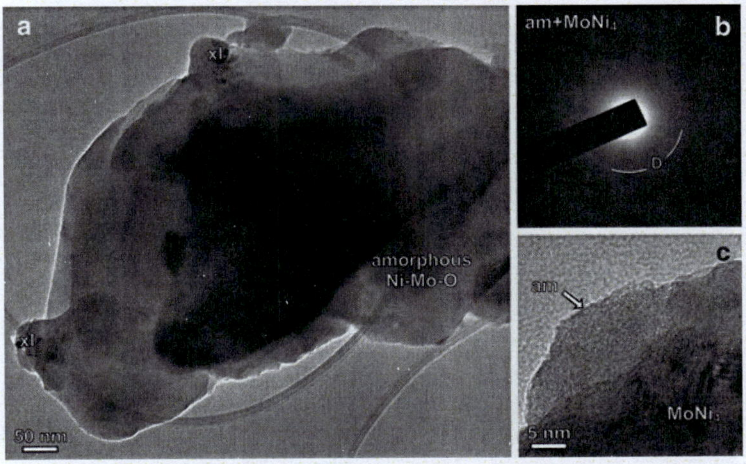

Fig. 5.61 (**a**) TEM image of the amorphous Mo–Ni–O particle detected in the powder electrodeposited at the Ni/Mo ratio 1/0.3. (**b**) SAED pattern recorded from this area (**a**) shows predominantly amorphous material (am) with weak reflections corresponding to $MoNi_4$ nanocrystals (x1). (**c**) High magnification of amorphous Mo–Ni–O particle with $MoNi_4$ nanocrystals (Reprinted from [123] with the permission of Elsevier.)

An area of prevailing $MoNi_4$ composition is shown in Fig. 5.63a, with corresponding SAED pattern in Fig. 5.63b. The strongest reflections were indexed to $D_1 = \{211\}$, $D_2 = \{002\}$, $D_3 = \{132\}$, and $D_4 = \{501\}$ [152]. In some areas we also find isolated crystals of NiO with a predominating cubic morphology and the edges truncated by rhombic dodecahedral faces. A close-up of a typical NiO nanocrystal is shown in Fig. 5.63c, showing well-resolved {200} lattice planes. The diffraction rings of the NiO phase cannot be seen most probably due to much stronger diffraction lines for the $MoNi_4$ and MoO_3 phases. Accordingly, its presence is determined by the measurement of D values, as shown in Fig. 5.63c [123].

It should also be stated that on both SAED patterns, and other SAED patterns, additional diffraction lines (more or less pronounced) are present, indicating that at the position of TEM analysis other phases are also present (actually at each position of TEM analysis mixture of different phases exists), but the best defined are

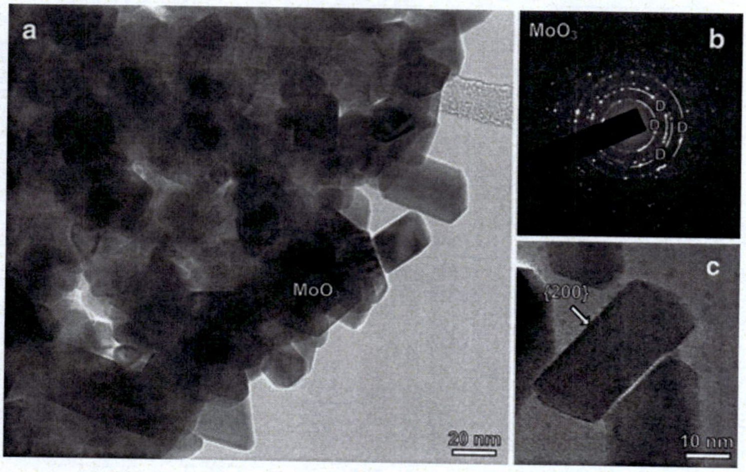

Fig. 5.62 (**a**) TEM image of euhedral MoO_3 nanocrystals in the sample with the Ni/Mo ratio 1/0.3. (**b**) Electron diffraction pattern from this area indicates the presence of monoclinic MoO_3 phase. The strongest SAED diffraction rings in the pattern correspond to and $D_1 = \{200\}$, $D_2 = \{211\}$, $D_3 = \{220\}$, and $D_4 = \{022\}$. (**c**) MoO_3 crystals are terminated with well-defined {200} dominating pinacoidal faces (Reprinted from [123] with the permission of Elsevier.)

the ones determined by the analysis of the SAED patterns. Hence, it could be concluded that the $MoNi_4$ and MoO_3 phases prevail in the powder with high Ni content (electrodeposited at the Ni/Mo ratio 1/0.3), while some amount of NiO phase has also been detected.

5.4.6.2 The TEM Analysis of Electrodeposited Mo-Rich Mo–Ni–O Powders

A general appearance of Mo-rich sample does not differ much from the Ni-rich sample. The TEM of typical amorphous cluster is shown in Fig. 5.64a. In the small crystalline areas (x1) traces of $MoNi_4$ and MoO_3 phases were detected and their diffraction patterns are presented in Fig. 5.64b, c, respectively. The diffraction lines in Fig. 5.64b are barely visible, indicating amorphous character of the

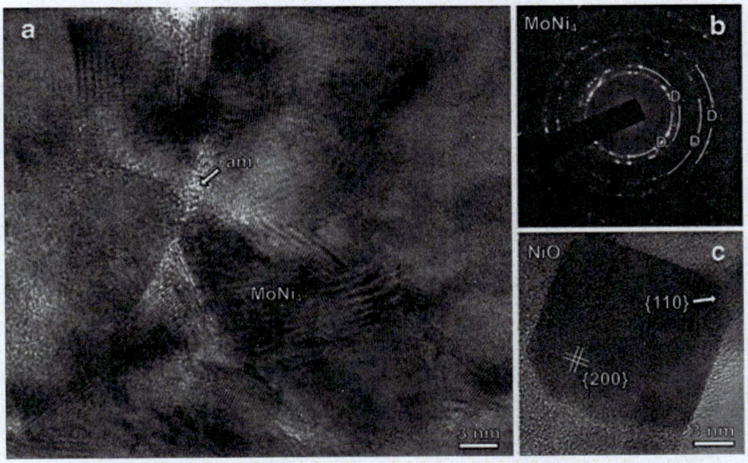

Fig. 5.63 (**a**) High-resolution TEM image of intergrown $MoNi_4$ nanocrystals. In the pockets of $MoNi_4$ grains there are remnants of uncrystallized amorphous phase (am). (**b**) SAED pattern shows rings of high intensity spots at reciprocal distances corresponding to $MoNi_4$ phase. The strongest reflections were indexed to $D_1 = \{211\}$, $D_2 = \{002\}$, $D_3 = \{132\}$, and $D_4 = \{501\}$. (**c**) Idiomorphic NiO particle (rock-salt structure) with well-resolved {200} lattice planes. The crystal is faceted with cube {001} and dodecahedral {110} faces (Reprinted from [123] with the permission of Elsevier.)

$MoNi_4$ phase, while much stronger diffraction lines for the MoO_3 phase (Fig. 5.64c) indicate its more pronounced crystallinity. In this sample the presence of NiO phase has not been detected.

It appears that in both samples $MoNi_4$ phase is more strongly bound to the amorphous phase than the oxide phases. This might be the reason why in our previous investigation [121, 122] this phase has clearly been detected by XRD in the recrystallized samples only after annealing at 600°C, while at all other temperatures a broad peak around $2\theta = 44°$ was only an indication of its existence.

It should be emphasized here that most of the phases detected in the recrystallized electrodeposited Mo–Ni–O powders [121, 122] ($MoNi_4$ and MoO_3) were also found in as-deposited Mo–Ni–O powders, indicating their formation during the process of alloy electrodeposition.

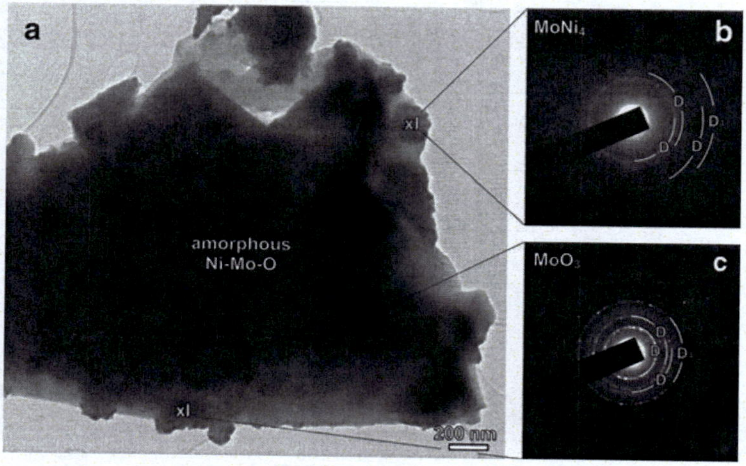

Fig. 5.64 (**a**) TEM image of the amorphous Mo–Ni–O cluster containing some crystalline areas in the powder electrodeposited at the Ni/Mo ratio 1/3. (**b**) SAED pattern from the area rich in $MoNi_4$ nanocrystals, similar to those in the Ni-rich sample, shown in this figure. (**c**) SAED pattern of the area containing MoO_3 nanocrystals, similar to those in the Ni-rich sample, shown in Fig. 5.63 (Reprinted from [123] with the permission of Elsevier.)

As stated earlier the most probable mechanism for Ni and Mo codeposition is the one reported by Podlaha and Landolt [117–120] after X-ray fluorescence analysis of the electrodeposited alloy. Their investigations were performed under controlled mass transport conditions (rotating cylinder electrode). The model assumes that the Ni electrodeposition occurs on the surface not covered by the molybdate ions as a reaction intermediate, by direct reduction of nickel species (all of them being complex of Ni^{2+} cations with the citrate anions), independently on the molybdate reaction which can occur only in the presence of nickel species [117–120]. The model of the Mo–Ni alloy electrodeposition is described by the following reduction reactions:

$$\mathrm{NiCit^{-}} + 2e^{-} \rightarrow \mathrm{Ni(s)} + \mathrm{Cit^{3-}}, \qquad (5.17)$$

$$MoO_4{}^{2-} + NiCit^- + 2H_2O + 2e^- \rightarrow [NiCitMoO_2]_{ads}{}^- + 4OH^- \tag{5.18}$$

$$[NiCitMoO_2]_{ads}{}^- + 2H_2O + 4e^- \rightarrow Mo(s) + NiCit^- + 4OH^- \tag{5.19}$$

$$H_2O + e^- \rightarrow \frac{1}{2}H_2 + OH^- \tag{5.20}$$

assuming that the Mo–Ni alloy electrodeposition is always accompanied by the simultaneous hydrogen evolution [reaction (5.20)]. This model has been recently confirmed, since the existence of adsorbed intermediate $[NiCitMoO_2]_{ads}{}^-$ by the in situ surface Raman spectroscopic studies has been detected [154]. However, it should be noted here that the whole mechanism has been based on the X-ray fluorescence analysis of the electrodeposited alloys, and from the content of Ni and Mo their partial current densities of electrodeposition were calculated and used for the analysis of the mechanism [117–120].

In our investigations Ni^{2+} ions were complexed with ammonia. Six complexes of nickel with ammonium anion can be formed (their stability constants are also given): [124]

$$Ni^{2+} + NH_3 \leftrightarrow [Ni(NH_3)]^{2+} \quad K_1 = 6.31 \times 10^2 \tag{5.21}$$

$$Ni^{2+} + 2NH_3 \leftrightarrow [Ni(NH_3)_2]^{2+} \quad K_2 = 2.51 \times 10^5 \tag{5.22}$$

$$Ni^{2+} + 3NH_3 \leftrightarrow [Ni(NH_3)_3]^{2+} \quad K_3 = 5.89 \times 10^6 \tag{5.23}$$

$$Ni^{2+} + 4NH_3 \leftrightarrow [Ni(NH_3)_4]^{2+} \quad K_4 = 9.12 \times 10^7 \tag{5.24}$$

$$Ni^{2+} + 5NH_3 \leftrightarrow [Ni(NH_3)_5]^{2+} \quad K_5 = 5.13 \times 10^8 \tag{5.25}$$

$$Ni^{2+} + 6NH_3 \leftrightarrow [Ni(NH_3)_6]^{2+} \quad K_6 = 5.49 \times 10^8 \tag{5.26}$$

Another complex which could be involved in the reaction mechanism is $[Ni(OH)_3]^-$ complex. Taking into account that the solution pH is 9, the concentration of OH^- ions is very low (of the order of 10^{-5} M), while the concentration of ammonium ions is 1.7 M. Due to the low dissociation constant of ammonium hydroxide, actual concentration of ammonia in such solution is 0.6 M [155]. According to the literature [156], the dominant complex of nickel is $[Ni(NH_3)_3]^{2+}$, formed by reaction (5.23), although its stability constant is not the highest. Hence, this complex is involved in the reaction of Mo–Ni–O powder deposition.

Taking into account that NiO, MoO_3, and $MoNi_4$ phases were detected in as-deposited samples by TEM, it is obvious that the mechanism of alloy electrodeposition similar to the one presented by equations (5.17)–(5.20) could be applied for the formation of $MoNi_4$ phase only, since it assumes complete reduction of both metal ions. By the analogy with the mechanism proposed by Podlaha and Landolt [117–120], we believe that the $MoNi_4$ phase could be formed by the following mechanism:

$$[Ni(NH_3)]^{2+} + 2e^- \rightarrow Ni(s) + 3NH_3 \tag{5.27}$$

$$\begin{aligned} &MoO_4{}^{2-} + [Ni(NH_3)_3]^{2+} + 2H_2O + 2e^- \\ &\quad \rightarrow [Ni(NH_3)_3MoO_2]_{ads}{}^{2+} + 4OH^- \end{aligned} \tag{5.28}$$

$$\begin{aligned} &[Ni(NH_3)_3MoO_2]_{ads}{}^{2+} + 2H_2O + 4e^- \\ &\quad \rightarrow Mo(s) + [Ni(NH_3)_3]^{2+} + 4OH^- \end{aligned} \tag{5.29}$$

The most probable mechanism for the MoO_3 phase formation is

$$\begin{aligned} &MoO_4{}^{2-} + [Ni(NH_3)_3]^{2+} + H_2O + 2e^- \\ &\quad \rightarrow MoO_3 + Ni(s) + 3NH_3 + 2OH^- \end{aligned} \tag{5.30}$$

The presence of NiO phase cannot be explained by the reduction of nickel complex. According to Palmer et al. [157] NiO could be formed by following chemical reactions:

$$Ni^{2+} + H_2O \rightarrow NiO + 2H^+ \tag{5.31}$$

$$[Ni(OH)]^+ \rightarrow NiO + H^+ \quad (5.32)$$

$$[Ni(OH)_2] \rightarrow NiO + H_2O \quad (5.33)$$

$$[Ni(OH)_3]^- + H^+ \rightarrow NiO + 2H_2O \quad (5.34)$$

Reactions (5.31), (5.32), and (5.34) are highly impossible since Ni^{2+}, $[Ni(OH)]^+$, and $[Ni(OH)_3]^-$ are not present in the solution, while the reaction (5.33) could be considered as possible (as explained below). Although at a given pH the concentration of the OH^- ions is very low (of the order of 10^{-5} M), due to massive hydrogen evolution during the process of powder electrodeposition, reaction (5.20), the concentration of the OH^- ions in the vicinity of the electrode surface significantly increases and the reaction between OH^- ions and $[Ni(NH_3)_6]^{2+}$ can occur producing $[Ni(OH)_2]$, since the stability constant for $[Ni(OH)_2]$ is two orders of magnitude higher than that for $[Ni(NH_3)_3]^{2+}$. Hence, $[Ni(OH)_2]$ could be formed by the reaction

$$[Ni(NH_3)_3]^{2+} + 2OH^- \rightarrow [Ni(OH)_2] + 3NH_3 \quad (5.35)$$

Formed $[Ni(OH)_2]$ becomes entrapped in the deposit to a certain extent. During the procedure of powder drying at 120°C, reaction (5.33) can take place. The "transition temperature at which nickel hydroxide dehydrates to form the oxide is *ca.* 75°C" [Eq. (5.33)] [157]. Thus, the most possible way for the NiO formation is dehydration of entrapped $[Ni(OH)_2]$ during the procedure of powder drying. In such a case smaller amount of the NiO phase than the amounts of other phases (MoO_3 and $MoNi_4$) should be expected in the deposit, as it was the case in the TEM analysis [123].

Hence, it seems that complete reduction of both metal ions occurs to some extent [reactions (5.27)–(5.29)] forming $MoNi_4$ phase, but the MoO_3 and NiO phases must be produced by additional reactions, (5.30), (5.33) and (5.35), respectively. It is important to note that the formation of metallic Ni(s) is a result of two reactions, (5.27) and (5.30), and that this could be the reason for a formation of $MoNi_4$ phase during complete reduction of both metal ions by reactions (5.27)–(5.29). Of course, it should be stated here that the most probable mechanism proposed by Podlaha and Landolt [117–120] is valid for the deposition of compact deposits from citrate containing solution (although there is an indication for the presence of oxygen in the compact deposits too [152–154]), while we were investigating powder electrodeposition from different ammonium sulfate solution. On the other side, it seems reasonable to assume that the electrodeposition mechanism does not depend on the potential (current) and that it should be the same for the deposition of compact and powderous deposits.

After these findings it is much easier to understand the process of solid state reaction that is supposed to take place during the annealing of these powders [121, 122]. The formation of $NiMoO_4$ phase should be the consequence of the reaction between NiO and MoO_3 in the solid state, which takes place at a temperature of about 600°C [150, 158, 159] and is more pronounced in the Ni-rich Mo–Ni–O powder [122].

5.5 Conclusions

The composition of the electrolyte, i.e., Ni^{2+}/Co^{2+} ions ratio, influences polarization curves for Co–Ni alloy powders electrodeposition, as well as alloy powder composition and morphology. It is also concluded that the shape of the polarization curves in all cases is practically defined by the shape of the polarization curve for hydrogen evolution. In the case of Ni powder cauliflower and spongy-type particles are detected, while Co powder contains only spongy-type particles. Both types of particles are present in the Co–Ni alloy powders obtained at Ni^{2+}/Co^{2+} ions ratios 4.00, 1.50, and 0.67, while only spongy-type particles are detected in the Co–Ni alloy powder obtained at Ni^{2+}/Co^{2+} ions ratio 0.25. In all cases of alloy powder electrodeposition anomalous type of codeposition is detected.

The morphology and composition of electrodeposited Fe–Ni alloy powders depend on the Ni/Fe ions concentration ratio in both electrolytes. Anomalous codeposition of Fe and Ni has been confirmed by the EDS analysis of alloy powders, being more pronounced in the electrolyte containing Fe(III) salts due to the possibility of $Fe(OH)_2$ formation. A common characteristic for all alloy powder samples was the presence of cone-shaped cavities and nodules. The possibility of the formation of single crystal $FeNi_3$ phase was found to exist in both electrolytes. It appears that it is better to use electrolyte with Fe(II) salts, independently of the supporting electrolyte, since the current efficiency for Fe–Ni powder electrodeposition is much higher than that from the solution containing Fe(III) salts.

After annealing in air at 400°C, 600°C, and 700°C for 3 h all alloy powders oxidized forming NiO, $NiFe_2O_4$, and Fe_2O_3 with the NiO phase disappearing in the samples with higher percentage of Fe. The $NiFe_2O_4$ phase was found to be dominant in the sample with the highest percentage of Fe annealed at 600°C. From the EDS and XRD analysis it could be concluded that the powders were not completely oxidized during annealing in air at 400°C and 600°C.

The morphology, chemical, and phase composition of as-deposited and annealed Mo–Ni–O powders depend on the Ni/Mo ions concentration ratio. XRD and SEM analysis revealed that all alloy powders are partially recrystallized already at 300°C, showing the presence of MoO_3 and $NiMoO_4$ phases. After annealing at 600°C complete recrystallization of all powders is confirmed by SEM analysis, showing the absence of any powder agglomerates and the presence of well-defined crystal grains. In the powder electrodeposited at the Ni/Mo ratio 1/0.3 XRD revealed the presence of only one phase, $NiMoO_4$, indicating that a single phase powder of the system Ni–Mo–O could be obtained by proper choice of electrodeposition parameters. At the same time the disappearance of MoO_3 phase for this sample confirmed that the phase transition from MoO_3 into $NiMoO_4$ phase occurred during annealing at 600°C. The TEM and XRD analysis revealed that two types of particles could be detected in both Ni-rich and Mo-rich as-deposited powders: amorphous and crystalline particles. Among the crystalline particles in both powders MoO_3 and $MoNi_4$ phases prevail, while in the Ni-rich powder NiO phase has also been detected. Based on these findings the corresponding mechanism of Mo–Ni–O powders electrodeposition has been proposed.

Acknowledgments This work was financially supported by the Ministry of Education and Science of the Republic of Serbia through the Projects No. 1806/2002, No. 142032G/2006, and No. 172054/2011.

The authors are indebted to D. Poleti from the Faculty of Technology and Metallurgy University of Belgrade, Serbia, for DSC-TGA analysis.

The authors are also indebted to V.M. Maksimović from the Institute of Nuclear Sciences—Vinča, Belgrade, Serbia, for the XRD analysis of all investigated systems.

Special thanks to A. Rečnik from the Jožef Stefan Institute, Ljubljana, Slovenia, for TEM analysis of as-deposited samples.

The authors also wish to express their gratitude to M.G. Pavlović, Institute of Electrochemistry ICTM, Belgrade, Serbia, for useful discussions in the case of the Co–Ni system.

References

1. Papers CT, Brabyn SM (1987) Met Powder Rep 42:863
2. Ekemar S, Lindholm L, Hartzell T (1982) Int J Refract H 1:37
3. Erol S, Nursel D (1999) J Adhes Sci Technol 13:679
4. Rehim SS, Halim AM, Osman MM (1985) J Appl Electrochem 15:107
5. Laughlin D, Lu B, Hsu Y, Zou J, Lambeth D (2000) IEEE Trans Magn 36:48
6. Vassal N, Salmon E, Fauvarque J (1999) J Electrochem Soc 146:20
7. Benvenuti F, Carlini C, Marchetti F, Marchionna M, Galletti A, Sbrana G (2001) J Organomet Chem 622:286
8. Tsay M, Chang F (2000) Appl Catal A Gen 203:15
9. Diskin A, Cunningham R, Ormerod R (1998) Catal Today 46:147
10. Kapoor S, Salunke HG, Tripathi AK, Kulshreshtha SK, Mittal JP (2000) Mater Res Bull 35:143
11. Ishihara T, Horiuchi N, Inoue T, Eguchi K, Takita Y, Arai H (1992) J Catal 136:232
12. Kikuko K, Teruh iko I (1998) Jpn Kokai Tokkyo Koho JP 10302790
13. Bianco A, Gusmano G, Montanari R, Montesperelli G, Traversa E (1994) Mater Lett 19:263
14. Bianco A, Gusmano G, Montanari R, Montesperelli G, Traversa E (1995) Thermochim Acta 269(270):117
15. Aymard L, Dumont B, Viau G (1996) J Alloys Compd 242:108
16. Huang J, Wu Y, Ye H (1996) Acta Mater 44:1201
17. Koltypin Y, Katabi G, Cao X, Prozorov R, Gedanken A (1966) J Non Cryst Solids 201:159
18. Kapoor S, Salunke H, Tripathi A, Kulshreshta S, Mittal J (2000) Mater Res Bull 35:143

19. Kurikka V, Gedanken A, Prozorov R, Revesz A, Lendvai J (2000) J Mater Res 15:332
20. Fievet F, Lagier J, Blin B, Meaudoin B, Figlarz M (1989) Solid State Ionics 32(33):198
21. Li Y, Li L, Liao H, Wang H, Qian Y (1999) J Mater Chem 9:2675
22. Degen A, Macek J (1999) Nanostruct Mater 12:225
23. Gibson KP (1995) Science 267:1338
24. Chen D, Wu S (2000) Chem Mater 12:1354
25. Zhang D-E, Ni X-M, Zhang X-J, Zheng H-G (2006) J Magn Magn Mater 302:290
26. Hayashi T, Ohno T, Yatsuda S, Uyeda R (1977) Jpn J Appl Phys 16:705
27. Dong L, Zhang Z, Jin S, Sun W, Chuang Y (1998) Nanostruct Mater 10:585
28. Bianco A, Gusmano G, Montanari R, Montesperelli G, Traversa E (1995) Thermochim Acta 269:117
29. Viau G, Ravel F, Acher O, Fiévet-Vincent F, Fiévet F (1995) J Magn Magn Mater 144:377
30. Gao X, Chen D, Dollimore D, Skrzypczak-Jankum E, Burckel P (1993) Thermochim Acta 220:75
31. Neddermann R, Binnewies M (1996) Z Anorg Allg Chem 622:17
32. Girirdin D, Maurer M (1990) Mater Res Bull 25:119
33. Xiaoli X, Zuoren N, Yabao J, Peiyun T, Shunlin S, Jie X, Tieyong Z (2008) J Alloys Compd 466:387
34. Jang HC, Ju SH, Kang YC (2009) J Alloys Compd 478:206
35. Abd El-Halim AM, Khalil RM (1986) Surf Coat Technol 27:103
36. Yur'ev BP, Golubkov LA (1969) Trudy—Leningradskii Politekhnicheskii Institut imeni M. I. Kalinina 14:269
37. Jović VD, Maksimović V, Pavlović MG, Popov KI (2006) J Solid State Electrochem 10:373
38. Jović VD, Jović BM, Pavlović MG, Maksimović V (2006) J Solid State Electrochem 10:959
39. Jović VD, Jović BM, Maksimović V, Pavlović MG (2007) Electrochim Acta 52:4254
40. Jović VD, Jović BM, Pavlović MG (2006) Electrochim Acta 51:5468
41. Jović VD, Maksimović V, Pavlović MG, Jović BM (2006) Mater Sci Forum 518:307
42. Maurice DR, Courtney TH (1990) Metall Mater Trans A 21A:289
43. Hamzaoui R, Elkedim O, Greneche JM, Gaffet E (2005) J Magn Magn Mater 294:e145
44. Hamzaoui R, Elkedim O, Gaffet E (2004) Mater Sci Eng A 381:363
45. Valderruten JF, Perez Alcazar GA, Greneche JM (2006) Phys B 384:316
46. Zhou PH, Deng LJ, Xie JL, Liang DF, Chen L, Zhao XQ (2005) J Magn Magn Mater 292:325
47. Kaloshkin SD, Tcherdyntsev VV, Tomilin IA (2001) Phys B 299:236
48. Tcherdyntsev VV, Kaloshkin SD, Tomilin LA, Shelekhov EV, Baldokhin YuV (1999) Nanostruct Mater 12:139

49. Baldokhin YuV, Tcherdyntsev VV, Kaloshkin SD, Kochetov GA, Pustov YuA (1999) J Magn Magn Mater 203:313
50. Baldokhin YV, Kolotyrkin PY, Petrov YI, Shafranovsky EA (1994) Phys Lett A 189:137
51. Schirmer B, Wuttig M (1999) Phys Rev B 60:945
52. Kuhrt C, Schultz L (1993) J Appl Phys 73:1975
53. Jartych E, Zurawicz JK, Oleszak D, Pekala M (2000) J Magn Magn Mater 208:221
54. Koohkana R, Sharafia S, Shokrollahib H, Janghorbanb K (2008) J Magn Magn Mater 320:1089
55. Pandey B, Verma HC (2007) J Phys Condens Matter 19:406207
56. Wang H, Liu Q, Zhang J, Hsu TY (Zuyao X) (2003) Nanotechnology 14:696
57. Schneeweissa O, Davida B, Zaka T, Zborilb R, Mashlanb M (2007) J Magn Magn Mater 310:e858
58. Song HB, Lee KJ, Kim KH, Oh ST, Lee SK, Choa YH (2010) J Nanosci Nanotechnol 10:106
59. Kim KH, Yu JH, Lee SB, Lee SK, Choa YH, Oh ST, Kim JR (2008) IEEE Trans Magn 44:3805
60. Kasagi T, Tsutaoka T, Hatakeyama K (1999) IEEE Trans Magn 35:3424
61. Oh ST, Joo MH, Choa YH, Kim KH, Lee SK (2010) Phys Scr T139:014050
62. Lačnjevac U, Jović BM, Jović VD (2009) Electrochim Acta 55:535
63. Lačnjevac U, Jović BM, Maksimović VM, Jović VD (2010) J Appl Electrochem 40:701
64. Zhelibo EP, Kravets NN, Gamarkin MYu, Remez SV (1995) Powder Metall Metal Ceram 34:113
65. Zhelibo EP, Kravets NN (1997) Powd Metall Metal Ceram 36:264
66. Chu CM (2003) J Chin Inst Eng 34:689
67. Morrish AH, Haneda KJ (1981) Appl Phys 52:2496
68. Ishino K, Narumiya Y (1987) Am Ceram Soc Bull 66:1469
69. Zhang Q, Itoh T, Abe M, Tamaura Y (1992) In: Yamaguchi T, Abe M (eds) Proceedings of the 6th international conference on ferrites. The Japan Society of Powder and Powder Metallurgy, Tokyo, p 481
70. Dube GR, Darshane YS (1993) J Mol Catal 79:285
71. Gopal Reddy CV, Manorama SV, Rao VJ (1999) Sens Actuators B Chem 55:90
72. Satyanarayana LK, Reddy KM, Manorama SV (2003) Mater Chem Phys 82:21
73. Abe M, Itoh T, Tamaura Y et al (1998) J Appl Phys 63:3774
74. Itoh T, Abe M, Sasao T et al (1989) IEEE Trans Magn 25:4230
75. Suran G, Heurtel A (1972) J Appl Phys 43:536
76. Naoe M, Yamanaka S (1970) Jpn J Appl Phys 9:293
77. Marshall DJ (1971) J Cryst Growth 9:305
78. Gibart P, Robbins M, Kane AB (1974) J Cryst Growth 24–25:166
79. Pulliam GR (1967) J Appl Phys 38:1120
80. Mee JE, Pulliam GR, Archer JL et al (1969) IEEE Trans Magn 5:717

81. Fitzgerald AG, Engin R (1974) Thin Solid Films 20:317
82. Itoh H, Takeda T, Naka S (1986) J Mater Sci 21:3677
83. Tsuchiya T, Yamashiro H, Sei T et al (1992) J Mater Sci 27:3645
84. Jung DS, Kang YC (2009) J Magn Magn Mater 321:619
85. Deschanres JL, Langlet M, Joubert JC (1990) J Magn Magn Mater 83:437
86. Lee PY, Ishizaka K, Suematsu H et al (2006) J Nanocryst Res 8:29
87. Sartale SD, Lokhande CD, Giersig M et al (2004) J Phys Condens Matter 16:773
88. Fang J, Shama N, Tung L et al (2003) J Appl Phys 93:7483
89. Ceylan A, Ozcan S, Ni C et al (2008) J Magn Magn Mater 320:857
90. Kinh VO, Chassaing E, Saurat M (1975) Electrodep Surf Treat 3:205
91. Yao SW, Zeng Y, Guo HT (1994) Surf Tech (Japan) 45:643
92. Friend WZ (1980) Corrosion of nickel and nickel-base alloys. Wiley Interscience, New York, p 248
93. Kriz JF, Shimada H, Yoshimura Y, Matsubayashi N, Nishijama A (1995) Fuel 74:1852
94. Astier MP, Dji G, Teichner SJ (1991) Appl Catal 72:321
95. Tsenta TE, Knyazheva VM, Svistunova TV, Kolotyrkin YM, Zakharin DS (1989) Prot Met 25:28
96. Beltowska-Lehman E (1990) J Appl Electrochem 20:132
97. Drown DE, Mahmood MN, Turner AK, Hall SM, Fogarty PO (1982) Int J Hydrogen Energy 7:405
98. Huot JY, Brossard L (1988) J Appl Electrochem 18:815
99. Conway BE, Bai L, Sattar MA (1987) Int J Hydrogen Energy 12:607
100. Raj IA, Vasu KI (1992) J Appl Electrochem 22:471
101. Conway BE, Bai L, Tessier DF (1984) J Electroanal Chem 161:39
102. Fan C, Piron DL, Paridis P (1994) Electrochim Acta 39:2715
103. Conway BE, Bai L (1985) J Chem Soc Faraday Trans I 81:1841
104. Raj IA, Kovenkatesan V (1988) Int J Hydrogen Energy 12:215
105. Fan C, Piron DL, Sleb A, Paridis P (1994) J Electrochem Soc 141:382
106. Divisek J, Schmotz H, Balej J (1989) J Appl Electrochem 19:519
107. Lasia A, Rami A (1990) J Electroanal Chem 294:123
108. Jakšić JM, Vojnović MV, Krstajić NV (2000) Electrochim Acta 45:4151
109. Gennero de Chialvo MR, Chialvo AC (1998) J Electroanal Chem 448:87
110. Jakšić MM (1989) Mater Chem Phys 22:1
111. Kedzierzawski P, Oleszak D, Janik-Czachor M (2001) Mater Sci Eng A300:105
112. Oleszak D, Portnoy VK, Matyja H (1999) Mater Sci Forum 312:345
113. De la Torre SD, Oleszak D, Kakitsuji A, Miyamoto K, Miyamoto H, Martinez SR, Almeraya CF, Martinez VA, Rois JD (2000) Mater Sci Eng A276:226
114. Goswami GL, Kumar S, Galun R, Mordike BL (2003) Lasers Eng 13:1
115. Bhattacharjee PP, Ray RK, Upadhyaya A (2006) Physica C449:116
116. Brenner A (1963) Electrodeposition of alloys. Priniciples and practice, vol 2. Academic, New York

117. Podlaha EJ, Landolt D (1996) J Electrochem Soc 143:885
118. Podlaha EJ, Landolt D (1996) J Electrochem Soc 143:893
119. Podlaha EJ, Landolt D (1997) J Electrochem Soc 144:1672
120. Marlot A, Kern P, Landolt D (2002) Electrochim Acta 48:29
121. Jović BM, Jović VD, Maksimović VM, Pavlović MG (2008) Electrochim Acta 53:4796
122. Lačnjevac U, Jović BM, Baščarević Z, Maksimović VM, Jović VD (2009) Electrochim Acta 54:3115
123. Jović VD, Jović BM, Lačnjevac U, Branković G, Bernik S, Rečnik A (2010) Electrochim Acta 55:4188
124. Dean JA (1985) Lange's handbook of chemistry, 13th edn. Mc-Graw Hill, New York, p 5
125. Despić AR, Jović VD (1995) In: White RE et al (eds) Modern aspects of electrochemistry, chap 2, vol 27. Plenum, New York
126. Horkans J (1979) J Electrochem Soc 126:1861; (1981) J Electrochem Soc 128:45
127. Jepson F, Meecham S, Salt FW (1955) Trans Inst Met Finish 32:160
128. Young CBF, Struyk C (1946) Trans Electrochem Soc 89:383
129. Schoch EP, Hirsch A (1907) Trans Am Electrochem Soc 11:135
130. Jović VD, Tošić N, Stojanović M (1997) J Electroanal Chem 420:43
131. Lönnberg B (1994) J Mater Sci 29:3224
132. Hansen M, Andrenko K (1958) Constitution of binary alloys. Mc-Graw Hill, New York
133. Calusaru A (1979) Electrodeposition of powders from solutions. Elsevier, New York
134. Wranglen G (1960) Electrochim Acta 2:1845
135. Kieling VC (1997) Surf Coat Technol 96:135
136. Bard AJ, Faulkner LR (2001) Electrochemical methods: fundamentals and applications, 2nd edn. Wiley, New York
137. Zhou XM, Wei XW (2009) Cryst Growth Des 9:7
138. Popov KI, Djokić SS, Grgur BN (2002) Fundamental aspects of electrometallurgy. Kluwer, New York
139. Chassaing E, Portal N, Levy AF, Wang G (2004) J Appl Electrochem 34:1085
140. Sanches LS, Domingues SH, Marino CEB, Mascaro LH (2004) Electrochem Commun 6:543
141. Donten M, Celsiulis H, Stojek Z (2005) Electrochim Acta 50:1405
142. Morgenstern T, Lienhardt JL, Reichelt W, Koenig U, Oppermann H (1993) Mater Sci Forum 133–136:627
143. Pejryd L (1985) Scand J Metall 14:268
144. Hussain OM, Ramana CV, Zaghib K, Julien CM (2006) In: Chowdari BVR et al (eds) Proceedings of the 10th Asian conference on solid state ionics: advanced materials for emerging technologies. World Scientific, River Edge, NJ, p 136
145. McCarron EM III (1986) J Chem Soc Chem Commun 336

146. Parise JB, McCarron EM III, Sleight W (1987) Mater Res Bull 22:803
147. Smith GW (1962) Acta Cryst 15:1054
148. Smith GW, Ibers JA (1965) Acta Cryst 19:269
149. Abrahams SC, Reddy JM (1965) J Chem Phys 43:2533
150. Sleight AW, Chamberland BL (1968) Inorg Chem 7:1672
151. Plyasova LM, Ivanchenko IYu, Andrushkevich MM, Buyanov RA, Itenberg ISh, Khramova GA, Karakchiev LG, Kustova GN, Stepanov GA, Tsailingold AL, Pilipenko FS (1973) Kinet Catal 14:1010
152. Harker D (1944) J Chem Phys 12:315
153. Parise JB et al (1991) J Solid State Chem 93:193
154. Zeng Y, Li Z, Ma M, Zhou S (2000) Electrochem Commun 2:36
155. Min X, Zhou M, Chai L, Wang Y, Shu Y (2009) Trans Nonferrous Met Soc China 19:1360
156. Younes O, Gileadi E (2002) J Electrochem Soc 149:C100
157. Palmer DA, Benezeth P, Wesolowski DJ (2005) In: Proceedings of the 14th international conference on the properties of water and steam, Kyoto, p 264
158. Trambouze Y, Colleuille Y, The TH, Hebd CR (1956) Seances Acad Sci Ser C 242:497
159. Yanushkevich TM, Zhukovskii VM, Ust'yantsev VM (1974) Russ J Inorg Chem 19:1056

Chapter 6
Electrochemical Synthesis of Dispersed Metallic Nanoparticles

Luca Magagnin and Paula Cojocaru

6.1 Introduction

Micro- and nanosized particles of noble and precious metals, with size uniformity, morphological homogeneity, and high purity, are a class of material of great interest for many applications in the field of nanotechnology, due to their unusual catalytic and physical–chemical properties, and for their possible exploitation in catalysis and advanced technology applications. Several techniques have been employed aiming at the fabrication of nanoparticles of different metals, e.g. the chemical reduction of the chloro-metallate anion, metal vapor synthesis, and electrochemical deposition [1–8]. Some other methods are being used for the preparation of gold nanoparticles like sol–gel, deposition–precipitation, and dip-coating techniques [9, 10]. Of these, recent studies have shown that electrodeposition is an easy and a fast procedure for the preparation of nanoparticles on a substrate [6, 11].

Electrochemical synthesis of metal powder has been traditionally used for mass production of micrometer size powder being primarily used in powder metallurgy as well as other engineering applications [12].

L. Magagnin (✉) • P. Cojocaru
Dip. Chimica, Materiali e Ing., Chimica G. Natta, Politecnico di Milano, Via Mancinelli, 7, 20131 Milano, Italy
e-mail: luca.magagnin@polimi.it; paula.cojocaru@polimi.it

Stojan S. Djokić (ed.), *Electrochemical Production of Metal Powders*,
Modern Aspects of Electrochemistry 54, DOI 10.1007/978-1-4614-2380-5_6,

The electrochemical synthesis of metal nanoparticles and other nanostructures is being currently considered as a potential route for the development of advanced functional materials and devices in different technology area, including chemical sensing, magnetic storage, micro-electronics, electroanalysis, and electrocatalysis [13–17].

6.2 Electrolytic Synthesis of Nanoparticles

Two major parameters control the chemical, physical, and electro-catalytic properties of the nanoparticles, namely, the particle size and shape (and the crystallographic orientation) of the nanometer scale material. The control of the size and geometry of the prepared nanoparticles remains a difficult aspect. To decrease dispersion in particle size during electrocrystallization, two important principles should be taken into account. First, the crystal seed formation should occur instantaneously thus preventing progressive nucleation. Second, the crystal growth has to be conducted at a slow rate, i.e., at low overvoltage. The importance and coherence of these fundamental principles has been recently elucidated by Penner et al. [18–22]. His research group successfully started extensive investigations on the electrodeposition of silver, gold, platinum, and cadmium nanoparticles on graphite surfaces [18–22]. The ideal method for transforming both principles which favor monodispersity into an experimental procedure can be best attained by applying the potentiostatic double-pulse technique. This method introduced by Scheludko and Todorova is based on an extremely short nucleation pulse of high cathodic polarization followed by a much longer growth pulse at low cathodic overvoltage [23, 24]. The ideal model situation is where nucleation only occurs within the first pulse and exclusive particle growth during the second pulse. The first pulse is more negative than E_{CRIT}, the critical potential for nucleation, and the second pulse is more positive than E_{CRIT} but more negative than the reversible potential (Fig. 6.1). The high cathodic amplitude of the first pulse is necessary to initiate nucleation. To achieve this, the overpotential of the second pulse must be low enough to inhibit generation of new nuclei. Under such circumstances, the original

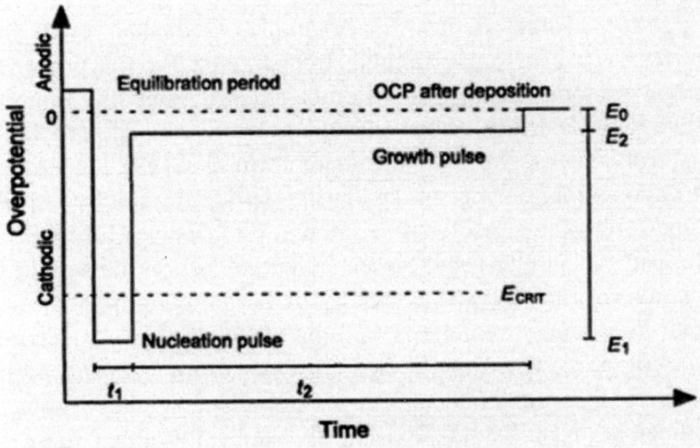

Fig. 6.1 Schematic representation of the potentiostatic double-pulse method. After an equilibration period, a nucleation pulse of potential E_1 is applied for a time t_1. To form nucleation sites, E_1 must be more negative than the critical nucleation potential E_{CRIT}. The first pulse is immediately followed by a growth pulse of potential E_2 for a time t_2. As E_2 is more positive than E_{CRIT}, no further nucleation on the substrate is possible (Reprinted from [24], Copyright 2002 with permission from Elsevier Science [24].)

distribution of stable nuclei formed during the first pulse should be preserved. With this method, the conflict between both optimal conditions for nucleation and growth is partially defused. This is due to the amount of small seed additionally nucleated at the higher polarization and resolved as soon as the potential is switched over to the lower polarization of the growth pulse. According to theory, the double-pulse procedure is an efficient way to control the particle size distribution of depositions.

Very recently, pulsed sonoelectrochemical synthesis, which was first introduced by Reisse et al. and involves alternating sonic and electric pulses, has been employed to obtain shape-controlled synthesis of nanoparticles [25–28]. The present understanding of the sonoelectrochemical mechanism describes an electrochemical deposition of a thin layer on the sonoelectrode during the electric pulse. The thin layer is then destroyed by the coming sonic pulse, to form jet-suspended nanoparticles in the solution. The electrochemical and sonic parameters, including the pulse duration, can be used to control the particle characteristics. For example, Gedanken et al. have reported the use of pulse sonoelectrochemistry for the preparation of silver nanoparticles with different shapes, including spherical, rodlike, and dendritic nanoparticles, and uncovered the crucial role of electrolyte composition in the shape formation [29]. According to their explanation, during sonication the particles are not completely detached from the sonoelectrode, so that the particles formed in the next pulse can further grow on seeds that are left on the surface.

To control the size and geometry of the prepared nanoparticles, another important parameter is the kind of substrate used. According to Markov et al. when metal crystallites or drops nucleate or grow onto a foreign substrate, in their immediate vicinity zones of zero nucleation probability exist [30]. Nucleation under these conditions occurs under the influence of the external electrical field. The growth of nuclei causes deformations in the field, resulting in an overpotential drop in the screened surface segments. When the overpotential is lower than the minimum value for nucleation, further nucleation is impossible. It was shown that for a hemispherical drop on a flat substrate, nucleation is impossible inside a zone of radius r, defined as

$$r = f\frac{E}{E - E_k}\rho, \tag{6.1}$$

where f is a numerical factor accounting for the deviation of geometry, E is cell voltage, and E_k is the critical voltage for nucleation. When $E > E_k$, the zone radius, r, approaches the effective value of the drop size. This explanation is valid only for processes under ohmic control, for example, for deposition of metals having high exchange current densities, as is the case with silver [31, 32]. Another possibility is the deposition of a metal onto its own surface as substrate. In this case, nucleation is also a prerequisite for the deposition process to start, but the critical overpotential value, E_k, is much lower than on a foreign substrate.

Recently, electrodeposition has been proved to provide a versatile method for producing metal nanoparticles and nanowires in the pores of polycarbonate, mica, and aluminum oxide templates [33–36]. In this case, the template imparts size control. Behm et al. obtained periodic arrays of size-similar metal islands by electrochemically decorating defects on the reconstructed gold (111) surface [37–39]. Kolb et al. transferred electrodeposited metal from the tip of an atomic force microscope to a metal surface [40–42]. Metal nanoparticles narrowly dispersed in size were thereby obtained. For example, hundreds of copper nanoparticles were positioned in a square array on a gold surface using this approach. The small size of these particles makes them ideal catalysts for many chemical as well as electrochemical applications [10, 11, 43]. For instance, Au nanoparticles were found to have extraordinary electrocatalytic properties towards the electrochemical reduction of molecular oxygen in acidic media, for the catalytic hydrogenation of unsaturated alcohols and aldehydes, as well as for the low-temperature oxidation of carbon monoxide [10, 11, 43]. Several other applications of Au nanoparticles are currently emerging to include reduction of NO with propene, CO, or H_2, removal of CO from H_2 streams, selective oxidation, e.g., epoxidation of olefins, as well as a selective hydrogenation of CO and CO_2 [44]. Additionally, recent studies have shown that these nanometer scale materials can be used to form quantum dots.

6.3 Sonoelectrochemical Synthesis of Nanoparticles

Sonoelectrochemistry is concerned with the coupling of power ultrasound to electrochemical systems to achieve and develop new processes [45, 46]. Ultrasounds are known to cause a multitude of complex phenomena such as turbulent flow, bubble oscillation and cavitation, generation of microjets, and shock waves [47, 48]. So far, many reports with empirical findings, predominantly the massive increase in the rate of mass transport at the electrode–solution interface, have appeared. Most experimental findings confirm that ultrasound effects are mainly related to mass transport and thermal effects. The mass transport model has been developed into some detail and it is now possible to qualitatively predict the mass transport parameters for certain cell designs and/or ultrasound sources depending on the effect of acoustic streaming [49].

The application of ultrasound irradiation to electrochemistry processes dates back to the 1930s [50]. However, in the last decade the expansion of the sonoelectrochemistry has become increasingly important. Since recently there is a growing interest of the application of the sonoelectrochemistry in environmental remediation and in the preparation of nanopowders. Reisse et al. have described a device for the production of metal powders using pulsed sonoelectrochemical reduction [51]. In those experiments a titanium probe (20 kHz) acts both as a cathode and as an ultrasound emitter [51, 52]. The electroactive part of the sonoelectrode is the planar circular surface at the bottom of the horn and the immersed cylindrical part into the electrolyte is covered by an isolating plastic jacket. The ultrasound probe is connected to a generator and a potentiostat using a pulse driver. The use of a three-electrode configuration (working, reference, and auxiliary electrodes) rather than just a cathode and an anode in a sonoelectrochemistry system was later proposed to apply a controlled potential to the sonoelectrode to get a better control of the process. In the majority of cases the processes have been carried out under galvanostatic conditions. Pulsed sonoelectrochemistry techniques, which use 20 kHz ultrasound horn both as a working electrode and as an ultrasound emitter, have been used to prepare nanopowders. The majority of nanomaterials produced by this method are pure metals. More recently the syntheses have been extended to include the preparation of nanosized metallic alloys, metal oxide semiconductors, and conductive polymers. The ultrasound pulse time and the current density are fundamental for controlling the process yield and particle size: a decrease in size can be achieved by decreasing the temperature, shortening the pulse duration, and increasing the current density and the ultrasound intensity. Nanoparticles aggregates can be formed, even if the shape and size of the nanoparticles can be tailored through the operational parameters and using suitable stabilizers [53].

The present importance of sonoelectrochemistry emerges from the wide range of its applications [54]. The first documented application of an ultrasound field was focused on electroplating. Nowadays, different nanomaterials are prepared by sonoelectrochemical methods. Nanoparticles of magnesium, tungsten, copper, palladium, silver, gold, PbS, copper oxide (I), Au–Ag alloys, and Co–Fe alloys have been synthesized by sonoelectrochemistry. In the same way, nanotubes of CdSe and TiO_2 and silver nanowires have also been synthetized [54–60].

6.4 Experimental Considerations for Electrolytic Synthesis of Nanoparticles

The development and the characterization of an electrochemical process for the synthesis of silver and gold particles from sulfite-based electrolytes are discussed in this section. The solution chemistry is selected in alternative to electrolyte composition inherited from chemical deposition processes and on the basis of its environmental benignity and reasonable chemical stability. The electrolyte composition was preliminary tested and, if needed, modified by adding appropriate components. The electrochemical deposition of silver nanoparticles was carried out from an electrolyte of the following base composition: $AgNO_3$ in the range from 10^{-3} to 0.1 M; $NaNO_3$ in the range from 0.05 to 0.5 M; Na_2SO_3 0.25 M, at pH 8.6. By slowly adding dilute silver nitrate solution to a stirred solution of sodium sulfite and sodium nitrate, precipitation of Ag_2S was avoided. Gold nanoparticles were electrodeposited from an electrolyte containing gold ammonium sulfite $(NH_4)_3Au(SO_3)_2$ 10 mM, Na_2SO_3 0.1 M, and ethylenediamine $C_2H_8N_2$ 10 mM, at pH 6.5. All solutions were prepared using reagent grade chemicals and double distilled water. Deposition tests were performed at room temperature (Ag) or at 55°C (Au). Electrochemical experiments were carried out in a conventional three electrode cell using as working electrode either an amorphous glassy carbon GC disc, apparent surface area 0.2826 cm^2, a titanium sheet or an indium–tin oxide (ITO) substrate, on which an apparent surface area of 1 cm^2 was defined by an insulating tape. The amorphous carbon electrode was polished by 1 μm diamond paste, ultrasonically washed in alcohol and distilled water and finally dried in a nitrogen stream. The titanium sheets were etched in dilute HF solution and polished with 1200 grit paper, ultrasonically washed in distilled water and air dried. A saturated calomel electrode (SCE) served as a reference electrode. The counter electrode was a platinum wire with a much larger surface area than the working electrode. The electrochemical characterization of metal particles deposition was carried out by cyclic voltammetry using a model 273A EG&G PAR potentiostat. The scan rate during voltammetric cycling was changed in the range from 50 to 300 mV s^{-1}. The formation of particles and their morphological features, characterized by scanning electron microscopy, was examined focusing on deposition on titanium substrate, in view of the possible exploitation of nanoparticles deposition as a method to modify and functionalize an electrodic material of widespread industrial use.

6.5 Deposition of Silver Nanoparticles

Cyclic voltammetry was first employed to define the potential region of the deposition process at changing Ag(I) concentration. The electrode potential was driven in the cathodic direction from 0.7 to −0.7 V at a scan rate of 100 mV s^{-1}. Figure 6.2 shows typical voltammetric curves at the glassy carbon electrode changing Ag(I) concentration in solution from 1 to 30 mM.

The peak potential for the cathodic reduction of the silver sulfite complex $[Ag(SO_3)_2]^{3-}$ shifts towards more positive value as the

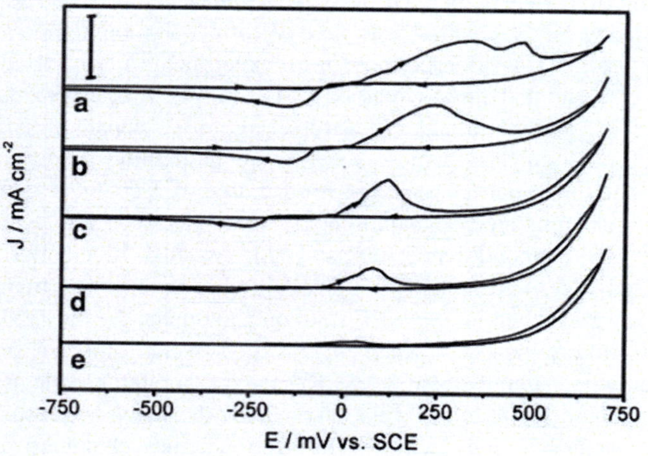

Fig. 6.2 Cyclic voltammograms from 0.7 to −0.7 V at glassy carbon electrode in electrolyte with changing Ag(I) concentration, Na_2SO_3 0.25 M, $NaNO_3$ 0.1 M, at pH 8.6; Ag(I): (**a**) 3×10^{-2} M, (**b**) 2×10^{-2} M, (**c**) 1×10^{-2} M, (**d**) 5×10^{-3} M, (**e**) 1×10^{-3} M. Scale bar 1 mA cm^{-2}

Ag(I) concentration increases, indicating the lowering of the nucleation overpotential as the silver concentration is increased. The formal potential of the complex may be estimated as about 0.02 V SCE by using the reported value of the stability constant (log $K \approx 9.0$) [61]. This is qualitatively confirmed by the voltammetric analysis, observing that, as the Ag(I) concentration increases, the potential value midway between the anodic and cathodic peaks approaches the value of the formal potential as given above. In each voltammogram, two distinct current peaks are clearly resolved. The first peak, which is distinctly seen for Ag(I) concentration in excess of 5 mM, is associated with the Ag(I) reduction process whereas the second peak represents the Ag dissolution process from the amorphous carbon substrate. For Ag(I) concentration 1 mM the voltammogram presents only a shallow wave during the cathodic semicycle, and no crossing of forward and reverse scan is observed, suggesting that nucleation is very limited in these conditions. The peak parameters, current density, and potential change significantly as the Ag(I) concentration increases: the potential of both peaks shift towards more positive values, as expected; besides, both peaks change in shape, showing broadening, and the crossing of the forward and reverse scan becomes more and more evident, as a result of increased nucleation intensity. The anodic current peak suddenly appearing in the voltammogram at Ag(I) concentration 0.03 M may be attributed to the oxidation of dithionite [62].

The above mentioned changes occurring in the silver deposition process at increasing Ag(I) concentration were further characterized. As shown in Fig. 6.3 in the case of Ag(I) 20 mM solution, a linear relationship between the peak current and the root square of the potential scan rate $v^{1/2}$ was found in the range 50–300 mV s^{-1}. In addition, as shown in Fig. 6.4, a linear relationship is found between the peak current density at a fixed potential scan rate and the concentration of Ag(I). These observations indicate that the cathodic reduction of the electroactive species is controlled by diffusive transport [63]. The relationship determined between peak current density and the Ag(I) concentration, as shown by the graph in Fig. 6.4, supports this conclusion.

A further step in this work was to test the effect of a supporting electrolyte on the electrodeposition of silver particles. To this aim,

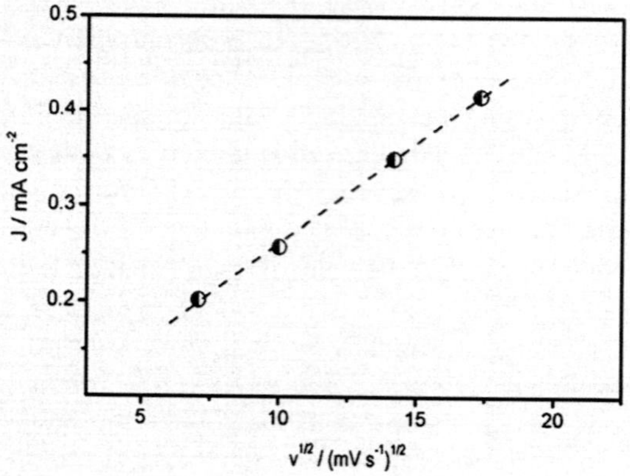

Fig. 6.3 Change of peak current density with potential scan rate at the glassy carbon electrode in Na_2SO_3 0.25 M, $NaNO_3$ 0.1 M, and Ag(I) 20 mM, at pH 8.6

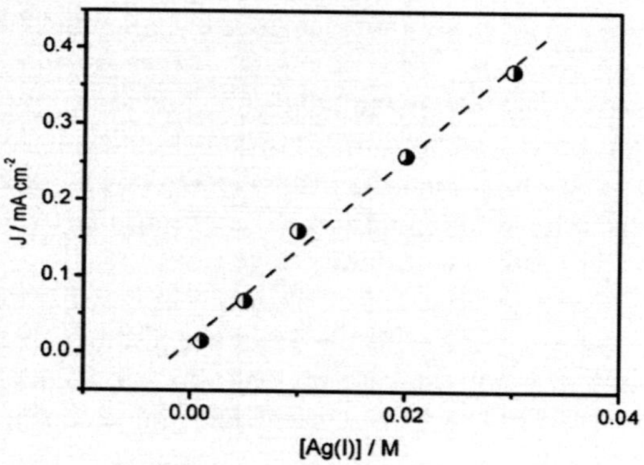

Fig. 6.4 Peak current density at scan rate 100 mV s^{-1} vs. Ag(I) concentration

sodium nitrate $NaNO_3$ was added in increasing concentration to the silver sulfite electrolyte and a cyclic voltammetry characterization (Fig. 6.5) at the glassy carbon electrode was performed. The current density and the potential of voltammetric peaks changed significantly as the sodium nitrate concentration was varied in the deposition bath. In Fig. 6.6 the trend of cathodic peak current density versus $NaNO_3$

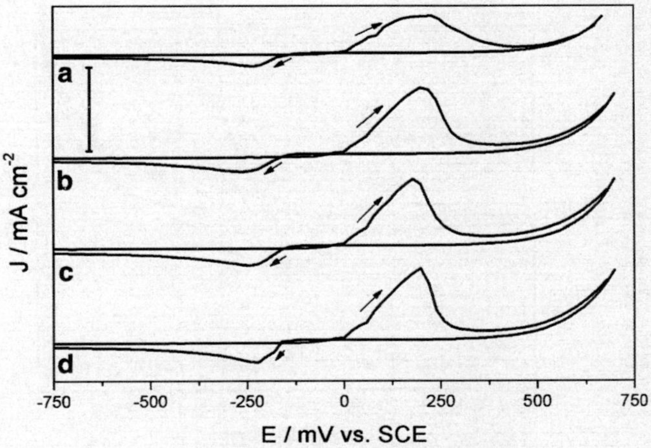

Fig. 6.5 Cyclic voltammograms from 0.7 to −0.7 V at glassy carbon electrode in electrolyte with changing $NaNO_3$ concentration, Na_2SO_3 0.25 M, Ag(I) 20 mM, at pH 8.6; $NaNO_3$ conc.: (**a**) 4×10^{-1} M, (**b**) 3×10^{-1} M, (**c**) 2×10^{-1} M, (**d**) 5×10^{-2} M. Scale bar 1 mA cm^{-2}

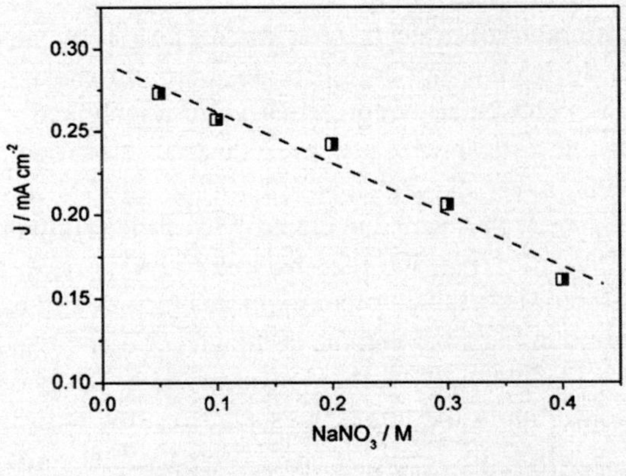

Fig. 6.6 Cathodic peak current density from cycling voltammograms (100 mV s^{-1}) at glassy carbon electrode versus $NaNO_3$ concentration. The electrolyte composition is the following: Ag(I) 20 mM, Na_2SO_3 0.25 M, $NaNO_3$ as reported, pH 8.6

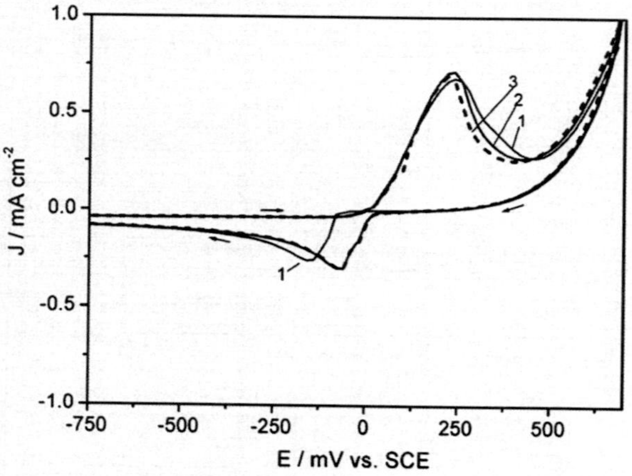

Fig. 6.7 A series of three consecutive cyclic voltammograms obtained for the vitreous carbon electrode in 20 mM Ag(I) + 0.25 M Na_2SO_3 + 0.1 M $NaNO_3$ bath, at a scan rate of 50 mV s^{-1}

concentration is shown. The change is linear with concentration emulating a diffusion limited relationship, which suggests that the main effect of sodium nitrate is upon the effective diffusion coefficient of the silver sulfite complex.

Three different substrate materials were tested for nanoparticles deposition: the glassy carbon electrode, titanium sheets, and ITO. The latter was chosen as a representative of a technical material of interest for its widespread use as industrial electrode substrate. Cycling voltammetry experiments from 0.7 to −0.7 V at a scan rate of 50 mV s^{-1} were performed in electrolyte of the following composition: Ag(I) 20 mM, Na_2SO_3 0.25 M, and $NaNO_3$ 0.1 M, at pH 8.6. A series of three successive voltammograms recorded in each condition is reported in Figs. 6.7 and 6.8 at the glassy carbon and titanium electrode, respectively. The behavior observed at the glassy carbon electrode conforms to the above presented results, with the further remark that after the first scan a remarkable shift towards more positive value of the nucleation potential is apparent. The voltammetric behavior observed at the titanium electrode does not show the standard features expected for a transient nucleation and growth process. In fact, during the first potential scan a steadily increasing cathodic current is observed, which is substantially

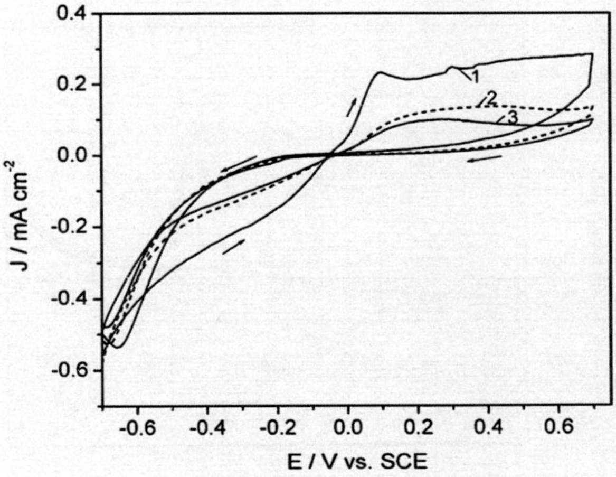

Fig. 6.8 A series of three consecutive cyclic voltammograms obtained at the titanium electrode in Ag(I) 20 mM, Na_2SO_3 0.25 M + $NaNO_3$ 0.1 M electrolyte at pH 8.6; scan rate 50 mV s^{-1}

reduced during the following cycling. In the anodic part of the reverse scan, the superposition of a current peak and a wave is observed during the first scan while upon cycling a single wave or a shallow broadened peak is consistently recorded. These features indicate that silver deposition on the titanium surface follows a quite different behavior compared to the glassy carbon electrode. Besides, the titanium surface shows evidence of inhibition upon cycling, as demonstrated by the current density decrease during both the cathodic and anodic part of the scan. The same type of study was performed with other concentration, obtaining similar results (Fig. 6.9).

Silver nanoparticles deposited on glassy carbon and titanium substrate are shown in Fig. 6.10. SEM images reported refer to relatively long pulse time duration, with one exception, to clearly visualize the deposition features. As expected, the glassy carbon surface performs much better compared to titanium sheets as long as the particle distribution is concerned. The glassy carbon electrode is inherently more fitted to a Volmer–Weber growth mode and in addition presents a highly polished surface allowing for a more uniform activity. The SEM image in Fig. 6.10 (top left) shows in fact that the surface is uniformly scattered with a dense array of particle of wide size

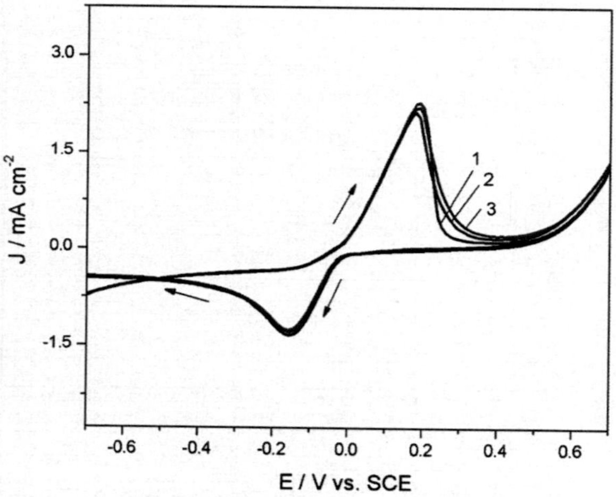

Fig. 6.9 A series of three consecutive cyclic voltammograms obtained at the ITO electrode in Ag(I) 20 mM, Na_2SO_3 0.25 M + $NaNO_3$ 0.1 M electrolyte at pH 8.6; scan rate 50 mV s^{-1}

distribution. Particle formation on titanium is characterized by the tendency to grow relatively large agglomerates as the potential pulse duration is increased, possibly also as a result of surface inhibition, as suggested by the above-presented voltammetric behavior.

This tendency towards clustering with deposition time is usually related to diffusion coupling between nearest-neighbor particles but, in the present case, it could also be triggered at the very stage of nucleation by the influence of the relatively high surface roughness of the substrate. Besides, it is worth noting that as the deposition time increases and particles agglomerates grow in size and number, the density of very fine particles, which are obtained at short pulse time, apparently lowers (Fig. 6.11).

The effect of the deposition potential is shown by the SEM micrographs presented in Fig. 6.12, also compared to image (b) in Fig. 6.10. Though not defined quantitatively, it is apparent that the increase of cathodic potential of a 50-ms pulse slightly reduces the tendency towards clustering and, as expected, increases the particle nucleation density.

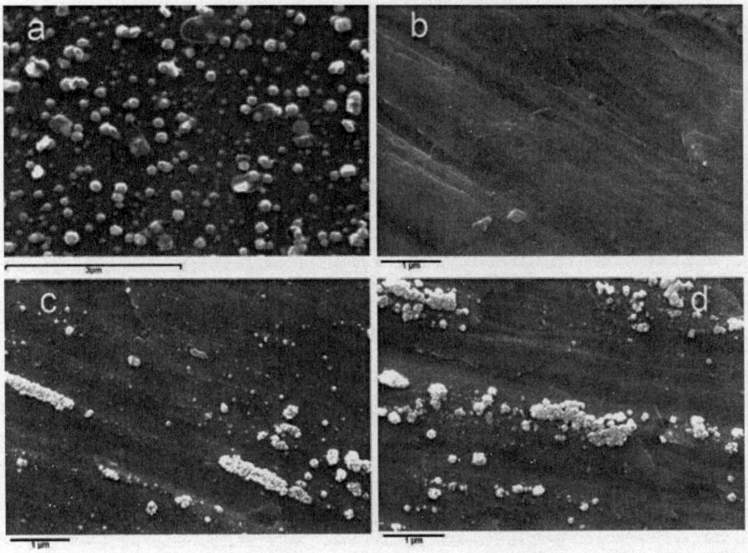

Fig. 6.10 Silver nanoparticles on glassy carbon electrode (**a**) (*top left*: −200 mV, 200 ms) and titanium substrate (*top right* and *bottom*: −200 mV); (**b**) 5 ms; (**c**) 50 ms; and (**d**) 100 ms from Ag(I) 20 mM, Na_2SO_3 0.25 M and $NaNO_3$ 0.1 M electrolyte at pH 8.6

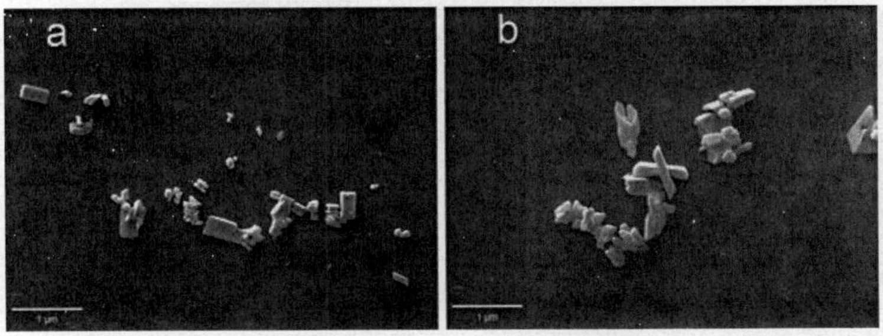

Fig. 6.11 Silver nanoparticles on ITO electrode: −200 mV; (**a**) 50 ms; (**b**) 100 ms from Ag(I) 20 mM, Na_2SO_3 0.25 M, and $NaNO_3$ 0.1 M electrolyte, at pH 8.6. The scale bar is 1 μm

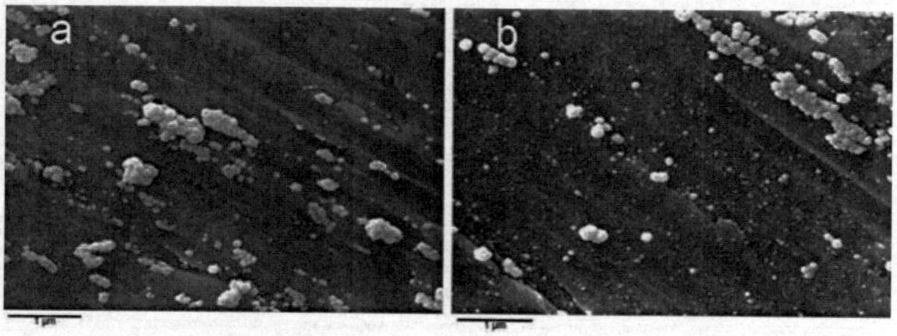

Fig. 6.12 Silver nanoparticles on titanium substrate from Ag(I) 20 mM, Na_2SO_3 0.25 M, and $NaNO_3$ 0.1 M electrolyte, at pH 8.6. Deposition potential: −400 (*left*), −600 mV (*right*). The scale bar is 1 μm

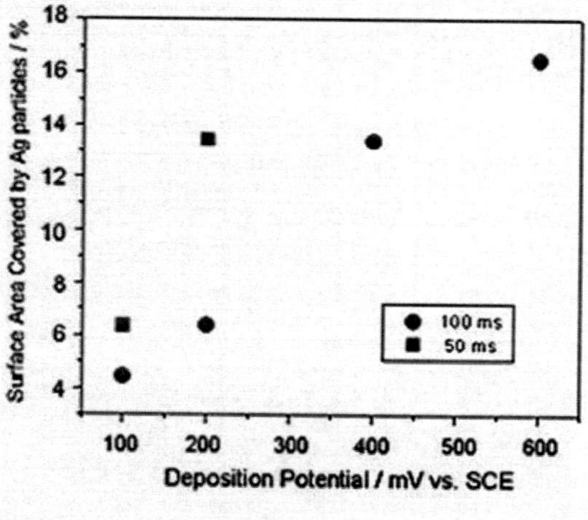

Fig. 6.13 Surface area of titanium electrode covered by silver particles versus the deposition potential for pulse duration of 50 and 100 ms. Electrolyte composition: Ag(I) 20 mM, Na_2SO_3 0.25 M, and $NaNO_3$ 0.1 M; pH 8.6

A pseudoquantitative assessment of the above-discussed effects of deposition potential and pulse duration on silver particles deposition on titanium can be made by evaluating the surface area covered by silver particles and aggregates as a function of these parameters. Some data are shown in the graph of Fig. 6.13 and they agree with the above observations: with the increase of deposition time at given potential (in the time range considered in this work and relevant to the objective of particles formation) the coverage by silver particle is reduced; with the increase of cathodic potential at a given pulse duration, the surface area covered by silver particles increases.

6.6 Deposition of Gold Nanoparticles

Gold deposition from the $(NH_4)_3Au(SO_3)_2$ 10 mM, Na_2SO_3 0.1 M, and $C_2H_8N_2$ 10 mM, solution at pH 6.5, shows the characteristics of a highly irreversible process. In Fig. 6.14 cyclic voltammograms at the glassy carbon electrode at 50 mV s^{-1} scan rate are reported. From the first scan it may be inferred that the discharge of the Au(I) complex at the glassy carbon electrode is highly inhibited since no distinct feature, that could be related to gold deposition, is clearly seen but a shallow current wave starting from about −300 mV. This is soon superseded by a second current wave starting at about −500 mV.

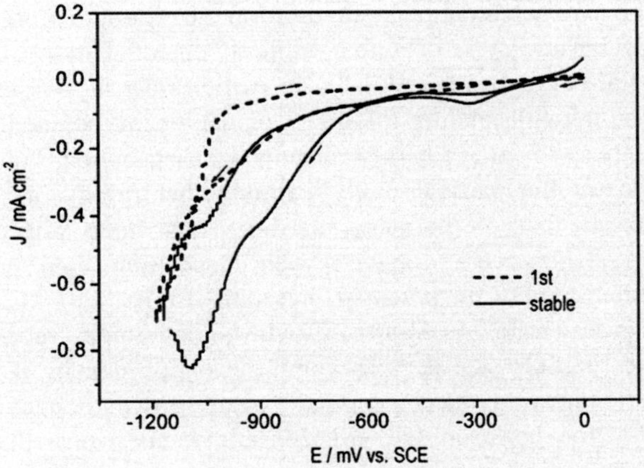

Fig. 6.14 Cyclic voltammograms at amorphous carbon electrode (first scan: *dashed line*; stable scan: *solid line*) in $(NH_4)_3Au(SO_3)_2$ 10 mM, Na_2SO_3 0.1 M, and ethylenediamine $C_2H_8N_2$ 10 mM, at pH 6.5; scan rate 50 mV s^{-1}, from 0 to −1.2 V

The latter can be attributed to the reduction of sulfite to dithionite ion, in agreement with early reports on the polarization behavior of Au(I) sulfite solution [64]. Upon cycling, a steady voltammogram is obtained where a distinct current peak for the discharge of the Au(I) sulfite complex appears at about −300 mV. This is followed by the reduction of sulfite to dithionite ion at the gold-plated surface at higher current density compared to the glassy carbon electrode. Similar features, though ill defined and poorly reproducible, could be recognized in voltammograms obtained by cycling at titanium electrode.

The voltammetric behavior of the gold deposition solution strongly suggests that particles formation may be controlled by a mechanism of progressive nucleation, as a result of kinetic control on the Au(I) sulfite complex discharge. However, the evidence from SEM observation of gold particles grown on titanium is not in line with this observation. SEM micrographs of gold particles deposited on titanium at changing pulse potential and 50 ms pulse duration are reported in Fig. 6.15. The nucleation density appears to be extremely sensitive to the deposition potential, possibly in connection with an inhibition effect of the side reaction of sulfite reduction. In fact, at −200 mV potential pulse, a uniform distribution of regularly sized gold particles is obtained, while at potential more negative than −400 mV a strong reduction in particles density is observed. However, notwithstanding the effect of deposition potential, SEM micrographs show that a highly uniform size distribution is obtained, even though the particle dimension is rather on the mesoscale (50–100 nm) than on the nanoscale dimension. Interestingly, quite different results were obtained when depositing gold particles on titanium sheets by potential sweeping from 0 to −1.2 V at changing scan rate. As shown by SEM micrographs reported in Fig. 6.16, particle size decreases and nucleation density increases as the potential scan rate is increased, but size distribution, as clearly shown by the insert in Fig. 6.16, is apparently worse than that obtained by a single potential pulse.

This behavior may be explained taking into account the influence of surface inhibition related to the electroreduction of sulfite, which could result in a twofold effect: inhibition of growth of already formed gold particles and nucleation of new particle on the titanium

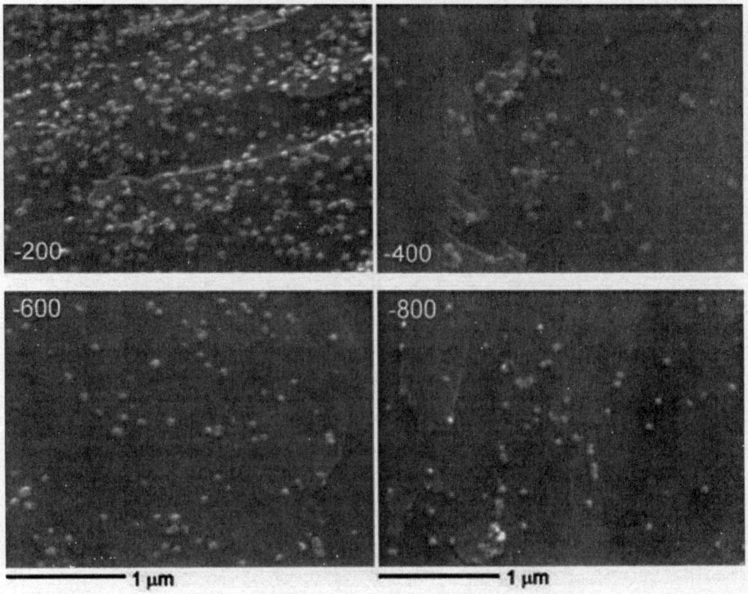

Fig. 6.15 SEM micrographs of gold particles on titanium deposited from $(NH_4)_3Au(SO_3)_2$ 10 mM, Na_2SO_3 0.1 M, and $C_2H_8N_2$ 10 mM, at pH 6.5, at changing potential, as indicated (vs. SCE) and 50 ms pulse duration. Scale bar is 1 µm

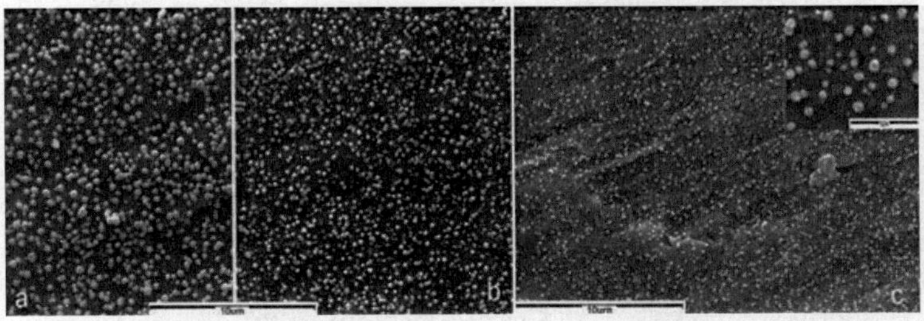

Fig. 6.16 SEM micrographs of Au particles on Ti deposited from Au(I) 10 mM, Na_2SO_3 0.1 M, and $C_2H_8N_2$ 10 mM solution, at pH 6.5, by potential sweeping from 0 to −1.2 at scan rate of 10 (**a**), 50 (**b**), and 200 mV s^{-1} (**c**). Scale bar is 10 µm; 1 µm in the *top-right* inset

surface as the deposition potential increases. Besides, it is also worth noting that particles size obtained by potential sweep is sensibly larger than that obtained by potential pulse deposition.

6.7 Galvanic Displacement of Nanoparticles on Silicon

The selective deposition of gold nanoclusters of controlled size on silicon surfaces by immersion plating in water-in-oil reversed micelle microemulsions is presented. Reversed micellar solutions are essentially dispersions of nanometer-sized water droplets in oil, stabilized by the presence of surfactant; they have long been used to synthesize metallic nanoclusters [65–67]. The conventional approach is based on the mixing and reacting of two different micellar solutions, e.g., one carrying a metal salt and the other carrying a reducing agent [68–70]. Nanocluster formation ensues upon micelle–micelle collision, material exchange between micelles, and subsequent reduction of the metal ions by the reducing agent. Here, a new strategy for directing nanocluster growth onto exposed silicon surfaces selectively has been adopted [71]. This selectivity is rooted in a galvanic displacement process. Galvanic displacement is a redox process, in which the substrate acts as the reducing agent for a species in solution. Thus, it achieves substrate selectivity, improved adhesion, and conformality, as demonstrated in recent thin film, interconnect, and MEMS applications. In these experiments, a water-based plating solution, containing hydrofluoric acid and a gold salt, is combined with sodium bis(2-ethylhexyl) sulfosuccinate (AOT, a surfactant) and heptane to produce a reversed micelle microemulsion. Metal cluster size is in good quantitative agreement with nominal micelle size over a wide range of values of the microemulsion parameter $R_{\mathrm{micelle}}(\mathrm{nm}) = 0.175R + 1.5$ (R = [water]/[AOT]). Figure 6.17a, b shows SEM images of gold and palladium nanoclusters on Si(100), respectively. A uniform coverage of the substrate is obtained [71].

The cluster-size distribution also appears fairly uniform; however, some larger aggregates can be observed and are attributed to cluster

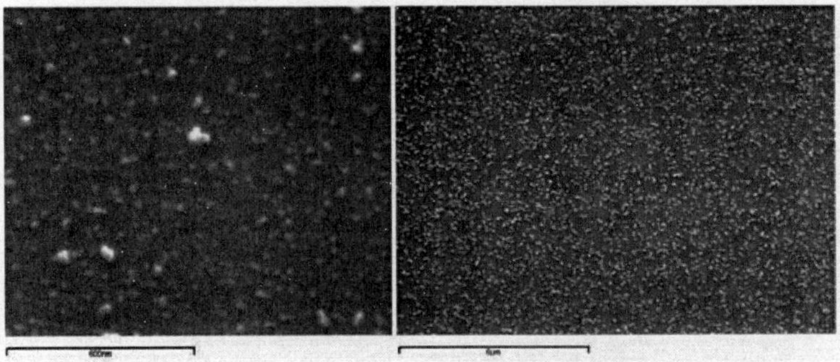

Fig. 6.17 SEM micrographs of gold (*left*) and palladium (*right*) clusters on silicon (*left*: $t = 15$ s, $R = 10$; *right*: $t = 600$ s, $R = 30$)

coalescence, especially for palladium. This phenomenon is well known from conventional studies of cluster formation in microemulsion systems, where cluster aggregates are in fact prevalent over single clusters. In the present case, it appears that cluster deposition by galvanic displacement from a single microemulsion is effective in reducing cluster aggregation.

6.8 Conclusions

Electrochemical synthesis of metal powder has been traditionally used for mass production of micrometer size powder being primarily used in powder metallurgy as well as other engineering applications. The electrochemical synthesis of metal nanoparticles and other nanostructures is being currently considered as a potential route for the development of advanced functional materials and devices in different technology area, including chemical sensing, magnetic storage, microelectronics, electroanalysis, and electrocatalysis. Nowadays, different nanomaterials are prepared by sonoelectrochemical methods. Silver and gold particles with size spanning from the nanoscale to the mesoscale were deposited from sulfite-based solutions. The proposed solution chemistry is reasonably stable and is shown to be a viable process for the electrochemical deposition of both Ag and Au particles. The voltammetric behavior of the electrolytes is studied at different substrates surface, GC, Ti and ITO, showing that kinetic control plays a role in the discharge of the sulfite complex of Ag(I) or Au(I). Silver particles of size below 50 nm were easily obtained on glassy carbon substrate by potential-controlled deposition achieving a high nucleation density. On the other hand, size distribution was not easily controllable, most likely as a consequence of the irreversible nature of the discharge reaction of the Ag(I) complex. Silver particles deposition on titanium showed low nucleation density and a strong tendency to form large particles clusters and agglomerates, mostly in connection with surface irregularities. Gold particles were successfully deposited by either a potential pulse or a potential sweeping technique, achieving good results in terms of nucleation density. This was observed in particular when depositing on titanium substrate, on which by using a single potential pulse Au particles could be deposited with high nucleation density and uniform size distribution, though particle dimension was rather on the mesoscale than on the nanoscale dimension.

A novel method for depositing gold clusters on silicon surfaces, conformally and selectively, has been demonstrated. It is based on immersion plating in reversed micellar microemulsions of a plating solution containing metal and fluoride ions. A good correlation has been found between gold cluster size and micelle radius.

References

1. Chen CW, Takezako T, Yamamoto K, Serizawa T, Akashi M (2000) Colloids Surf A 169:107
2. Chen W, Cai WP, Liang CH, Zhang LD (2001) Mater Res Bull 36:335
3. Brust M, Walker M, Bethell D, Schiffrin DJ, Whyman R (1994) J Chem Soc Chem Commun 801
4. Magnusson M, Deppet K, Malm J-O, Bovin J-O, Samuelson L (1999) Nanostruct Mater 12:45
5. Collier PJ, Iggo JA, Whyman R (1999) J Mol Catal A Chem 146:149
6. Finot MO, Braybrook GD, McDermott MT (1999) J Electroanal Chem 466:234
7. Finot MO, McDermott MT (2000) J Electroanal Chem 488:125
8. Guerin S, Attard GS (2001) Electrochem Commun 3:544
9. Boccuzzi F, Chiorino A, Manzola M (2001) Mater Sci Eng C 15:215

10. Schimpf S, Lucas M, Mohr C, Rodemerck U, Bruckner A, Radnik J, Hofmeister H, Claus P (2002) Catal Today 72:63
11. El-Deab MS, Ohsaka T (2002) Electrochim Acta 47:4255
12. Popov KI, Pavlovic MG (1993) In: White RE, Conway BE, Bockris JO'M (eds) Modern aspects of electrochemistry, vol 24. Plenum, New York, p 299
13. Walter EC, Penner RM, Liu H, Ng KH, Zach MP, Favier F (2002) Surf Interface Anal 34:409
14. Sun M, Zangari G, Shamsuzzoha M, Metzger RM (2001) Appl Phys Lett 78:2964
15. Walter EC, Ng K, Zach MP, Penner RM, Favier F (2002) Microelectron Eng 61(62):555
16. Welch CM, Compton RG (2006) Anal Bioanal Chem 384:601
17. Arvia AJ, Salvarezza RC, Triaca WE (2004) J New Mater Electrochem Syst 7:133
18. Zoval V, Stiger RM, Biernacki P, Penner RM (1996) J Phys Chem 100:837
19. Zoval JV, Biernacki P, Penner RM (1996) Anal Chem 6:1585
20. Liu H, Penner RM (2000) J Phys Chem B 104:9131
21. Zoval JV, Lee J, Gorer S, Penner RM (1998) J Phys Chem B 102:1166
22. Anderson M, Gorer S, Penner RM (1997) J Phys Chem B 101:5895
23. Scheludko A, Todorova M (1952) Bull Acad Bulg Sci Phys 3:61
24. Ueda M, Dietz H, Anders A, Kneppe H, Meixner A, Plieth W (2002) Electrochem Acta 48:377
25. Reisse J, Francois H, Vandercammen J et al (1994) Electrochim Acta 39:37
26. Reisse J, Caulier T, Deckerkheer C et al (1996) Ultrason Sonochem 3:S147
27. Durant A, Delplancke JL, Libert V, Reisse J (1999) Eur J Org Chem 11:2845
28. Delplancke JL, Dille J et al (2000) Chem Mater 12:946
29. Zhu J, Liu S, Palchik O, Koltypin Yu, Gedanken A (2000) Langmuir 16:6396
30. Markov I, Boynov A, Toschev S (1973) Electrochim Acta 18:377
31. Gerisher H (1958) Z Electrochem 62:256
32. Price PB, Vermilyea DA (1958) J Chem Phys 28:720
33. Almawlawi D, Liu CZ, Moskovits M (1994) J Mater Res 9:1014
34. Liu KI, Chien CL, Searson PC, YuZhang K (1998) Appl Phys Lett 73:2222
35. Hulteen JC, Martin CR (1997) J Mater Chem 7:1075
36. Martin CR (1994) Science 266:1961
37. Strbac S, Maroun F, Magnussen OM, Behm RJ (2001) J Electroanal Chem 500:479
38. Strbac S, Magnussen OM, Behm R (1999) J Phys Rev Lett 83:3246
39. Moller FA, Magnussen OM, Behm R (1996) J Phys Rev Lett 77:5249
40. Kolb DM, Engelmann GE, Ziegler JC (2000) Solid State Ionics 131:69
41. Kolb DM, Engelmann GE, Ziegler JC (2000) Angew Chem Int Ed 39:1123
42. Ziegler JC, Engelmann GE, Kolb DM (1999) Z Phys Chem 208:151
43. Haruta M (1997) Catal Today 36:153
44. Maximov M et al (2000) Microelectron Eng 51:61
45. Yeager E, Hovorka F (1953) J Acoust Soc Am 25:443
46. Lickiss PD, McGrath VE (1996) Chem Br 32:47

47. Lighthill J (1978) Waves in fluids. Cambridge University Press, Cambridge, p 1221
48. Suslick KS (1990) Science 247:1439
49. Cornpton RG, Eklund JC, Marken F (1997) Electroanalysis 9:509
50. Moriguchi N (1934) J Chem Soc Jpn 55:749
51. Reisse J, Francois H, Vandercammen J, Fabre O, Kirsch-de Mesmaeker A, Maerschalk C, Delplancke JL (1994) Electrochim Acta 39:37
52. Durant A, Delplancke JL, Winand R, Reisse J (1995) Tetrahedron Lett 36:4257
53. Sáez V, Mason TJ (2009) Molecules 14:4284
54. Yeager E, Kovorka F, Dereska J (1957) J Acoust Soc Am 29:769
55. Saterlay AJ, Wilkins SJ, Goeting ChH, Foord JS, Compton RG, Marken F (2000) J Solid State Electrochem 4:383
56. Pollet B, Lorimer JP, Phull SS, Hihn JY (2000) Ultrason Sonochem 7:69
57. Goeting ChH, Foord JS, Marken F, Compton RG (1999) Diam Relat Mater 8:824
58. Han HM, Phillips GJ, Mikhalovsky SV, FitzGerald S, Lloyd AW (2008) J Mater Sci Mater Med 19:1787
59. Murali KR, Sasindran P (2004) J Mater Sci 39:6347
60. González-García J, Deseada Esclapez M, Bonete P, Vargas Hernández Y, Gaete Garretón L, Sáez V (2010) Ultrasonics 50:318
61. NIST Standard Reference Database 46, NIST Critically selected stability constants of metal complexes: Version 8.0
62. Hemmingsen T (1992) Electrochim Acta 37:2775
63. Bard AJ, Faulkner LR (1980) Electrochemical methods, fundamentals and applications. Wiley, New York
64. Derivaz J-P, Resin A, Losi S (1977) Surf Technol 5:369
65. Landfester K (2001) Adv Mater 13:765
66. Fletcher PDI, Howe AM, Robinson BH (1987) J Chem Soc Faraday Trans 83:985
67. Eastoe J, Fragneto G, Robinson BH, Towey TF, Heenan RK, Leng FJ (1992) J Chem Soc Faraday Trans 88:461
68. Arcoleo V, Cavallaro G, La Manna G, Turco Liveri V (1995) Thermochim Acta 254:111
69. Aliotta F, Arcoleo V, Buccoleri S, La Manna G, Turco Liveri V (1995) Thermochim Acta 265:15
70. Arcoleo V, Turco Liveri V (1996) Chem Phys Lett 258:223
71. Magagnin L, Bertani V, Cavallotti PL, Maboudian R, Carraro C (2002) Microelectron Eng 64:479

Chapter 7
Production of Metallic Powders from Aqueous Solutions Without an External Current Source

Stojan S. Djokić

7.1 Introduction

The production of metallic powders without an external current source has many advantages compared to electrochemical (electrolytic) methods. Generally speaking, all the powders that can be produced electrolytically from aqueous solutions can also be obtained using chemical methods, i.e. without an external current source. Classical examples of metallic powders produced from aqueous solutions by chemical methods on an industrial scale via the so-called hydrometallurgical processes include nickel and copper.

The hydrometallurgical production of metal powders is based on the reduction of metal ions with hydrogen gas from their aqueous solutions, usually prepared under pressure and elevated temperatures.

Metallic powders can also be successfully produced without an external current source from aqueous solutions via electroless deposition. There are two clearly distinguished types of electroless deposition (a) galvanic displacement deposition and (b) autocatalytic deposition [1].

S.S. Djokić (✉)
Elchem Consulting Ltd., Edmonton, AB, Canada T5X 6B3
e-mail: sdjokic@telus.net

Stojan S. Djokić (ed.), *Electrochemical Production of Metal Powders*, Modern Aspects of Electrochemistry 54, DOI 10.1007/978-1-4614-2380-5_7,

In the galvanic deposition, the ions of more noble metals are reduced by a less noble metal. In this way the deposition of powders of a more noble metal proceeds at the surface of a less noble metal which is, in turn, dissolved.

In the autocatalytic deposition from homogenous solutions, the formation of powders into the bulk solution is usually seen when the concentration of the reducing agent is too high or at elevated temperatures [1]. While the deposition of thin smooth films proceeds only onto catalytic or previously activated metallic or non-metallic surfaces, the formation of powders is observed both on the activated surfaces and into bulk solution. The growth of particles is obviously catalyzed by the metal being deposited, which is the same as the autocatalytic deposition.

Hydrometallurgical processes, i.e. production of Ni, Co, Cu, or similar metals, are well established in practice. Similarly, the electrodeposition of metallic powders has been extensively studied over the last century. A significant portion of this book is devoted to electrodeposition of metallic powders. The study of electroless deposition of metal powders has to a certain extent been neglected. The formation of metallic powders occurring during electroless deposition of metallic powders is considerably less investigated than its electrolytic counterpart.

The aim of this chapter is to review the deposition of metallic powders from aqueous solutions without an application of the external current.

7.2 Production of Metallic Powders with Hydrogen Reduction

The precipitation of metal powders from aqueous solutions via hydrogen reduction of their respective ions dates back to the nineteenth century [2]. The knowledge of metal precipitation at high temperature and pressure in an autoclave has significantly advanced in the first half of the twentieth century and was successfully applied worldwide in the industry for the production of many metal powders.

Examples of the application of autoclave processes in the production of metal powders with hydrogen reduction at high temperature and pressure include nickel, cobalt, copper, noble metals, and less common metals [2].

The description of details of production of metal powders with hydrogen is beyond the scope of this chapter. Only the basic and most important concepts of this technology are briefly discussed.

The reduction of metal ions with hydrogen takes place at elevated temperature and pressure, according to the following reaction:

$$M^{2+} + H_2 \rightarrow M + 2H^+. \tag{7.1}$$

Reaction (7.1) is an overall reaction involving the reduction of metal ions:

$$M^{2+} + 2e^- \rightarrow M \tag{7.2}$$

and the oxidation of hydrogen gas to its ions:

$$H_2 \rightarrow 2H^+ + 2e^-. \tag{7.3}$$

The process is carried out in an autoclave, usually at higher temperatures and pressures as well as controlled pH to enhance the reaction kinetics and to produce the powder of the desired specification.

The nucleation, crystal growth, agglomeration, particle size, and morphology of powders are controlled by the conditions, including additives such as complexing agents, buffering agents, and stabilizers. The precipitation of metal powders by hydrogen has been used in the commercial production in industrial settings for over fifty years [3]. On an industrial scale, nickel is produced from ammoniacal solutions by hydrogen reduction under pressure at elevated temperatures in mechanically agitated autoclaves [3–6].

Ammoniacal solutions form soluble complexes, considering the example of nickel which can be described by a generic formula $[Ni(NH_3)_n]^{2+}$, where n is in the range from 2 to 6. The deposition of nickel on nucleating sites is usually described by the following reaction:

$$[Ni(NH_3)_n]SO_4 + H_2 \rightarrow Ni + (HH_4)_2SO_4 + (n-2)NH_3. \quad (7.4)$$

A prerequisite in nickel powder production is the formation of fine particles (sites) of apparent density in the nucleation run. The number of nucleation sites should be as large as possible to achieve the best reduction results. Different surface morphology, shape, and size of powders, can be obtained by varying the composition of the feed solution and by adding inorganic or organic compounds which are able to catalyze or modify the reaction mechanisms.

In the reduction of Ni(II) ion with hydrogen catalysts such as platinic chloride, ferrous sulfate, palladium chloride, and silver sulfate, the number of active sites can significantly increase, resulting in a finer subgrain size [7]. The subgrains were spherical or block-like indicating a uniform growth rate in all directions.

A catalyst is used only in the nucleation run and its concentration of about 0.1 g/L is usually restricted as a practical minimum [3]. Although an increase in the amount of the catalyst within the reasonable limits is beneficial, excessive additions should be avoided due to their inclusion into the final product.

Metallic nickel is known as a hydrogenation catalyst in the reduction of Ni(II) ions with hydrogen. Since metallic nickel is the product of the reduction the reaction is autocatalytic. The rate constant is approximately directly proportional to the amount of nickel powder. The diminishing effect at higher amounts of nickel powder is attributed to the difficulty in maintaining a homogenous suspension with a large number of nucleation sites [8].

Ferrous sulfate is the most frequently used catalyst for nickel precipitation via hydrogen reduction reaction. A possible explanation for this is that Fe(II) ions form ferrous sulfate which will hydrolyze in the presence of ammonia according to the following reaction:

$$FeSO_4 + 2NH_4OH \rightarrow Fe(OH)_2 + (NH_4)_2SO_4. \quad (7.5)$$

The ferrous hydroxide, $Fe(OH)_2$, nuclei may initiate the reduction of Ni(II) ions to metallic nickel. The disadvantage of this process is the incorporation of Fe into the final product. In the current practice some of the $FeSO_4$ is replaced with other compounds which results in a lower content of Fe in the final nickel powder. Other compounds used or recommended as a seed for Ni powder precipitation include chromous sulfate, manganese salts, and ammonium carbonate [3].

It is believed that ammonium carbonate promotes the formation of a fine suspension of solid basic nickel carbonate. Fine particles of the basic nickel carbonate, in turn, act as a catalyst and provide the nuclei on which nickel metal precipitates. Very fine nickel powders can be produced using this method.

The presence of different additives which act as activating or passivating agents significantly influences the morphology of nickel powders. In general terms, activators (e.g., antraquinone, sodium boron hydride, etc. in small amounts) produce smooth spherical particles [9, 10]. On the other hand, when passivating agents are used, preferential deposition will occur at sites not covered by the additives, which results in the formation of irregular particles. For a successful commercial nickel powder production it is necessary to control the neutralization, buffering, and nucleation procedures during the hydrogen reduction process [3].

Similarly, cobalt powder can be produced from cobalt(III) hexamine sulfate ($[Co(NH_3)_6]_2(SO_4)_3$) [2] or $Co(OH)_2$ [11].

Copper is another example which can be successfully produced using the so-called hydrometallurgical processes. The reduction of Cu(II) with hydrogen from acidic and ammoniacal solutions is achieved under pressure in the absence of a nucleating agent [12–14].

The generic reactions for the precipitation of copper powders can be presented with the following equations:

(a) Acidic solutions:

$$Cu^{2+} + H_2 \rightarrow Cu + 2H^+. \tag{7.6}$$

(b) Ammoniacal solutions:

$$Cu^{2+} + 2NH_3 + H_2 \rightarrow Cu + 2NH_4^+. \tag{7.7}$$

The production of copper powder from leach solutions is used on a commercial scale. These processes are carried out at temperatures above 150°C and under high pressures of hydrogen gas.

The reduction of Cu(II) to Cu^o with hydrogen can proceed even from high acidic solutions [15]. Arsenic, antimony, and bismuth present in the liquor have been found to catalyze reduction. The rate of reduction of Cu(II) is proportional to the square root of the partial pressure of hydrogen.

The reduction of other metallic ions with hydrogen under pressure and at elevated temperatures is probably possible; however, the most important practical applications of this technology are for now related to the production of nickel, cobalt, and copper.

7.3 Production of Metallic Powders via Electroless Deposition

Besides hydrometallurgical means, where the reduction of ions, as described in the previous sections, is achieved by hydrogen under pressure and elevated temperature, metallic powders can be successfully produced from an aqueous solution without the application of the external current either via galvanic displacement (cementation) or using an appropriate reducing agent.

The production of metallic powders through hydrometallurgical or electrodeposition routes has extensively been studied over the last century. The formation of metallic powders during electroless deposition of films of metals or alloys is observed when the so-called bath instability takes place [1, 16, 17]. As such, the electroless deposition was far less investigated than its electrolytic counterpart. The bath instability is usually seen in the experimental conditions at elevated temperatures or when the concentration of the reducing agent is too high.

In this section, the production of metallic powders via galvanic displacement or cementation as well as via electroless deposition using different reducing agents is discussed.

7.3.1 Production of Metallic Powders by the Galvanic Displacement Reaction

The galvanic displacement reaction is also known as cementation or immersion plating. In the galvanic displacement reaction, a less noble metal (powder, foil, and any other form or shape of the substrate) is immersed into an aqueous solution containing the ions of a more noble metal. As a consequence, the more noble metal as a thin film or as powder particle will be produced.

In a simple manner, the galvanic displacement can be described by the following reactions:

1. Dissolution or oxidation of a less noble metal, M_1, in an appropriate solution containing $M_2^{z_2+}$ ions of a more noble metal M_2:

$$M_1 \rightarrow M_1^{z_1+} + z_1e^-. \tag{7.8}$$

2. Reduction of $M_2^{z_2+}$ ions and a subsequent deposition of a more noble metal (M_2):

$$M_2^{z_2+} + z_2 \rightarrow M_2. \tag{7.9}$$

A combination of reactions (7.8) and (7.9) leads to

$$\frac{z_2}{z_1}M_1 + M_2^{z_2+} \rightarrow \frac{z_2}{z_1}M_1^{z_1+} + M_2. \tag{7.10}$$

Reaction (7.10) represents a generic description of the deposition of more noble metals via galvanic displacement. In this way, many metals can be produced as thin films or powders.

Theoretically, there are many examples of the galvanic displacement reaction. However, practical importance of this type of deposition is related to the following systems: Fe/Cu, Zn/Cu, Ni/Au, Al/Cu, Al/Ag, and similar.

The cementation of copper on iron has attracted interest of particular importance because this simple process has been used since the sixteenth century for the recovery of copper from mine waters by contacting dilute process streams with iron scrap [18]. Similarly, aluminum as an alternative to iron can be used for the recovery of copper [19, 20].

In an ideal thermodynamic case, as soon as the surface of the less noble metal is covered with a film of more noble metal the deposition should stop. However, as experimentally observed, this is not the case. In the galvanic displacement deposition not only thicker coatings but also powders can be obtained.

The formation of powders during the galvanic deposition is attributed to the porosity of noble metal deposited onto less noble metal. In terms of electroless deposition, the less noble metal can be considered as the reducing agent of ions of a more noble metal. At the beginning, i.e., immediately upon the immersion, the rate of deposition of the more noble metal is the fastest, since the whole surface and the largest number of active sites are available for the reaction. A similar conclusion is applicable to the rate of dissolution of the less noble metal. At the beginning, the rate of dissolution or oxidation of the less noble metal is the fastest. Of course, as it will be shown later in this section, the rate of dissolution depends on the electrolyte composition, pH, temperature, etc. As the surface of the less noble metal is covered with the more noble metal, the rate of deposition progressively decreases, because smaller number of sites are in contact with the solution.

Due to porosity or uneven coating, the deposition of the more noble metal will continue as long as the surface of the less noble metal is in contact with solution. On the other hand, when the less noble metal is completely consumed, the electrons cannot be produced as a consequence of its dissolution [Eq. (7.8)]. In this case, the deposition stops. In another scenario, when the amount of less noble metal is larger than that required for the stoichiometric reduction of ions of a more positive metal, the deposition stops, although some secondary processes may occur, leading to complete dissolution of the less noble metal.

Various types of surface morphology and different shapes or sizes of powders can be produced using this technique of deposition. Additionally, via galvanic displacement reaction metallic powders can be produced on any type of a less noble metal substrate. This less noble metal could be powder, flat, or any other type of surface.

The growth of the particles, their shape, and surface morphology depend on the conditions such as the type of the electrolyte, metal being deposited, temperature, pH and time etc. Some examples of the powders produced via displacement reaction follow.

The SEM images of Ag powder produced from a solution containing 0.01 M $AgNO_3$ dissolved in a diluted NH_4OH solution at pH of about 10 (a) or from a 0.01 M $AgNO_3$ and 0.5 M citric acid solution at pH of about 2 (b) by the galvanic deposition on an Al foil are presented in Fig. 7.1 [21].

In the alkaline solutions (pH 10) on flat aluminium substrate globular powders of silver were produced (a), while in acidic solution (pH 2) dendritic structures are observed (b).

The production of silver powder from an acidic solution using the aluminium substrate can be explained by the following reaction:

$$Ag^+ + 2H^+ + Al \rightarrow Ag + Al^{3+} + H_2. \quad (7.11)$$

Similarly, in alkaline solutions containing $[Ag(NH_3)_2]^+$ ions, silver powder is produced onto aluminium foil via galvanic deposition according to the reaction

$$\begin{aligned} &[Ag(NH_3)_2]^+ + Al + 2OH^- + 2H_2O \\ &\qquad \rightarrow Ag + [Al(OH)_4]^- + H_2 + 2NH_3. \end{aligned} \quad (7.12)$$

It is worth to note that in both acidic and alkaline conditions the simultaneous hydrogen evolution reaction takes place due to aluminium dissolution, which is experimentally seen as bubble formation.

In general terms, the experimental observations suggest that the surface of the less noble metal *is not* uniformly covered with the more noble metal, but rather the film of more noble metal is porous. The existence of pores allows the electrolyte to penetrate through and get into contact with the less noble metal. Under these conditions, when in contact with an appropriate electrolyte, the less noble metal, M_1, will dissolve according to reaction (7.8).

The produced electrons according to reaction (7.8) are further transferred and used for the reduction of ions of the more noble

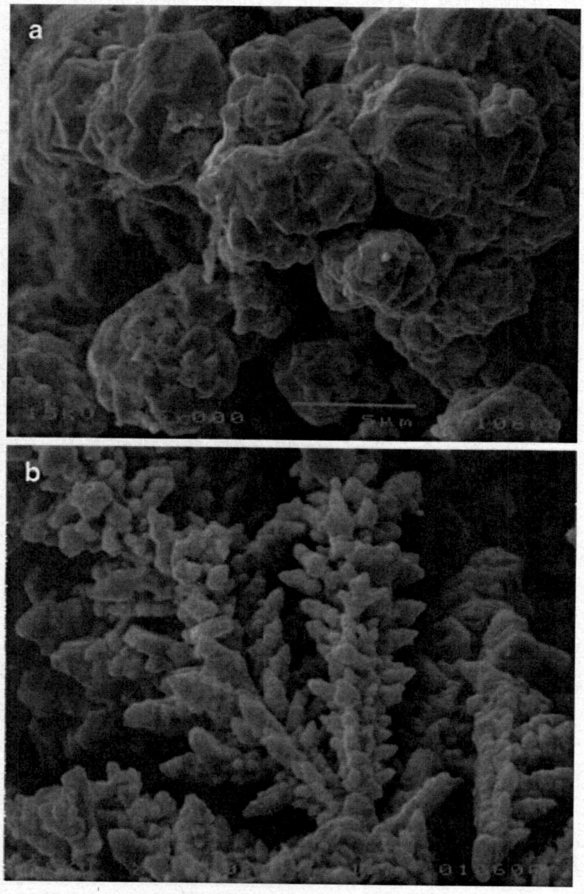

Fig. 7.1 Ag powders produced via galvanic deposition from (**a**) 0.01 M $AgNO_3$ dissolved in diluted NH_4OH (pH 10, 22°C) and from (**b**) 0.01 M $AgNO_3$ dissolved in 0.5 M citric acid solution (pH 2, 22°C) on an aluminium foil [21] (Reprinted with permission from ECS—The Electrochemical Society.)

metal, as described by the reaction (7.9). In this way the growth of the film or powders of the more noble metal occurs further. Due to the porosity of the deposited metal, particles can be obtained via galvanic deposition on powdery or flat substrates.

It seems that in some systems the galvanic deposition produces powders when the simultaneous hydrogen evolution reaction takes place:

$$2H^+ + 2e^- \rightarrow H_2. \tag{7.13}$$

This is demonstrated for the examples of Al/Ag (Fig. 7.1) or Al/Cu [3] in alkaline solutions. The simultaneous hydrogen evolution (as bubbles) may significantly change the hydrodynamic conditions at the surface at which the deposition of a more noble metal takes place. Changes in the hydrodynamic conditions may lead to growth of particles of various shapes.

For other systems, where the simultaneous hydrogen evolution is not evident, it seems that to produce powders by this type of deposition it is necessary that the surface of the less noble metal is not completely covered by the more noble metal. An incomplete surface coverage of the less noble metal may form porous (spongy) deposits with a poor adhesion, thus leading to the formation of various shapes of powders.

To show the formation of Ag powder under the conditions where hydrogen evolution does not occur, the following examples are presented in this section.

SEM images with two magnifications of silver powder produced from a solution containing $AgNO_3$ and citric acid (pH ~2) on a Cu substrate are given in Fig. 7.2. As shown in this figure, the silver particles produced from an acidic solution are dendritic.

The deposition of silver powder on copper substrates from an $AgNO_3$ solution containing citrate ions (acidic media) can be described by the following reaction:

$$Cu + 2Ag^+ \rightarrow Cu^{2+} + 2Ag. \tag{7.14}$$

SEM images of silver powders produced from an $Ag(NH_3)_2{}^+$ solution on copper substrate are presented in Fig. 7.3. The particles of this silver powder, as seen from this figure, are globular. The diameter of individual particles is estimated at 1 μm or less.

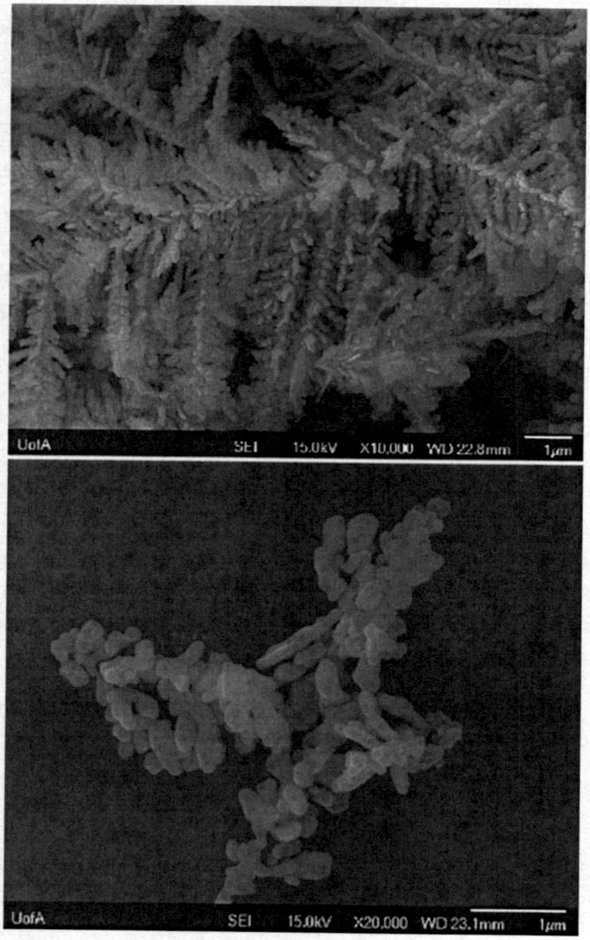

Fig. 7.2 Ag powders produced via galvanic deposition from 0.01 M $AgNO_3$ dissolved in 0.5 M citric acid solution (pH 2, 22°C) on a copper substrate

In the alkaline solutions, the formation of silver powders on copper substrate is described by the reaction:

$$Cu + 2Ag(NH_3)_2^+ \rightarrow Cu(NH_3)_4^{2+} + 2Ag. \quad (7.15)$$

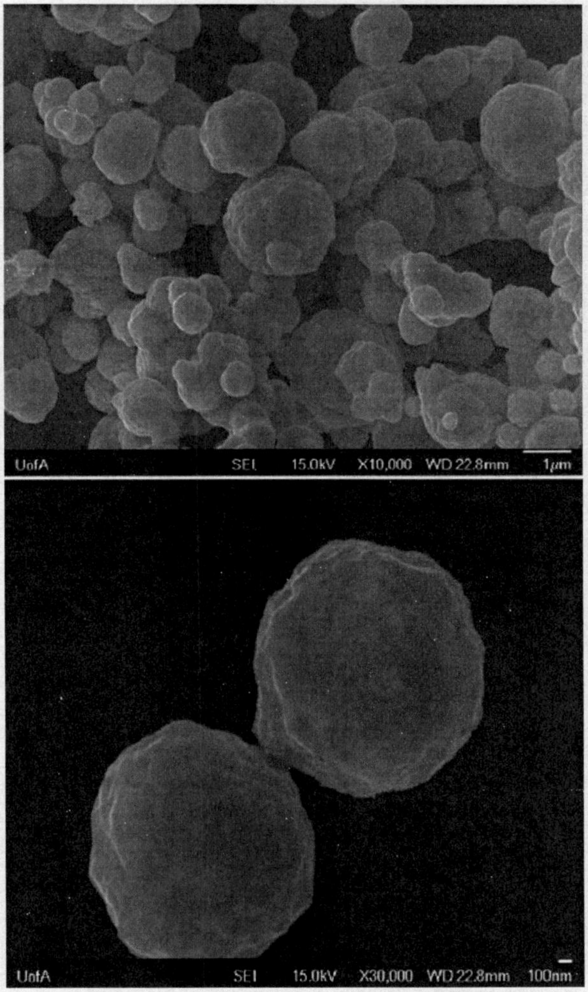

Fig. 7.3 Ag powders produced via galvanic deposition from 0.01 M $AgNO_3$ dissolved in diluted NH_4OH (pH 10, 22°C) on a copper substrate

Notably, in both cases there is no simultaneous hydrogen evolution as shown by reactions (7.14) and (7.15). The evolution of hydrogen gas bubbles was not observed in the experimental settings.

However, based on the experimental observations, in acidic solutions during the galvanic displacement deposition of silver powders a dendritic surface morphology is obtained on both Al and Cu as presented in Figs. 7.1a and 7.2. On the other hand, when the deposition is carried out from alkaline solutions, globular particles of silver were obtained on Al and Cu substrates as shown in Figs. 7.1b and 7.3. These results suggest that pH rather than the nature of the substrate in the case of galvanic deposition influences the surface morphology of the deposited powders.

The results so far do not give an explicit explanation of crystal growth and powder formation via galvanic displacement reaction. As such, this aspect definitely requires additional studies for the future advancement of this technology.

7.3.2 *Electroless Deposition of Metallic Powders Using Different Reducing Agents*

Many metallic powders can be successfully deposited from aqueous solutions using reducing agents such as formaldehyde, hypophosphite, borohydride, and dimethylamine borane (DMAB). All mentioned reducing agents have been used in electroless deposition of thin uniform and continuous films of various metals [21].

SEM micrographs of Cu powders obtained by electroless deposition with formaldehyde as the reducing agent are presented in Fig. 7.4.

This Cu powder was produced from an alkaline $CuSO_4$ solution containing sodium potassium tartrate as a complexing agent of Cu(II) ions. The reduction of Cu(II) complexed ions with formaldehyde is described by the following simple reaction:

$$Cu^{2+} + HCOH + 2OH^{-} \rightarrow Cu + HCOOH + H_2O. \qquad (7.16)$$

Fig. 7.4 SEM micrographs of Cu powders produced from an alkaline $CuSO_4$ solution containing sodium potassium tartrate as a complexing agent of Cu(II) ions and formaldehyde as a reducing agent [21] (Reprinted with permission from ECS—The Electrochemical Society.)

Electroless deposition of copper accompanies the simultaneous hydrogen evolution as it is presented with the reaction below:

$$2\text{HCOH} + 2\text{OH}^- + \text{Cu}^{2+} \rightarrow 2\text{HCOOH} + \text{H}_2 + \text{Cu}. \tag{7.17}$$

A combination of Eqs. (7.16) and (7.17) leads to:

$$2\text{Cu}^{2+} + 3\text{HCOH} + 4\text{OH}^- \rightarrow 2\text{Cu} + 3\text{HCOOH} + \text{H}_2\text{O} + \text{H}_2. \tag{7.18}$$

It is obvious from Fig. 7.4 that Cu powders are agglomerated. The individual particles are predominantly larger than 1 μm and they have a very rough surface morphology.

Similarly, other metallic powders can be produced by the electroless deposition when *bath instability* takes place [1, 16, 17]. A few examples showing Ni powder produced with hydrazine, Co powder produced with hydrazine or hypophosphite, and Ag powder produced with formaldehyde are presented in Fig. 7.5. Hypophosphite, hydrazine, and formaldehyde were used as reducing agents of the respective ions.

Based on the results presented in Fig. 7.5a, the produced nickel particles are agglomerated globules with a rough surface morphology. The size of these particles, according to the SEM image, is estimated to be between 100 and 300 nm. Silver powders obtained from an alkaline solution containing $[\text{Ag(NH}_3)_2]^+$ and formaldehyde as reducing agents represent agglomerates composed of globular silver particles with the size less than 1 μm as illustrated in the SEM image (Fig. 7.5b). The shapes of Co particles produced by the reduction with hypophosphite (Fig. 7.5c) and hydrazine (Fig. 7.5d) are quite different. While the Co powders deposited via the reaction with hydrazine represent the 3D dendrites, particles produced with hypophosphite are platelets. In both cases the size of the majority of individual particles is estimated to be more than 1 μm. The difference in the shape of Co particles can be attributed to their various chemical compositions. It is known that a reduction of Co(II) ions with hypophosphite leads to the incorporation of phosphorous and the formation of amorphous deposits [1], while the reduction with hydrazine produces pure Co [16].

Fig. 7.5 Powders of (**a**) nickel produced by electroless deposition with hydrazine, (**b**) silver produced with formaldehyde, (**c**) cobalt produced with hypophosphite, and (**d**) cobalt produced with hydrazine [21] (Reprinted with permission from ECS—The Electrochemical Society.)

The results presented in Figs. 7.4 and 7.5 show that the shape, size, and morphology of the particles are significantly influenced by the reducing agent as well as by the nature of the metal deposited.

Ascorbic acid, as an environmentally friendly reducing agent of various metallic ions (e.g. silver, gold, and copper), has particularly attracted researchers in the field of electroless deposition [1]. Using ascorbic acid as the reducing agent, Ag, Cu, Au, Pd, etc. powders were produced by electroless deposition from alkaline or acidic solutions [21].

SEM micrographs of silver particles produced from a solution of $AgNO_3$ in water (a) and from an alkaline $[Ag(NH_3)_2]^+$ solution (b) with ascorbic acid as the reducing agent are shown in Fig. 7.6. As these images show, there is a significant difference in the shapes of silver particles produced in acidic or alkaline solutions using the

Fig. 7.6 Particles of silver produced from (**a**) a solution of $AgNO_3$ in water and (**b**) an alkaline $[Ag(NH_3)_2]^+$ solution with ascorbic acid as a reducing agent [21] (Reprinted with permission from ECS—The Electrochemical Society.)

same reducing agent. In both acidic (Fig. 7.6a) and alkaline (Fig. 7.6b) solutions the agglomerated powders were obtained. In the acidic solutions platelets or geometric crystals of individual particles of silver with a majority being 1 μm in size or more were found. On the other hand, in the alkaline solutions spherical-like particles with a rough surface morphology were obtained. The size of these spherical particles is less than 1 μm, with the majority being in the size range 300–500 nm. The difference in the shape of produced powders can be attributed to various reaction mechanisms in alkaline and acidic conditions.

SEM micrographs of Cu, Au, and Pd powders are presented in Fig. 7.7. As shown in Fig. 7.7a, copper represents agglomerated geometric crystals with a wide range in size from 500 nm to more than 2–3 μm. Gold powder produced with ascorbic acid (Fig. 7.7b) represents an agglomerate composed of 3D dendrites, smaller than 500 nm. This result is in a way different from the results published before [23], where the gold particles were found to be starlike. The differences in the shape of gold particles are attributed to various conditions including different solutions used in experiments. The palladium powder presented in Fig. 7.7c consists of agglomerates of Pd particles. It seems that the individual Pd particles are spherical with a rough surface morphology. The size of the individual particles is estimated at about 300–600 nm. This result is different than the previously reported images of Pd particles [24] obtained with a reduction of Pd(II) ions with hypophosphite. The reduction of Pd(II) ions with hypophosphite leads to the production of smooth spherical particles [24]. The differences in the surface morphology of Pd particles are attributable to various reducing agents.

Importantly, the morphologies of electrolessly produced powders illustrated in this chapter are quite different and depend on the nature of the reducing agent used in the experiments. This can be seen clearly with a comparison of the SEM results presented in Figs. 7.4–7.7.

As suggested above for the case of cobalt, the difference in the shapes of powders is attributed to the various chemical compositions of the powder itself. On the other hand, the composition of the powder is determined by the nature of the reducing agent, e.g., hydrazine or

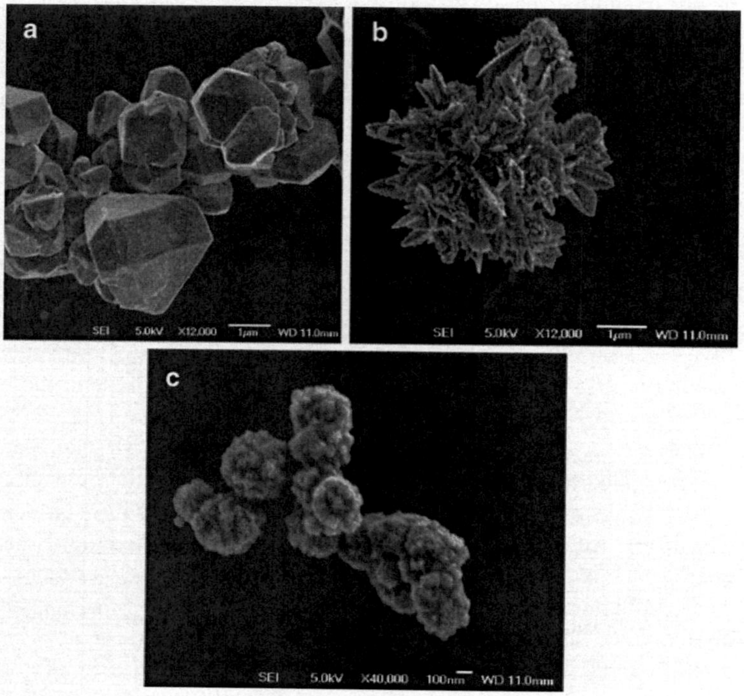

Fig. 7.7 SEM micrographs of (**a**) Cu, (**b**) Au, and (**c**) Pd powders, produced from $[Cu(NH3)_4]^{2+}$, $AuCl_3$, and $[Pd(NH_3)_2]^{2+}$ solutions, respectively, using ascorbic acid as a reducing agent [21] (Reprinted with permission from ECS—The Electrochemical Society.)

hypophosphite. Using the same reducing agent, but different metallic ions, e.g., Co^{2+} or Ni^{2+}, different shapes of powders can be obtained, as illustrated in Fig. 7.5a, d. This suggests that not only the reducing agent but also the nature of the metal, i.e., Ni or Co, would influence the surface morphology of the deposited powder. In addition, as demonstrated in Fig. 7.6 for the example of silver using the same reducing agent (ascorbic acid), pH may have a significant influence on the shape of electroless deposited powders.

It is obvious that the further investigations are required to find a more sophisticated correlation between the shape of the particles and the nature of the reducing agent. Although the electroless deposition is frequently used in practice, the fundamental aspects governing this process are not yet established. In the following section the mechanisms of electroless deposition are discussed.

7.4 Discussion of Mechanistic Aspects of Electroless Deposition

The electroless deposition of a metal M from a homogenous liquid solution, using a reducing agent R^{n-}, is described by the following reaction:

$$M^{z+} + R^{n-} \rightarrow M + R^{z-n}. \tag{7.19}$$

In the electroless deposition, the hydrolysis phenomena, as has already been published [1, 16, 17], plays a very significant role in the reduction of metallic ions from aqueous solutions. Due to an increase in pH, the following reaction may occur:

$$M^{z+} + xOH^{-} \rightarrow M(OH)_x^{(z-x)+}. \tag{7.20}$$

The hydrolyzed metallic species ($M(OH)_x^{(z-x)+}$) are further reduced in the presence of an appropriate reducing agent (R^{n-}) to a metal according to the reaction

$$M(OH)_x^{(z-x)+} + R^{n-} \rightarrow M + R(OH)_x^{(z-x-n)+}. \tag{7.21}$$

An increased concentration of the reducing agent, pH, and an elevated temperature significantly contribute to the hydrolysis of the metal ions causing the formation of hydroxyl complexes, hydroxides, and even oxides [1, 16, 17]. These hydroxides and/or oxides, formed into the bulk solution due to hydrolysis, serve as sites which are further reduced to the metallic state with an appropriate reducing agent. Since various sizes of particles can be produced, it is obvious that the further growth of the metallic powder is autocatalytic. Furthermore, since the formation of hydrolyzed metallic species is significantly determined by pH, it seems that the degree of hydrolysis can significantly influence the shape of powder particles.

The concept that the formation of hydroxides or oxides in the bulk solution leads to their further reduction to the metallic state is documented for Co [16] or Cu and Ag [21]. Oxide powders, i.e., Ag_2O, Cu_2O, and CuO were used as starting materials for the production of Ag and Cu powders [21]. Slurries were simply prepared by a dispersion of solid oxide powders in distilled water. The dispersions were vigorously stirred with a magnetic stirrer and an adequate amount of ascorbic acid which was used as the reducing agent was added. The mixing was continued until metallic powders (e.g. Ag and Cu) were obtained after approximately 30 min, which was visually observed.

XRD patterns of silver powders produced from the slurry of Ag_2O in water and using the ascorbic acid as a reducing agent are presented in Fig. 7.8a, b. Both patterns (a) and (b) in Fig. 7.8 clearly show Ag peaks, suggesting that Ag_2O is reduced to elemental silver. The reduction of Ag_2O to the elemental silver using the ascorbic acid can be presented as

$$Ag_2O + H_2O \rightarrow 2Ag^+ + 2OH^-, \tag{7.22}$$

$$Ag^+ + OH^- \rightarrow AgOH, \tag{7.23}$$

$$2AgOH + C_6H_8O_6 \rightarrow 2Ag + C_6H_6O_6 + 2H_2O. \tag{7.24}$$

XRD patterns of the powders produced by the reduction of the aqueous slurries of Cu_2O and CuO with the ascorbic acid are presented in Fig. 7.9. It is very clear from the patterns in Fig. 7.9 that only Cu peaks were identified. No other impurities were found.

The reduction of Cu_2O and CuO to the metallic state is explained by the following reactions:

$$Cu_2O + H_2O \rightarrow 2Cu^+ + 2OH^-, \tag{7.25}$$

$$Cu^+ + OH^- \rightarrow 2CuOH, \tag{7.26}$$

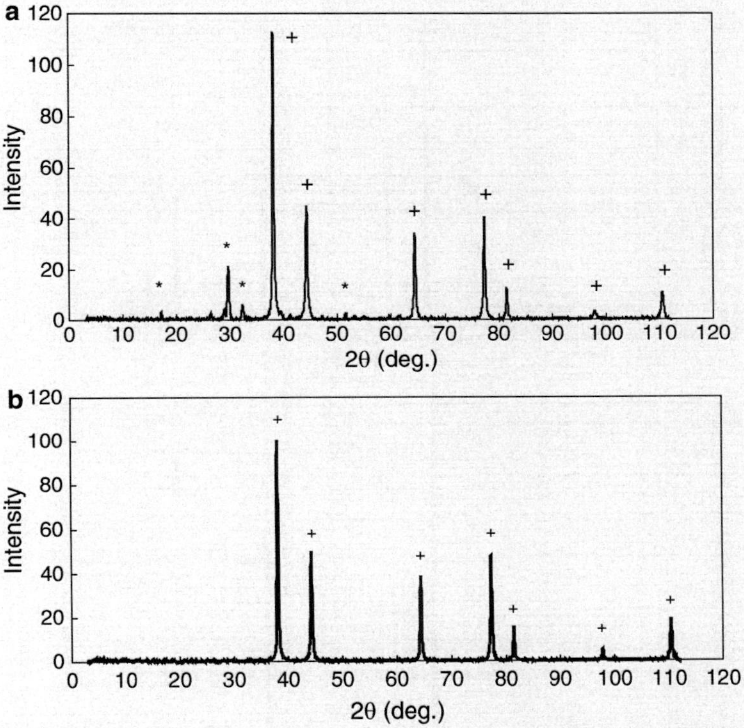

Fig. 7.8 XRD patterns of (**a**) coarse and (**b**) fine silver powders produced from a suspension of Ag_2O and ascorbic acid as a reducing agent (*plus* Ag, *asterisk* $Ag_2C_2O_4$) [21] (Reprinted with permission from ECS—The Electrochemical Society.)

$$2CuOH + C_6H_8O_6 \rightarrow 2Cu + C_6H_6O_6 + 2H_2O, \qquad (7.27)$$

and

$$CuO + H_2O \rightarrow Cu^{2+} + 2OH^-, \qquad (7.28)$$

$$Cu^{2+} + 2OH^- \rightarrow Cu(OH)_2, \qquad (7.29)$$

$$Cu(OH)_2 + C_6H_8O_6 \rightarrow Cu + C_6H_6O_6 + 2H_2O. \qquad (7.30)$$

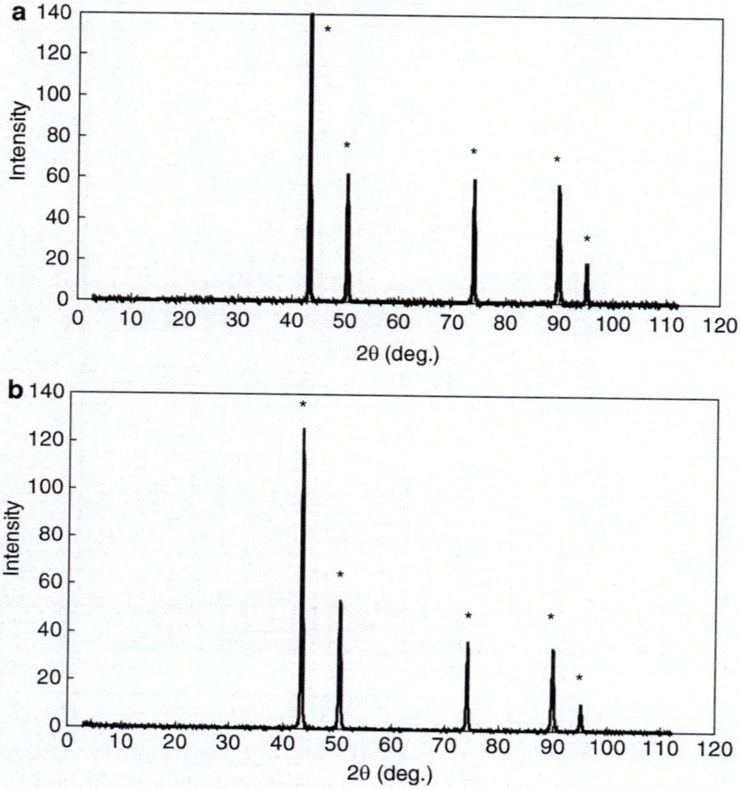

Fig. 7.9 XRD patterns of copper powders produced from suspensions of (**a**) Cu_2O and (**b**) CuO and ascorbic acid as a reducing agent (*asterisk* Cu) [21] (Reprinted with permission from ECS—The Electrochemical Society.)

The SEM images of silver and copper powders produced by the reduction of the respective oxides with ascorbic acid are presented in Fig. 7.10. In both cases, i.e. when the starting material was Cu_2O or CuO, similar shapes of Cu particles were obtained. The results presented in Fig. 7.10a, b show that the Cu particles obtained via the reduction of Cu_2O or CuO with ascorbic acid are geometric crystals with sizes from 500 nm to about 2 μm. Silver powder (Fig. 7.10c, d), produced via the reduction of Ag_2O with ascorbic

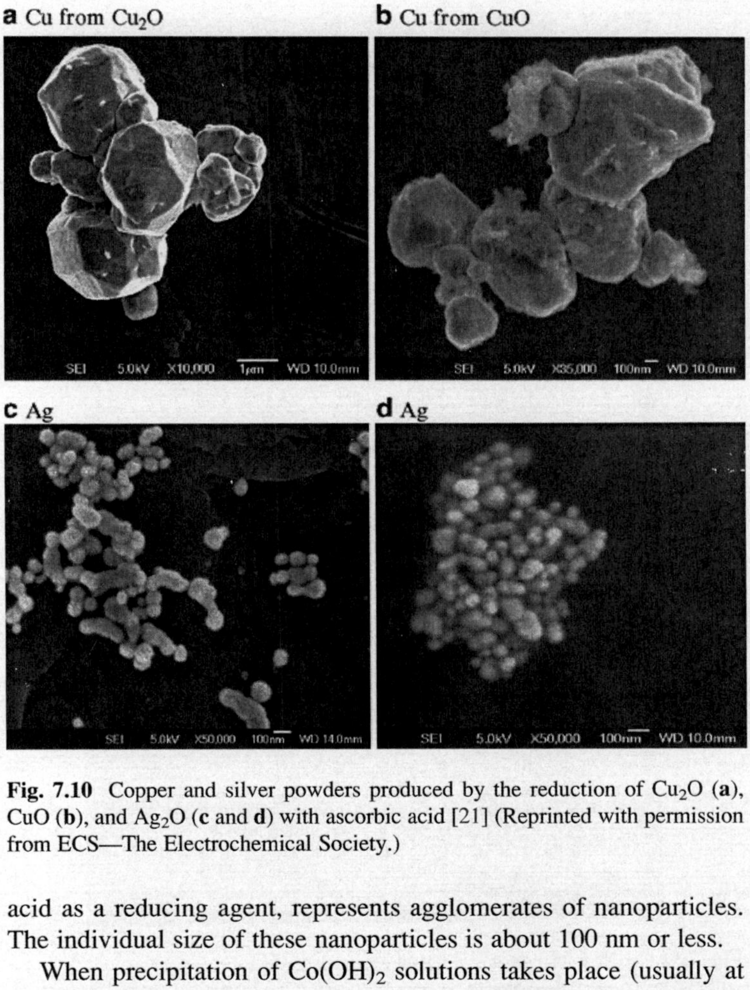

Fig. 7.10 Copper and silver powders produced by the reduction of Cu_2O (**a**), CuO (**b**), and Ag_2O (**c** and **d**) with ascorbic acid [21] (Reprinted with permission from ECS—The Electrochemical Society.)

acid as a reducing agent, represents agglomerates of nanoparticles. The individual size of these nanoparticles is about 100 nm or less.

When precipitation of $Co(OH)_2$ solutions takes place (usually at higher pH), the reduction of Cu(II) to metallic cobalt was observed even at room temperature, immediately upon the addition of hydrazine [16]. Consequently, cobalt powder was produced. A test experiment was carried out as described below to prove the observation that the deposition of cobalt metal by the reduction of cobalt hydroxide with hydrazine is possible. Cobalt hydroxide was precipitated by mixing aqueous solutions of $CoCl_2$ and NaOH. Immediately upon

Fig. 7.11 SEM micrographs of Co powder obtained by a reduction of $Co(OH)_2$ with hydrazine (**a**) 30 s after hydrazine was added, and (**b**) 10 min after hydrazine was added [16] (Reprinted with permission from ECS—The Electrochemical Society.)

the addition of hydrazine to such a prepared slurry, the reduction of ions to metal occurred even at room temperature without the presence of a metallic substrate. Micrographs of the Co powder product are presented in Fig. 7.11.

7.5 Electroless Deposition of Metal Powder Nanoparticles

Due to developments in the field of nanotechnology, electroless deposition has attracted many researchers in the last few decades.

Metallic nanoparticles can successfully be produced via a galvanic displacement reaction on suitable substrates in the same manner as described in Sect. 7.3.1. In this case, due to poor adhesion, the metal nanoparticles can easily be separated from the substrate. It seems that the duration of the process should be reduced at a reasonable minimum to avoid the growth and production of particles larger than a "nanometer" range.

Similarly, using the electroless deposition and an appropriate reducing agent in a way as described in Sect. 7.3.2, metallic nanoparticles can be successfully produced. Again, to produce metallic nanoparticles, attention should be paid to keep the process as short as possible to avoid the growth of particles larger than "nanometer" range via autocatalytic deposition.

In the electroless deposition, the growth of particles can be successfully avoided by a proper control of deposition parameters including pH, temperature, time, composition of the solution including concentrations of the metal ion, and a reducing agent. The conditions for the powder formation via electroless deposition are described in this chapter.

The powder formation occurs when the concentration of the reducing agent (e.g., formaldehyde, hydrazine, borohydride, hypophosphite, and similar) is too high and at elevated temperature. Consequently, to produce nanoparticles of a specific metal or alloy it is necessary to keep the concentration of the reducing agent at a level which would be sufficient to start the deposition in the bulk solution. Too high concentrations of the reducing agent would unavoidably lead to growth and a consequent production of larger particles. Of course, based on this scenario the concentration of metal ions to be reduced with an appropriate reducing agent should be kept at a minimum to produce nanoparticles.

Another approach that should further be investigated is the addition of compounds which can immediately suppress deposition and the growth of metal particles if the target is to produce nanoparticles. In this case, the compounds suppressing the growth of metal particles will depend on the metal itself, reducing agent, pH, temperature, etc. Accordingly, every individual system would require future studies to produce a specific shape and size of the desired nanoparticles. At present, most of the nanoparticles produced via galvanic displacement or electroless deposition using appropriate reducing agent are agglomerated. Methods to keep the produced individual nanoparticles in a suspension to avoid agglomeration should further be investigated as well.

If the field of nanotechnology is aimed to grow in the future where metallic nanoparticles find realistic applications, both galvanic displacement and electroless deposition with reducing agents open a tremendous window for future studies of the kinetics and mechanisms of these processes. This approach could lead to the successful production of desired sizes and shapes of the required nanoparticles.

7.6 Conclusions

Electroless deposition can successfully be used for the production of metallic powders with various sizes and shapes. A traditional approach in the production of metallic powders from aqueous solutions without the application of an external current, e.g., Ni, Co etc., is based on the reduction of the respective metal ions with hydrogen at elevated temperatures and under pressure. These hydrometallurgical processes are well established in industry.

Galvanic displacement deposition of metallic powders is achieved as a consequence of the porosity of a more positive metal or due to a simultaneous hydrogen evolution reaction. Many metallic powders can be produced using this simple process.

From homogenous solutions, many metallic powders can be produced when the instability of the solution takes place. This is seen when the concentration of the reducing agent is too high or at elevated temperatures. Many metallic powders, e.g., Cu, Ni, Co, Ag, Au, Pd, etc., using appropriate reducing agents can be obtained from homogenous aqueous solutions. The observations suggest that the hydrolysis of metallic ions is crucial in the deposition of metallic powders via electroless deposition from homogenous aqueous solutions. It was shown that if starting sources of metallic ions are oxides (e.g., Ag_2O, CuO, and Cu_2O), using an appropriate reducing agent (ascorbic acid), metallic powders (Ag or Cu) can be successfully produced. Similarly, when the $Co(OH)_2$ is precipitated as a result of hydrolysis, an addition of hydrazine to this slurry leads to the successful production of Co powder.

References

1. Djokić SS (2002) Electroless deposition of metals and alloys. In: Conway BE, White RE, Bockris JO'M (eds) Modern aspects of electrochemistry, chap 2, no. 35. Kluwer Academic/Plenum, New York, p 51
2. Agrawal A, Kumar V, Pandey BD, Sahu KK (2006) Mater Res Bull 41:879
3. Saarinen T, Lindfors L, Tugleberg S (1998) Hydrometallurgy 47:309
4. Weir DR, Sefton VB (1979) International laterite symposium. AIME, New Orleans, NY, p 325
5. Wajszcuk RA, Charewicz WA (1994) Hydrometallurgy 35:99
6. Schaufelberger FA, McGauley PJ (1957) Process for recovering copper by chemical precipitation. US Patent 2,796,342
7. Mackiw VN, Kunda W, Evans DJT (1966) Effect of addition of agents on the properties of nickel powders, in modern developments in powder metallurgy. Plenum, New York, p 15
8. Mackiw VN, Kunda W, Lin WC (1957) Trans AIME 209:786
9. Yamasaki N, Huanzhen L (1993) Metall Trans B 24B:557
10. Chou EECJ, Cooperrider MW (1985) Production of cobalt and nickel powder. US Patent 4,545,814
11. Kim DJ, Chung HS, Yu K (2002) Mater Res Bull 37:2067
12. Schuafelberger FA (1956) J Met 8:695
13. Evans DJI (1968) Advances in extractive metallurgy. Institution of Mining and Metallurgy/Elsevier, New York, p 831
14. Macgregor ER, Halpern J (1958) Trans AIME 214:244
15. Wodka J, Chareweiz WA, Janicki T (1999) Hydrometallurgy 52:177
16. Djokić SS (1997) J Electrochem Soc 144:2358
17. Djokić SS, Cavallotti PL (2010) Electroless deposition: theory and applications. In: Djokić SS (ed) Electrodeposition: theory and practice, modern aspects of electrochemistry, chap 6, no 48. Springer, New York, pp 251–289
18. Nadkarni RM, Jelden CF, Bowles KC, Flanders HE, Wadsworth ME (1967) Trans TMS-AIME 239:239

19. Mur LE, Annamalai V (1978) Metall Trans 9B:515
20. MacKinnon DJ, Ingraham TR (1970) Can Met Quart 9:443
21. Djokić SS, Djokić NS (2011) J Electrochem Soc 158:D204
22. Djokić SS (1996) J Electrochem Soc 143:1300
23. Djokić SS (2000) In: Landolt D, Maltosz M, Sato Y (eds) Fundamentals of electrochemical deposition and dissolution, PV 99-33, The Electrochemical Society Proceedings, Pennington, NJ, p 38
24. Djokić SS (1999) Plat Surf Finish 86:104

Index

Stojan S. Djokić (ed.), *Electrochemical Production of Metal Powders*,
Modern Aspects of Electrochemistry 54, DOI 10.1007/978-1-4614-2380-5,